LCA Compendium – The Complete World of Life Cycle Assessment

Series editors

Walter Klöpffer, LCA Consult & Review, Frankfurt am Main, Germany
Mary Ann Curran, BAMAC, Ltd., LCA & Sustainability Consultant, Cincinnati, OH, USA

Aims and Scope

Life Cycle Assessment (LCA) has become the recognized instrument to assess the ecological burdens and human health impacts connected with the complete life cycle (creation, use, end-of-life) of products, processes and activities, enabling the assessor to model the entire system from which products are derived or in which processes and activities operate. Due to the steady, world-wide growth of the field of LCA, the wealth of information produced in journals, reports, books and electronic media has made it difficult for readers to stay abreast of activity and recent developments in the field. This led to the realization of the need for a comprehensive and authoritative publication.

The *LCA Compendium* Book Series will discuss the main drivers in LCA (SETAC, UNEP/SETAC Life Cycle Initiative, etc.), the strengths and limitations of LCA, the LCA phases as defined by ISO standards, specific applications of LCA, Life Cycle Management (LCM) and Life Cycle Sustainability Assessment (LCSA). Further volumes, which are closely related to these themes will cover examples of exemplary LCA studies ordered according to the importance of the fields of application. They will also present new insights and new developments and will keep the whole work current. The aim of the series is to provide a well-structured treatise of the field of LCA to give orientation and guidance through detailed descriptions on all steps necessary to conduct an LCA study according to the state of the art and in full agreement with the standards.

The *LCA Compendium* Book Series anticipates publishing volumes on the following themes:

- Background and Future Prospects in Life Cycle Assessment (published in March 2014)
- Goal and Scope Definition in Life Cycle Assessment
- Life Cycle Inventory Analysis (LCI)
- Life Cycle Impact Assessment (LCIA)
- Interpretation, Critical Review and Reporting in Life Cycle Assessment
- Applications of Life Cycle Assessment
- Special Types of Life Cycle Assessment
- Life Cycle Management (LCM)
- Life Cycle Sustainability Assessment (LCSA)
- Life Cycle Assessment Worldwide

More information about this series at http://www.springer.com/series/11776

Michael Z. Hauschild • Mark A.J. Huijbregts
Editors

Life Cycle Impact Assessment

Springer

Editors
Michael Z. Hauschild
Division for Quantitative Sustainability Assessment, Department of Management Engineering
Technical University of Denmark (DTU)
Lyngby, Denmark

Mark A.J. Huijbregts
Department of Environmental Science
Institute for Water and Wetland Research
Radboud University
Nijmegen, The Netherlands

ISSN 2214-3505 ISSN 2214-3513 (electronic)
LCA Compendium – The Complete World of Life Cycle Assessment
ISBN 978-94-017-9743-6 ISBN 978-94-017-9744-3 (eBook)
DOI 10.1007/978-94-017-9744-3

Library of Congress Control Number: 2015932962

Springer Dordrecht Heidelberg New York London

Printed on acid-free paper

Springer Science+Business Media B.V. Dordrecht is part of Springer Science+Business Media (www.springer.com)

Preface

Life Cycle Assessment (LCA) is an instrument to assess the impacts on the environment and on human health connected with the complete life cycle (creation, use, end-of-life) of products, processes and activities. It enables the practitioner to model the entire system from which products are derived or in which processes and activities operate.

Life cycle impact assessment (LCIA) is used in LCA to establish a linkage between the inventory of elementary flows for the system of the product or process and its potential environmental impacts. The aim of this book of the *LCA Compendium* is to give the reader a thorough insight into LCIA presenting the history, the state of the art, the existing problems and research needs, and the foreseeable future. It starts with an introduction of fundamental characteristics and principles of LCIA (Chap. 1) followed by the selection of impact categories and classification of inventory flows (Chap. 2). It continues with an in-depth description and discussion of the following impact categories (Chaps. 3, 4, 5, 6, 7, 8, 9, 10, 11, 12, and 13):

- Abiotic resource use
- Acidification
- Climate change
- Ecotoxicity
- Eutrophication
- Human toxicity
- Land use
- Particulate matter formation
- Photochemical ozone formation
- Stratospheric ozone depletion
- Water use

These chapters on the individual impact categories share a common structure presenting the characteristics of the impact category in terms of:

- Impact pathways
- Affected areas of protection
- Contributing inventory flows
- Geographical scale
- Spatial and temporal variability

This is followed by a presentation and discussion of different midpoint and endpoint characterisation methods for the relevant impact category with discussions of metrics, model uncertainties and new developments, and concluded by a discussion of research needs.

The chapters on characterisation are followed by chapters on the two optional LCIA steps which are normalisation (Chap. 14) and weighting (Chap. 15).

The 15 chapters have been elaborated by leading experts in the field to provide the reader a qualified and up-to-date insight into the vast and important field of LCIA. Each chapter functions as a self-containing unit, simultaneously playing its individual role in the overall concept of this book. The editors have taken care to avoid unwanted repetitions, especially regarding the common principles of LCIA presented in the introductory chapter.

We thank all authors contributing to this volume, bringing their insights to the benefit of the reader. We would also like to express our sincere gratitude to our colleagues Mikolaj Owsianiak and Morten Ryberg for their valuable comments and thorough work in the final editing of this book and to Almut B. Heinrich, managing editor for the *LCA Compendium*, for her tremendous help in getting the volume ready for print.

Lyngby, Denmark — Michael Z. Hauschild
Nijmegen, The Netherlands — Mark A.J. Huijbregts

Contents

Contributors

Rodrigo A.F. Alvarenga Department of Sustainable Organic Chemistry and Technology, Faculty of Bioscience Engineering, Ghent University, Ghent, Belgium

Laura de Baan Institute of Environmental Engineering, ETH Zurich, Zurich, Switzerland

Jo Dewulf Department of Sustainable Organic Chemistry and Technology, Faculty of Bioscience Engineering, Ghent University, Ghent, Belgium

Peter Fantke Division for Quantitative Sustainability Assessment, Department of Management Engineering, Technical University of Denmark (DTU), Lyngby, Denmark

Jeroen B. Guinée Department of Industrial Ecology, Institute of Environmental Sciences (CML), Leiden University, Leiden, The Netherlands

Michael Z. Hauschild Division for Quantitative Sustainability Assessment, Department of Management Engineering, Technical University of Denmark (DTU), Lyngby, Denmark

Andrew D. Henderson Division of Epidemiology, Human Genetics and Environmental Sciences, School of Public Health, The University of Texas Health Science Center at Houston, Houston, TX, USA

Mark A.J. Huijbregts Department of Environmental Science, Institute for Water and Wetland Research, Radboud University, Nijmegen, The Netherlands

Sebastien Humbert Quantis, EPFL Innovation Park, Lausanne, Switzerland

Norihiro Itsubo Department of Environmental Studies, Tokyo City University, Yokohama, Tsuzuki-ku, Yokohama, Japan

Olivier Jolliet School of Public Health, Department of Environmental Health Sciences, University of Michigan, Ann Arbor, MI, USA

Joe L. Lane School of Chemical Engineering, The University of Queensland, Brisbane, QLD, Australia

Alexis Laurent Division for Quantitative Sustainability Assessment, Department of Management Engineering, Technical University of Denmark (DTU), Lyngby, Denmark

Annie Levasseur CIRAIG, Department of Chemical Engineering, École Polytechnique de Montréal, Montréal, QC, Canada

Llorenç Milà i Canals Division of Technology, Industry and Economics, United Nations Environment Programme, Paris, France

Stephan Pfister Institute of Environmental Engineering (IfU), ETH Zurich, Zurich, Switzerland

Philipp Preiss University of Stuttgart, Stuttgart, Germany

Ralph K. Rosenbaum IRSTEA, UMR ITAP, ELSA-PACT – Industrial Chair for Environmental and Social Sustainability Assessment, Montpellier Cedex 5, France

Pierre-Olivier Roy CIRAIG, Department of Chemical Engineering, École Polytechnique de Montréal, Montréal, QC, Canada

Pilar Swart Department of Sustainable Organic Chemistry and Technology, Faculty of Bioscience Engineering, Ghent University, Ghent, Belgium

Rosalie van Zelm Department of Environmental Science, Institute for Water and Wetland Research, Radboud University, Nijmegen, The Netherlands

Chapter 1
Introducing Life Cycle Impact Assessment

Michael Z. Hauschild and Mark A.J. Huijbregts

Abstract This chapter serves as an introduction to the presentation of the many aspects of life cycle impact assessment (LCIA) in this volume of the book series 'LCA Compendium'. It starts with a brief historical overview of the development of life cycle impact assessment driven by numerous national LCIA methodology projects and presents the international scientific discussions and methodological consensus attempts in consecutive working groups under the auspices of the Society of Environmental Toxicology and Chemistry (SETAC) as well as the UNEP/ SETAC Life Cycle Initiative, and the (almost) parallel standardisation activities under the International Organisation for Standardisation (ISO). A brief introduction is given on the purpose and structure of LCIA. As a common background for the 11 chapters dealing with the characterisation modelling of the most common impact categories, the chapter concludes with an introduction of the general principles and features of characterisation.

Keywords Decision support • ISO standards • LCA • Life cycle assessment • LCIA • Life cycle impact assessment • LCIA history • Spatial differentiation

1 Introduction

The inventory for a product system provides quantification of exchanges between the processes of the product system and the environment. Such an inventory can contain a very large number of substance emissions (>1,000) as well as input of resources (>100). The environmental relevance can differ dramatically between the different emissions and resource extractions. Since the main goal of a life cycle

M.Z. Hauschild (✉)
Division for Quantitative Sustainability Assessment, Department of Management Engineering, Technical University of Denmark (DTU), Produktionstorvet, Building 424, 2800 Lyngby, Denmark
e-mail: mzha@dtu.dk

M.A.J. Huijbregts
Department of Environmental Science, Institute for Water and Wetland Research, Radboud University, Nijmegen, The Netherlands

M.Z. Hauschild, M.A.J. Huijbregts (eds.), *Life Cycle Impact Assessment*, LCA Compendium – The Complete World of Life Cycle Assessment, DOI 10.1007/978-94-017-9744-3_1

assessment (LCA) study is often to decide which of the compared alternative products is preferable from an environmental perspective or where in the life cycle of a product we find the largest environmental impacts, such a comprehensive inventory provides limited decision support. It is the purpose of the life cycle impact assessment (LCIA) to aid the interpretation by translating the inventory of elementary flows into a profile with a limited but covering number of environmental impact scores, representing the product system's impact on, e.g., global warming or acidification, which can help provide answers to the questions posed in the goal definition of the study and support decision making.

1.1 Historical Overview

LCA has its early roots in the late 1960s and 1970s where the results of a study were reported as resource and emission profiles (e.g. Hunt et al. 1974), and no quantitative assessment of the associated impacts on environment or resources was performed (Huppes and Curran 2012). The 1980s saw an increase in the use of LCA in particular for packaging studies (e.g. BUS 1984; Franke 1984; Lundholm and Sundström 1985), and several LCA-related scientific working groups were formed under the auspices of the Society of Environmental Toxicology and Chemistry (SETAC) in the early 1990s (Klöpffer 2006). To increase the interpretability of LCA results, the need for an assessment of environmental impacts of the inventory results became apparent. The first methods for assessment of environmental impacts in LCA were published in the early 1990s with prominent examples as the Swiss Ecoscarcity (or Ecopoints) methodology (Ahbe et al. 1990) and the CML 1992 methodology (Heijungs et al. 1992).

In 1993 a standardisation process was launched for LCA under the International Organisation for Standardisation, ISO, resulting in a common framework and fundamental principles, including a specific standard addressing LCIA (ISO 14040 1997). Together with the standards describing life cycle inventory modelling (ISO 14041 1998) and interpretation (ISO 14043 2000). In 2006, a standard for the LCIA phase (ISO 14042 2000) was merged into the ISO 14044 standard (ISO 14044 2006) which, together with an updated version of the basic LCA standard (ISO 14040 2006), constitutes the present standards for LCA.

In parallel to the development of the ISO-standard, a large number of national projects, most notably in Canada, Denmark, Japan, The Netherlands, Sweden, Switzerland and the United States, specifically focused on the further development of LCIA methodologies [e.g. EDIP (Wenzel et al. 1997), CML 2002 (Guinée et al. 2002), Ecoindicator 99 (Goedkoop and Spriensma 2000), EDIP2003 (Hauschild and Potting 2005), EPS (Steen 1999a, b), IMPACT2002+ (Jolliet et al. 2003), TRACI (Bare et al. 2003), LIME (e.g. Itsubo et al. 2004), LUCAS (Toffoletto et al. 2007) and ReCiPe (Goedkoop et al. 2009)]. Most of the early methodologies built on midpoint methods meaning that the indicator for an impact category was chosen at some intermediary point in the underlying impact pathway (e.g. CML 2002, EDIP, TRACI). The number of impact categories included in these methodologies varied, but was typically higher than 10 and included impacts like global warming, stratospheric ozone depletion, human toxicity and acidification.

In parallel, the work on developing endpoint methods evolved (Eco-indicator 99, EPS) directing the impact assessment against indicators located at the very end of the impact pathway, such as human health, ecosystem quality and resource scarcity. Endpoint modelling was also inspired by parallel work on monetisation of the environmental impacts of energy technologies (e.g. Spadaro and Rabl 2002).

The main advantage of the midpoint indicators is considered their relatively strong scientific robustness, while endpoint indicators are considered to be easier to interpret (Bare et al. 2000). More recently, the midpoint and endpoint approaches have been combined into one LCIA methodology to enhance consistency in the impact pathway modelling (LIME, ReCiPe, IMPACT2002+).

The ISO 14044 standard brought standardisation of basic principles, but the standard did not specify which LCIA methods to apply in practice but gave the practitioner the advice to apply methods of general and international scientific acceptance. None of the methods mentioned above can be said to enjoy such a general acceptance, and the LCA practitioner was thus left with limited guidance on the choice between a large number of different LCIA methods which can give very different results and conclusions from the LCA study (Dreyer et al. 2003; Pant et al. 2004).

Since the early 90s and in parallel with the development of the ISO standards several attempts have been made to harmonise the LCIA field to avoid the confusion that arises when several different methods can be applied. This work has, in particular, been undertaken by consecutive SETAC working groups both in Europe and North America, later followed by task forces under the UNEP/SETAC Life Cycle Initiative that in 2002 started to work on identification of a recommended best practice (see Klöpffer 2006). While these activities have resulted in some consensus on the principles for best approaches (see, e.g., Udo de Haes et al. 2002), they have not resulted in a uniform globally accepted set of characterisation methods, except for a few impact categories. Promising results of this process have been:

- Consensus on the need to merge midpoint and endpoint models in a consistent framework to combine the advantages of both concepts (Bare et al. 2000).
- Development of a generic set of quality criteria for assessing characterisation methods, and application of these criteria on the most widely used impact assessment methods (Margni et al. 2007; Udo de Haes et al. 2002).
- For the toxicity-related impact categories, a consensus among model developers on what can be characterised as best practice, including the development of a scientific consensus model for characterisation of toxic impacts on human health and ecosystems (Hauschild et al. 2008; Rosenbaum et al. 2008, www.usetox.org).

As a recent development, the European Commission created recommendations of best practice for LCA and LCIA under the International Life Cycle Data system (ILCD) to support its Environmental Footprint guidelines. For the recommendations on LCIA practices, a systematic procedure was followed starting with the development of consistent criteria for good characterisation modelling practice. The criteria were then applied to existing LCIA methods to arrive at recommendations of the best among existing methods for a wide range of impact categories at both midpoint and endpoint. The resulting recommendations were vetted in extensive expert and stakeholder hearings, and the outcome of this process now specifies the LCIA methods to be applied in the European Environmental Footprinting work (Hauschild et al. 2013). More recently, the UNEP/SETAC Life Cycle Initiative launched a flagship project to develop global recommendations of best practice for a number of impact categories, following a scientific consensus-oriented approach (Jolliet et al. 2014).

While significant scientific progress has been made over the years and international standardisation and scientific consensus building have been going on for decades, the science of LCIA modelling is still under development. The focus areas include improvement of the endpoint modelling for all the impacts that are covered by LCIA, and development of common metrics for particularly the natural ecosystem endpoint to allow aggregation of the impact from many midpoint categories based on modelling of their ecosystem damage. Focus is also on development of characterisation methods for regional impact categories that can be applied in a consistent way in regions all over the world, and on accounting for the often very important differences in sensitivities at local or regional scale that are crucial to take into account when modelling environmental impacts.

2 Purpose and Structure

Life cycle impact assessment is the third phase of LCA according to the ISO 14040 standard, following the goal and scope definition and the inventory analysis and preceding the interpretation phase. It has the purpose to translate the elementary flows from the life cycle inventory into their potential contributions to the environmental impacts that are considered in the LCA and thus to support the interpretation phase where the questions posed in the goal definition are answered.

According to the ISO 14040 standard, LCIA consists of five steps as illustrated in Fig. 1.1:

Selection of impact categories, category indicators and characterisation models (selection)
This is where the impacts to be addressed are selected in accordance with the goal of the study and a method chosen for each impact category.

Assignment of LCI results to the selected impact categories (classification)
Here, the elementary flows of the inventory, such as resource consumption and emissions into air or water are assigned to the relevant impact categories among those selected under step 1, according to their ability to contribute to different environmental problems.

These two steps are further explained in Chap. 2 of this volume.

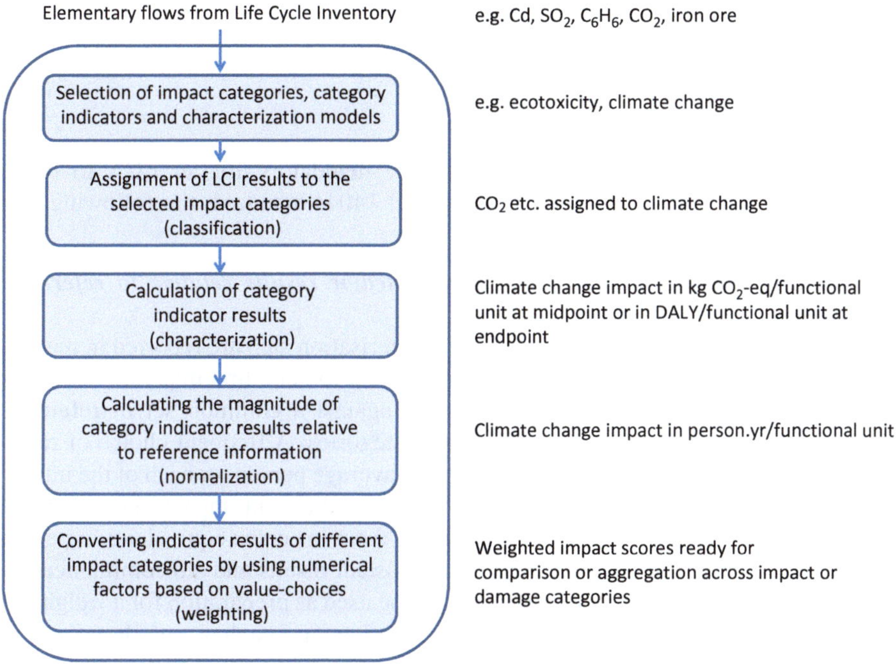

Fig. 1.1 The five steps of life cycle impact assessment

Calculation of category indicator results (characterisation)

For each elementary flow assigned to an impact category, the amount is multiplied with a so called characterisation factor. The characterisation factor for an elementary flow gives a quantitative representation of its importance for a specific impact category. The resulting indicator score is expressed in a metric common to all contributions within the impact category, for instance in terms of kg CO_2-equivalents for the contribution of a methane emission coming from the production of 1 kg of rice to global warming. The indicator scores for all the elementary flows that contribute to a specific impact category are summed to arrive at a total impact score for that impact category.

General principles of characterisation modelling are presented in Sect. 3, while a detailed description of the state-of-the-art in the modelling is given per impact category in Chaps. 3, 4, 5, 6, 7, 8, 9, 10, 11, 12 and 13:

- Abiotic resource use
- Acidification
- Climate change
- Ecotoxicity
- Eutrophication
- Human toxicity
- Land use

- Particulate matter formation
- Photochemical ozone formation
- Stratospheric ozone depletion
- Water use

While the first three steps of the LCIA are mandatory for an LCA to be in compliance with the requirements of the ISO 14044 standard, the following two steps are optional:

Calculating the magnitude of category indicator results relative to reference information (normalisation)
The indicator scores resulting from the characterisation step are reported in metrics that differ between the impact categories. In order to get a first impression of their relative magnitudes, they can be expressed against a common set of reference information that is available for all impact categories. A frequent choice of reference information is the annual impact from an average person for each of the impact categories. Normalisation thus brings scores for the different impact categories on a common scale, in the given case by expressing them in person equivalents or person years, which may support checking for inconsistent results and aid communication of the results. In addition, normalisation may be used as preparation for a weighting of the indicator results. Different normalisation approaches are described in Chap. 14.

Converting indicator results of different impact categories by using numerical factors based on value-choices (weighting)
In order to support a final comparison of indicator results across impact categories, a weighting can be performed, applying weighting factors that are based on value choices and represent the importance assigned to each of the impact categories. Weighting may be needed to conclude on studies where trade-offs exist between the results for the different impact categories, such as the comparison between fossil fuels with a main contribution to global warming and biofuels with main contributions to land use and water use impacts (Herrmann et al. 2013). Following weighting, the results can be aggregated across the impact categories to a single score in order to ease the interpretation. Weighting approaches are described in Chap. 15.

3 Characterisation Principles

As a common background for Chaps. 3, 4, 5, 6, 7, 8, 9, 10, 11, 12 and 13 that deal with characterisation modelling for individual impact categories, an introduction is given to general conditions and principles for the characterisation modelling in the following sections.

3.1 Characterisation Framework

Characterisation factors are calculated using characterisation models and relate or translate the elementary flow into its impact on the chosen indicator for the impact category. Udo de Haes et al. (2002) proposed a framework for calculation of characterisation factors (CF) according to which the CF is expressed as the product of a fate factor (FF), an exposure factor (XF) for the exposure of sensitive targets in the receiving environment and an effect factor (EF) expressing the effects of the exposure on the targets for the impact category.

$$CF = FF \cdot XF \cdot EF \tag{1.1}$$

This generic framework has since been applied with some modifications in most of the emission-related impact categories, with contents, metrics and meanings of the three factors that vary according to the impact category.

3.2 Impact Scores

Characterisation factors are expressed per unit of the elementary flow and their application is relatively straightforward:

$$IS_{j,i,k,l} = Q_{i,k,l} \cdot CF_{j,i,k,l} \tag{1.2}$$

Where $IS_{j,i,k,l}$ is the contribution from elementary flow *i* extracted at location *k* or emitted to environmental compartment *l* at location *k* to the indicator score for impact category *j*

$Q_{i,k,l}$ is the quantity of elementary flow *i* extracted at location *k* or emitted to compartment *l* at location *k* (from the inventory)

$CF_{j,i,k,l}$ is the characterisation factor under impact category *j* for elementary flow *i* extracted at location *k* or emitted to compartment *l* at location *k*

Equation 1.2 reveals two fundamental assumptions in characterisation:

- Characterisation factors depend on intrinsic properties of the elementary flow and for some impact categories also on the location of emission or extraction of the flow.
- The impact is proportional to the emitted quantity – the modelling of the impact from the product system is thus linear.

Following characterisation of the individual flows from the inventory, the results may be summed for each impact category according to:

$$IS_j = \Sigma_i \Sigma_k \Sigma_l IS_{j,i,k,l} \tag{1.3}$$

The degree of site dependency varies among the impact categories from none for the truly global impact categories like climate change and stratospheric ozone depletion to a high sensitivity to local or regional conditions for e.g. water use impacts or acidification. The need for spatial differentiation is discussed for each impact category in Chaps. 3, 4, 5, 6, 7, 8, 9, 10, 11, 12 and 13.

3.3 Midpoint Versus Endpoint Indicators

Characterisation applies models of the impact pathway leading from the elementary flows that are reported in the inventory through a sequence of causally related impacts to the Areas of Protection (AoPs) that are relevant for the impact category as illustrated in Fig. 1.2 for global warming. Areas of Protection were previously also referred to as 'Safeguard Subjects' (Udo de Haes et al. 2002). They represent the aspects that we care about and that we want the LCA to reveal the potential of the studied product system to damage. While different definitions of the areas of protection exist (ISO 14044 2006; Udo de Haes et al. 1999a, b; Steen 1999a, b), the focus in this volume is on the following three: Human health, Natural environment and Natural Resources (Hauschild et al. 2013).

According to ISO 14044, the indicator that is chosen to represent an impact category can be located anywhere along the impact pathway linking inventory data through consecutive environmental impacts to the damage that they cause on the AoPs. Midpoint indicators are located somewhere along the impact pathway, ideally at the point after which the mechanism is identical for all flows assigned to that impact category (Goedkoop et al. 2009). With this location of the midpoint indicator, the flows will have different midpoint characterisation factors, while their mid-to-endpoint characterisation factor is the same. Characterisation at the

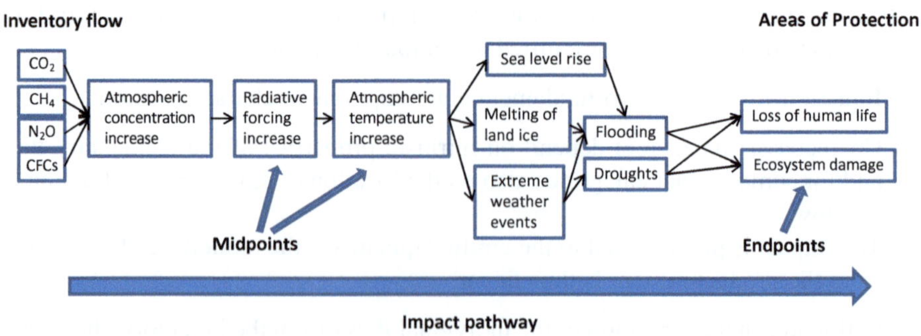

Fig. 1.2 Simplified impact pathway for global warming connecting elementary flows from the inventory to the areas of protection

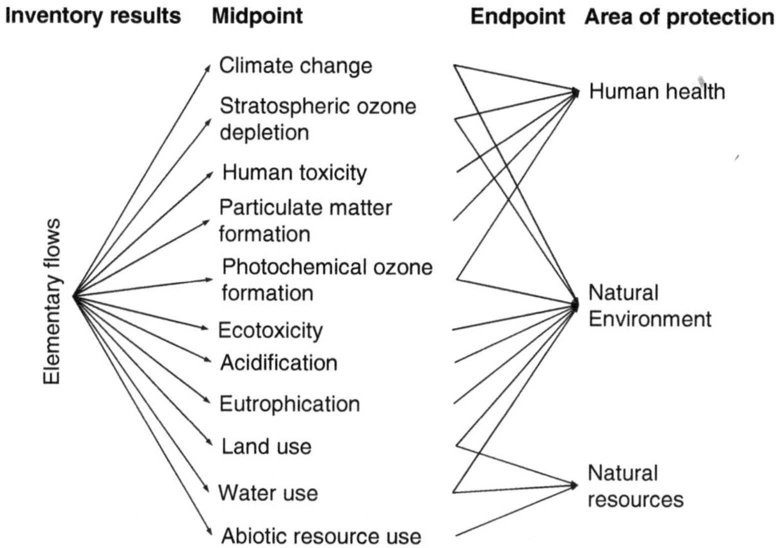

Fig. 1.3 Framework of midpoint impact categories covered in this volume illustrating their relation to the areas of protection (Adapted from Hauschild et al. 2013)

endpoint level requires modelling of the whole impact pathway to the point where the impacted entities are the areas of protection. Endpoint characterisation modelling is sometimes also referred to as 'damage modelling'. Figure 1.3 illustrates the framework of midpoint and endpoint categories for the impacts discussed later in this volume.

Life cycle impact assessment supports a holistic perspective on the environmental impacts of the product system and in principle attempts to model any impact from the system which can be expected to damage one or more areas of protection. This is in accordance with the requirements of the ISO standard that "the selection of impact categories shall reflect a comprehensive set of environmental issues related to the product system being studied, taking the goal and scope into consideration" (ISO 14044 2006).

The impact categories discussed in Chaps. 3, 4, 5, 6, 7, 8, 9, 10, 11, 12 and 13 are defined by their midpoint indicator but for each of them characterisation can be performed both at midpoint and endpoint level. The two approaches are complementary in that the midpoint characterisation has a stronger relation to the elementary flows and a lower modelling uncertainty, while the endpoint characterisation has a stronger relation to the areas of protection and hence a better information on the environmental relevance of the characterised flows.

3.4 Potential Impacts

The life cycle inventory presents elementary flows which have been attributed to an arbitrary functional unit, for instance, the consumption of 1 l of milk. The elementary flows of the inventory are generally aggregated over time (and space). The impacts expressed by the calculated indicator scores hence represent a sum of scores from elementary flows which occurred years ago and elementary flows which are predicted to occur sometime in the future. Furthermore, the impacts affect different ecosystems in different parts of the world depending on where the processes are located. Environmental effects arise, however, at a specific point in time and space. In LCA we have no way of knowing the simultaneous emissions from other processes, outside the product system, which expose the same ecosystem, and we typically have very limited information about the background concentration of other substances in the ecosystems that are affected by the flows from our product system. It is thus difficult to interpret the impacts which are modelled in LCIA in terms of real effects on the environment. They should rather be seen as relative performance indicators which can be the basis of comparisons and optimisation of the system or product. Product systems are fictitious entities that we cannot monitor in the real world, and characterisation models applied in LCIA are therefore difficult to validate. Their validity is typically based on their derivation from scientifically peer reviewed and accepted environmental models which are adopted to operate within the restrictions posed by the boundary conditions of LCA.

3.5 Best Estimates

In order to give a comprehensive view on the environmental impacts associated with the product system under study, LCIA must cover a widely encompassing suite of environmental impacts and support comparison between them. In order to ensure a fair comparison across the impact categories, the characterisation modelling has to aim for the same degree of realism for every category to avoid a bias and get the relative proportions of the impact categories right. This is ensured by aiming for best estimates in the modelling of all impacts, meaning that precautionary assumptions and conservative estimates are typically avoided (Hauschild and Pennington 2002).

3.6 Spatial Differentiation

Traditionally characterisation modelling assessed the behaviour of an elementary flow in a unit world, assuming a global set of standard conditions for the emission or resource extraction. However, global standard conditions can hide large and often unknown variations in the actual exposure that can be expected of sensitive parts of the environment. These differences in the sensitivity of the receiving environment can have a very strong influence on the resulting impact that may dwarf the influence of those intrinsic properties of the substance that are represented in the site-generic characterisation modelling (Potting and Hauschild 1997; Bare et al. 2003). A practical reason for disregarding variations in the modelling of the fate, exposure and effect has been ignorance about the location of processes in the product system, but it is often possible to differentiate the characterisation modelling according to local or regional specificities and derive site-dependent characterisation factors which, for instance, depend on the country of emission or extraction (Potting and Hauschild 2006).

3.7 Linear, Average or Marginal Modelling

The model approach to derive a characterisation factor can differ. The standard in LCIA modeling is to assess *marginal* changes (i.e. the influence of raising the background concentration/pressure on the impact indicator by a minimal amount). The advantages of the marginal approach are that it aims to realistically describe the influence of a change in pollution load for that specific situation and that it promotes emission changes with the highest reduction efficiency, i.e., where the slope of the cause-effect curve is steepest (Huijbregts et al. 2011). If the background concentration is not known, a *linear* approach can be followed as a simplified alternative. For instance, linearity between the concentration that affects 50 % of the species and the zero-effect concentration is assumed in this approach. A third approach is using an *average* effect factor. In this case the modelling depends on the background concentration, but rather than taking the derivative of the curve at that point, the average effect decrease per unit of concentration decrease towards zero-effect is used. Following this average approach, LCIA would focus on reaching the preferable state of the environment defined by society, and not on marginal changes. The advantage of the average approach is that it adopts a long-term perspective, focusing on ultimate environmental aims by society (Huijbregts et al. 2011). Figure 1.4 visualises the three approaches.

3.8 Uncertainty

Uncertainties are inevitable in LCIA. In general, three types of uncertainties can be distinguished: parameter uncertainty, model uncertainty and uncertainty from value choices (Huijbregts 1998).

Parameter uncertainty refers to uncertainty in the data required as input in the models applied for LCIA. Uncertainty in these data arises from measurement

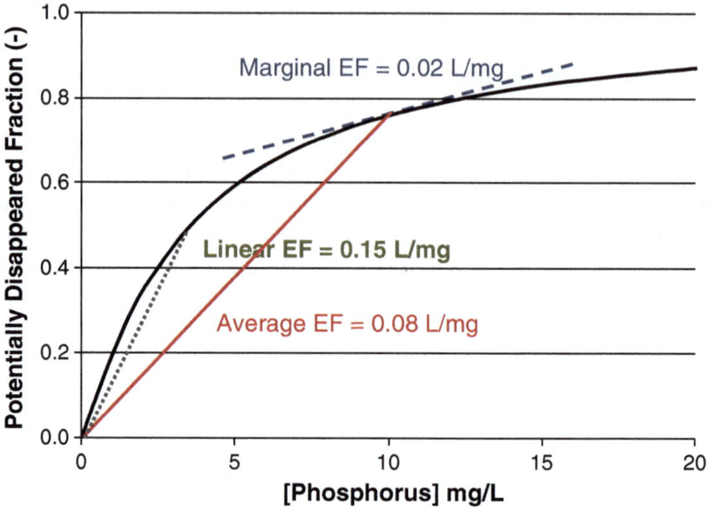

Fig. 1.4 Derivation of effect factors (EF) following a linear approach, marginal approach and an average approach, for the impact of total phosphorus concentrations on freshwater macro-invertebrate diversity with a logistic response curve $PDF = 1/(1 + 4.07 \cdot Cp^{-1.11})$ and working point of 10 mg/l (Derived from Huijbregts et al. 2011)

errors, analytical imprecision, and limited sample size. Statistical uncertainty in parameter values can be quantified via probabilistic simulation (Van Zelm and Huijbregts 2013). To perform a probabilistic simulation, input parameters need to be specified as uncertainty distributions. The method varies all the parameters at random, but the variation is restricted by the given uncertainty distribution for each parameter. Various parameter distributions, such as uniform or log-normal distributions, can be used. Repeated calculations produce a distribution of the predicted output values, reflecting the influence of the combined parameter uncertainties (Huijbregts 1998).

Model uncertainty is defined as uncertainty about the relations and mechanisms being studied (Huijbregts 1998). For instance, in the impact assessment it is commonly assumed that ecological processes respond in a linear manner to environmental interventions, while this may not be the case in reality. Model uncertainty can be made operational with the help of discrete choice analysis. In such an analysis, the choices that need to be made are identified. Subsequently, the various options to deal with every choice included are chosen to finally calculate results for each combination of choices (Van Zelm and Huijbregts 2013). Uncertainties from value choices are driven by personal beliefs and values that reflect what we care about, without any science being involved (Hertwich et al. 2000). A typical example is the equity of different age groups or species. Discrete choice analysis can also be used to investigate the uncertainties related to choices that reflect different personal values.

To avoid an unmanageable large number of choice combinations, the Cultural Theory has been used to cluster model assumptions and value choices in a logic and internally consistent way within LCIA (e.g., Hofstetter 1998; De Schryver et al. 2011). The Cultural Theory distinguishes five perspectives from which people perceive the world and behave in it. Three of these are generally used in LCIA: the individualist, the hierarchist, and the egalitarian perspective (Hofstetter 1998; Hofstetter et al. 2000). Each perspective reflects a hypothetical stakeholder or decision maker with a specific set of preferences and contextual values that explains his or her view on society and nature (De Schryver et al. 2011). De Schryver et al. (2013) showed that the value choices mainly responsible for the differences in results among perspectives are the choice of time horizon and inclusion of highly uncertain effects in the modelling. They also showed that value choices and different model assumptions in LCIA can modify the conclusions of an LCA and thus the practical implication of decisions based on the results of an LCA.

4 Reading Guide to This Volume in the Book Series 'LCA Compendium'

This volume of the LCA compendium is structured according to the ISO framework for LCIA, starting with the Selection of impact categories and classification of inventory flows in Chap. 2 and continuing with the description of characterisation modelling in separate chapters for each of the 11 impact categories mentioned in Chap. 2. The Chaps. 3, 4, 5, 6, 7, 8, 9, 10, 11, 12 and 13 share a common structure with a presentation of the characteristics of the impact category (impact pathway, affected AoPs, contributing substances, geographical scale and spatial and temporal variability) followed by a presentation and discussion of different midpoint and endpoint characterisation methods for the impact category with discussion of metrics, uncertainties and new developments, and concluded by a discussion of research needs. The chapters on characterisation are followed by chapters on the two optional steps, which are Normalisation in Chap. 14 and Weighting in Chap. 15.

The chapters have been written to function as self-containing units but at the same time avoid repetition of the common principles of LCIA that have been presented in this introductory chapter.

References

Ahbe S, Braunschweig A, Müller-Wenk R (1990) Methodik für Ökobilanzen auf der Basis ökologischer Optimierung, vol 133. Bundesamt für Umwelt, Wald und Landschaft (BUWAL), Bern

Bare JC, Hofstetter P, Pennington DW, Udo de Haes HA (2000) Life cycle impact assessment midpoints vs. endpoints: the sacrifices and the benefits. Int J Life Cycle Assess 5(5):319–326

Bare JC, Norris GA, Pennington DW, McKone TE (2003) The tool for the reduction and assessment of chemical and other environmental impacts. J Ind Ecol 6(3–4):49–78

Bundesamt für Umweltschutz (BUS) (1984) Ökobilanzen von Packstoffen, Schriftenreihe Umweltschutz Nr. 24. Bern

de Haes HA U, Finnveden G, Goedkoop M, Hauschild MZ, Hertwich E, Hofstetter P, Klöpffer W, Krewitt W, Lindeijer E, Jolliet O, Mueller-Wenk R, Olsen S, Pennington D, Potting J, Steen B (eds) (2002) Life cycle impact assessment: striving towards best practice. SETAC Press, Pensacola. ISBN 1-880611-54-6

De Schryver A, van Zelm R, Humbert S, Pfister S, McKone TE, Huijbregts MAJ (2011) Value choices in life cycle impact assessment of stressors causing human health damage. J Ind Ecol 15(5):796–815

De Schryver A, Humbert S, Huijbregts MAJ (2013) The influence of value choices in life cycle impact assessment of stressors causing human health damage. Int J Life Cycle Assess 18:698–706

Dreyer LC, Niemann AL, Hauschild MZ (2003) Comparison of three different LCIA methods: EDIP97, CML2001 and eco-indicator 99. Does it matter which one you choose? Int J Life Cycle Assess 8(4):191–200

Franke M (1984) Umweltauswirkungen durch Getränkeverpackungen – Systematik zur Ermittlung der Umweltauswirkungen von komplexen Prozessen am Beispiel von Einweg-und Mehrweg-Getränkebehältern. TechnischeUniversität Berlin, Institut für Technischen Umweltschutz, Berlin

Goedkoop MJ, Spriensma R (2000) Eco-indicator 99, a damage oriented method for life cycle impact assessment, methodology report (update April 2000)

Goedkoop M, Heijungs R, Huijbregts MAJ, De Schryver A, Struijs J, Van Zelm R (2009) ReCiPe 2008 A life cycle impact assessment method which comprises harmonised category indicators at the midpoint and the endpoint level. Report I: characterisation, 1st edn. 6 Jan 2009. http://www.lcia-recipe.net

Guinée JB (ed), Gorrée M, Heijungs R, Huppes G, Kleijn R, de Koning A van Oers L, Wegener Sleeswijk A, Suh S, Udo de Haes HA, de Bruijn JA, van Duin R and Huijbregts MAJ (2002) Handbook on Life Cycle Assessment: operational guide to the ISO standards. Eco-efficiency in industry and science. Kluwer Academic Publishers, Dordrecht (Hardbound, ISBN 1-4020-0228-9; Paperback, ISBN 1-4020-0557-1)

Hauschild MZ, Pennington D (2002) Indicators for ecotoxicity in life cycle impact assessment. Chapter 6. In: de Haes HA U, Finnveden G, Goedkoop M, Hauschild MZ, Hertwich E, Hofstetter P, Klöpffer W, Krewitt W, Lindeijer E, Jolliet O, Mueller-Wenk R, Olsen S, Pennington D, Potting J, Steen B (eds) Life cycle impact assessment: striving towards best practice. SETAC Press, Pensacola. ISBN 1-880611-54-6

Hauschild M, Potting J (2005) Spatial differentiation in life cycle impact assessment – the EDIP2003 methodology. Environmental News No. 80. The Danish Ministry of the Environment, Environmental Protection Agency, Copenhagen

Hauschild MZ, Huijbregts M, Jolliet O, MacLeod M, Margni M, van de Meent D, Rosenbaum RK, McKone T (2008) Building a model based on scientific consensus for life cycle impact assessment of chemicals: the search for harmony and parsimony. Environ Sci Technol 42 (19):7032–7037

Hauschild MZ, Goedkoop M, Guinée J, Heijungs R, Huijbregts M, Jolliet O, Margni M, De Schryver A, Humbert S, Laurent A, Sala S, Pant R (2013) Identifying best existing practice for characterization modelling in life cycle impact assessment. Int J Life Cycle Assess 18(3):683–697

Heijungs R, Guinée J, Huppes G, Lankreijer RM, Udo de Haes HA, Wegener Sleeswijk A, Ansems AMM, Eggels PG, van Duin R, Goede HP (1992) Environmental life cycle assessment of products: guide and backgrounds, october 1992. Centre of Environmental Science, Leiden University, Leiden

Herrmann IT, Jørgensen A, Bruun S, Hauschild MZ (2013) Potential for optimized production and use of rapeseed biodiesel. Based on a comprehensive real-time LCA case study in Denmark with multiple pathways. Int J Life Cycle Assess 18(2):418–430

Hertwich E, Hammitt JK, Pease W (2000) A theoretical foundation for life-cycle assessment: recognizing the role of values in environmental decision making. J Ind Ecol 4:13–28

Hofstetter P (ed) (1998) Perspectives in life cycle impact assessment. A structured approach to combine models of the technosphere, ecosphere and valuesphere. Kluwer Academic Publishers, Dordrecht

Hofstetter P, Baumgartner T, Scholz RW (2000) Modeling the valuesphere and the ecosphere: integrating the decision makers' perspectives into life cycle assessment. Int J Life Cycle Assess 5:161–175

Huijbregts MAJ (1998) Application of uncertainty and variability in LCA. Part I: a general framework for the analysis of uncertainty and variability in life cycle assessment. Int J Life Cycle Assess 3:273–280

Huijbregts MAJ, Helweg S, Hertwich E (2011) Do we need a paradigm shift in life cycle impact assessment? Environ Sci Technol 45:3833–3834

Hunt RG, Franklin WE, Welch RO, Cross JA, Woodall AE (1974) Resource and environmental profile analysis of nine beverage container alternatives, United States Environmental Protection Agency (US EPA), Office of Solid Waste Management Programs, EPA/530/SW-91c, Washington, DC

Huppes G, Curran MA (2012) Environmental life cycle assessment: background and perspective. In: Curran MA (ed) (2012) Life cycle assessment handbook – A guide for environmentally sustainable products. Scrivener Publishing, Salem, MA, and Wiley, Hoboken

ISO 14040 (1997) Environmental management – life cycle assessment – principles and framework. International Standards Organization, Geneva

ISO 14040 (2006) Environmental management – life cycle assessment – principles and framework. International Standards Organization, Geneva

ISO 14041 (1998) Environmental management – life cycle assessment – goal and scope definition and inventory analysis. International Standards Organization, Geneva

ISO 14042 (2000) Environmental management – life cycle assessment – life cycle impact assessment. International Standards Organization, Geneva

ISO 14043 (2000) Environmental management – life cycle assessment – life cycle interpretation. International Standards Organization, Geneva

ISO 14044 (2006) Environmental management – life cycle assessment – requirements and guidelines. International Standards Organization, Geneva

Itsubo N, Sakagami M, Washida T, Kokubu K, Inaba A (2004) Weighting across safeguard subjects for LCIA through the application of conjoint analysis. Int J Life Cycle Assess 9:196–205

Jolliet O, Margni M, Charles R, Humbert S, Payet J, Rebitzer G, Rosenbaum R (2003) IMPACT 2002+: a new life cycle impact assessment methodology. Int J Life Cycle Assess 8:324–330

Jolliet O, Frischknecht R, Bare J, Boulay A-M, Bulle C, Fantke P, Gheewala S, Hauschild M, Itsubo N, Margni M, McKone TE, Mila I, Canals L, Posthuma L, Prado-Lopez V, Ridoutt B, Sonnemann G, Rosenbaum RK, Seager T, Struijs J, Van Zelm R, Vigon B, Weisbrod A (2014) Global guidance on environmental life cycle impact assessment indicators: findings of the scoping phase. Int J Life Cycle Assess 19(4):962–967

Klöpffer W (2006) The role of SETAC in the development of LCA. Int J Life Cycle Assess 11 (Special Issue 1):116–122

Lundholm MP, Sundström G (1985) Resource and environmental impact of Tetra Brik carton and refillable and non-refillable glass bottles, Tetra Brik aseptic environmental profile. AB Tetra Pak, Malmö

Margni M, Gloria T, Bare J, Seppälä J, Steen B, Struijs J, Toffoletto L, Jolliet O (2007) Guidance on how to move from current practice to recommended practice in life cycle impact assessment: UNEP/SETAC Life Cycle Initiative

Pant R, Van Hoof G, Schowanek D, Feijtel TCJ, de Koning A, Hauschild M, Pennington DW, Olsen SI, Rosenbaum R (2004) Comparison between three different LCIA methods for aquatic ecotoxicity and a product environmental risk assessment – insights from a detergent case study within OMNIITOX. Int J Life Cycle Assess 9:295–306

Potting J, Hauschild MZ (1997) Predicted environmental impact and expected occurrence of actual environmental impact. Part 2: spatial differentiation in life-cycle assessment via the site-dependent characterisation of environmental impact from emissions. Int J Life Cycle Assess 2(4):209–216

Potting J, Hauschild M (2006) Spatial differentiation in life cycle impact assessment – a decade of method development to increase the environmental realism of LCIA. Int J Life Cycle Assess 11 (Special Issue 1):11–13

Rosenbaum RK, Bachmann TM, Gold LS, Huijbregts MAJ, Jolliet O, Juraske R, Köhler A, Larsen HF, MacLeod M, Margni M, McKone TE, Payet J, Schuhmacher M, van de Meent D, Hauschild MZ (2008) USEtox – the UNEP-SETAC toxicity model: recommended characterization factors for human toxicity and freshwater ecotoxicity in life cycle impact assessment. Int J Life Cycle Assess 13:532–546

Spadaro JV, Rabl A (2002) Air pollution damage estimates: the cost per kilogram of pollutant. Int J Risk Assess Manage 3(1):75–98

Steen B (1999a) A systematic approach to environmental priority strategies in product development (EPS). Version 2000-general system characteristics; CPM report 1999:4, Chalmers University of Technology, Gothenburg

Steen B (1999b) A systematic approach to environmental priority strategies in product development (EPS). Version 2000-Models and data of the default method; CPM report 1999:5, Chalmers University of Technology, Gothenburg

Toffoletto L, Bulle C, Godin J, Reid C, Deschênes L (2007) LUCAS – a new LCIA method used for a Canadian-specific context. Int J Life Cycle Assess 12:93–102

Udo de Haes HA, Jolliet O, Finnveden G, Hauschild M, Krewitt W, Müller-Wenk R (1999a) Best available practice regarding impact categories and category indicators in life cycle impact assessment (1). Int J Life Cycle Assess 4:66–74

Udo de Haes HA, Jolliet O, Finnveden G, Hauschild M, Krewitt W, Müller-Wenk R (1999b) Best available practice regarding impact categories and category indicators in life cycle impact assessment (2). Int J Life Cycle Assess 4:167–174

Van Zelm R, Huijbregts MAJ (2013) Quantifying the trade-off between parameter and model structure uncertainty in life cycle impact assessment. Environ Sci Technol 47:9274–9280

Wenzel H, Hauschild MZ and Alting L (1997) Environmental assessment of products, vol 1, Methodology, tools and case studies in product development, 544 p. Chapman & Hall, London, and Kluwer Academic Publishers, Hingham. ISBN 0 412 80800 5

Chapter 2
Selection of Impact Categories and Classification of LCI Results to Impact Categories

Jeroen B. Guinée

Abstract This chapter concerns 'selection of impact categories' and 'assignment of LCI results to selected impact categories (classification)'. These elements are the first two mandatory elements of Life Cycle Impact Assessment (LCIA). They have largely been developed during the 1990s. In practice these mandatory steps are often performed using default lists of impact categories and default lists of inventory items classified to these default impact categories as part of LCA handbooks, guides and software tools. Despite these default lists, it is still important to pay sufficient attention to both these steps in any LCA case study. Every practitioner of LCA will always need to justify the completeness of default lists of impact categories and default classification lists for their study. In addition, the handling of missing information needs to be reported explicitly and transparently, and needs to be taken into consideration when developing conclusions and recommendations for the study at stake. After the 1990s, the attention to selection of impact categories and classification in LCA methodology studies and papers has been limited. Still, there are issues that deserve further attention from LCA method developers such as the harmonisation of naming impact categories while distinguishing or not between names for midpoint impact categories and names for endpoint category indicators, keeping default lists of impact categories manageable, and the classification of inventory results that relate to more than one impact category.

Keywords Assignment to impact categories • Classification • Impact categories • LCA • LCIA • Life cycle assessment • Life cycle impact assessment • Selection of impact categories

J.B. Guinée (✉)
Department of Industrial Ecology, Institute of Environmental Sciences (CML), Leiden University, P.O. Box 918, 2300 RA Leiden, The Netherlands
e-mail: guinee@cml.leidenuniv.nl

M.Z. Hauschild, M.A.J. Huijbregts (eds.), *Life Cycle Impact Assessment*, LCA Compendium – The Complete World of Life Cycle Assessment, DOI 10.1007/978-94-017-9744-3_2

1 Introduction

According to ISO 14040 (1997, 2006a), the first two mandatory elements of LCIA concern:

- selection of impact categories, category indicators and characterisation models;
- assignment of LCI results to the selected impact categories (classification);

The result of these two elements is that LCI results like resource uses and emissions are assigned to different impact categories such as for example acidification; see Fig. 2.1.

From the early 1980s until today, LCIA rapidly evolved from a simple first impact assessment method, where airborne and waterborne emissions were divided by semi-politically set limits for those emissions and aggregated into so-called critical volumes of air and critical volumes of water, to full-fledged fate-exposure-impact based assessment methods of today. The developments in the field of selection of impact categories and classification on the other hand were much less spectacular. In Sect. 2, we will first briefly discuss the history of these two mandatory elements. Subsequently, we will discuss the purpose of selection and classification (Sect. 3), the choice of impact categories (Sect. 4), how LCI results are assigned to impact categories (Sect. 5), and finally we will discuss some potential research needs and expected future developments (Sect. 6).

Similarities and differences between LCA handbooks, guides and other method proposals will be discussed within particularly Sects. 3, 4 and 5.

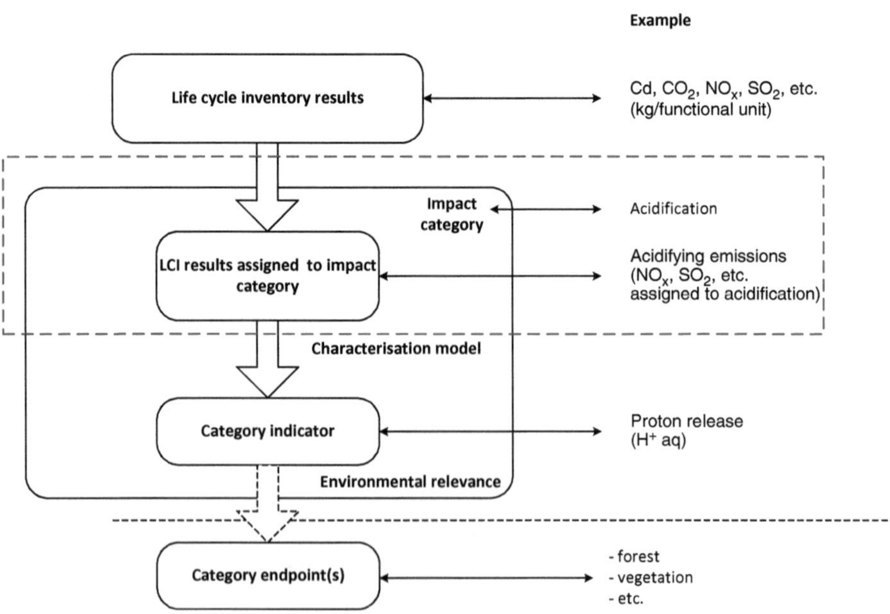

Fig. 2.1 The conceptual framework for defining category indicators (Slightly adapted from: ISO 14042 2000)

2 History of Selection of Impact Categories and Classification

ISO does not provide explicit definitions for selection of impact categories and classification. One may say that for this, ISO built on the preparatory work done by Udo de Haes (1996) who wrote in his report that selection of impact categories addresses the topic of defining impact categories: "the types of impact of the given interventions (elementary flows) are identified and a number of relevant impact categories are defined which cover impacts as much as possible". According to Udo de Haes (1996) classification addresses "the assignment of the environmental interventions (LCI results) to the defined categories" and this is basically also the definition that ISO 14042 adopted implicitly (ISO 14042 2000).

Although the topics covered were the same, selection of impact categories and classification used to be referred to as just 'classification' in the 1990s. The first time the term 'classification' was introduced with this meaning was as part of the work on the first Dutch LCA Guide (Heijungs et al. 1992). They defined classification as "the third component of a life cycle assessment in which the contribution made by the environmental interventions to the potential environmental effects is determined through model-based calculations". Note that classification at that time thus included selection of impact categories and classification, but also selection of category indicators and characterisation models and the characterisation itself, although the terminology for all of these elements was also different at that time.

One of the first workshops on LCA was held in Leuven (de Smet 1990) in 1990 and shortly after LCA as a topic was embraced by the Society of Environmental Toxicology and Chemistry (SETAC). SETAC started playing a leading and coordinating role in bringing LCA practitioners, users and scientists together to collaborate on the continuous improvement and harmonisation of LCA framework, terminology and methodology and a first workshop on LCA was held in Leiden, The Netherlands in 1992 (Anonymous 1992a). The workshop agreed to a three-step approach to the Classification phase: "(1) a classification based on main processes of the relevant effect chains, (2) the definition of units to measure these classes of effects and (3) an aggregation of the effects in terms of these units" (Udo de Haes 1992; Guinée 1992). Also as part of this workshop both Guinée (1992) and Baumann et al. (1992) proposed a list of environmental problems or environmental effects (later on referred to as 'impact categories'), which at the workshop were merged into a first common list of 'headings for classification' (impact categories; Udo de Haes 1992). This first common list was largely (with minor adaptations) adopted in the LCA Guide by Heijungs et al. (1992) and a short list was adopted in the LCA book of the Nordic Council (Anonymous 1992b) later on.

Shortly after the SETAC-Europe Leiden workshop, SETAC-US organised a similar exercise at the SETAC Sandestin workshop Florida in 1992 (Fava et al. 1993). The Sandestin workshop participants adopted the three-step approach to classification as agreed upon in the Leiden workshop, but proposed different names for these steps (classification, characterisation and valuation) and a new name for the Classification phase: Life Cycle Impact Assessment. This workshop also introduced the term 'stressor', i.e. sets of conditions that may lead to human health, ecological and resource depletion impacts. In addition the workshop introduced the term 'impact categories' but it referred to what was later called 'areas of protection': ecological health, human health, resource depletion and social welfare. Main question then was how this division was related to the common list of 'headings for classification' discussed during the Leiden workshop (Udo de Haes 1992).

In 1993, the SETAC Code of Practice (CoP, Consoli et al. 1993) combined the results from the Sandestin and Leiden workshops. One of the main aims of the CoP was to define a common methodological framework in order to streamline further methodological discussions and progress. The CoP adopted the Sandestin term 'Life Cycle Impact Assessment' (LCIA) for this phase of LCA and the Sandestin terms for its three steps: classification, characterisation and valuation. The other problem that had to be solved in the Code of Practice was the apparent contrast between the protection areas of the Sandestin workshop 1992 (Fava et al. 1993) and the common list of 'headings for classification' discussed during the Leiden workshop (Udo de Haes 1992). The CoP concluded that this was indeed an apparent contrast and not a principal one. Integration of protection areas and 'headings for classification' seemed possible by defining a matrix with one axis representing the protection areas and the other axis representing the impact categories (see Table 2.1). Note that in the CoP in this way the definition of impact categories as adopted by the participants of the Sandestin workshop was changed into 'general areas of protection', and that the Leiden 'headings of classification' was changed into 'impact categories'. A default list of relevant impact categories for LCA studies was not drafted in the CoP. Last but not least, the CoP also provided a

Table 2.1 Relationship between general areas of protection and specific impact categories

	General areas of protection		
Specific impact categories (examples)	Resource	Human health	Ecological health
Resource depletion			
Depletion of abiotic resources	+		
Depletion of biotic resources	+		
Pollution			
Global warming		(+)	+
Ozone depletion		(+)	(+)
Human toxicity		+	
Ecotoxicity		(+)	+
Photochemical oxidant formation		+	+
Acidification		(+)	+
Eutrophication			+
Degradation of ecosystems and landscape			
Land use			+

From Consoli et al. (1993)

slightly modified definition for the classification step: "the classification is the step in which the data from the inventory analysis [...] are grouped together into a number of impact categories".

Developments, of course, continued after the CoP, but changes to the CoP remained marginal until 1996. As mentioned above, Udo de Haes (1996) distinguished for the first time between selection of impact categories and classification and this was then adopted by ISO 14040–14044 (2006a, b) series of LCA Standards. The ISO Standards provided the requirements for these two steps, and these were copied into most LCA handbooks and guides afterwards without any significant modifications.

ISO describes procedures rather than specific default lists, methodologies or models for LCIA, implying that any impact category, methodology or model is acceptable as long as it satisfies the general ISO criteria. This left sufficient space for further elaboration of particularly two main topics related to 'selection of impact categories' and 'classification':

- The choice of impact categories including
  - the way of defining relevant impact categories for an LCA study: should we do that case by case or should we aim for a default list of impact categories that is basically valid for all LCA studies?
  - the difference between midpoint and endpoint approaches.
- The assignment of LCI results to impact categories including topics as
  - what to do with inventory results that cannot be assigned (yet) to impact categories, and
  - how to handle inventory results that relate to more than one impact category.

These topics will be further discussed in Sects. 4 and 5, respectively. First we will briefly discuss the purpose of selection of impact categories.

3 The Purpose of Selection of Impact Categories and Classification

In the impact assessment phase the results of the inventory analysis are transposed into contributions to relevant impact categories. To this end, relevant impact categories must be identified. This can be done case by case or can be facilitated by defining a default list of impact categories, with a possible distinction between 'baseline' impact categories, 'study-specific' impact categories and 'other' impact categories. The purpose of the ISO-element 'Selection of impact categories' is to compel LCA practitioners to explicitly select those categories relevant to the goal of their particular study. LCA practitioners generally do this, mostly as part of the Goal and Scope definition.

In the classification, the inventory results are assigned on a purely qualitative basis to the various pre-selected impact categories. Again, this can be done on a case by case basis or can be facilitated by defining a default list of elementary flows, for which characterisation factors have previously been derived. The classification step then involves no actual work as these elementary flows have already been assigned to the various impact categories. In the case of other elementary flows the practitioner will have to adopt an appropriate procedure of his own. The purpose of the ISO-element 'Classification' is to compel LCA practitioners to explicitly assign inventory results to impact categories. For some impact categories with a limited number of contributing flows, such as global warming/climate change, stratospheric ozone depletion, acidification and eutrophication, assigning inventory results to impact categories has in practice been solved. For other impact categories with a huge number of contributing flows, such as the toxicity-related impact categories, the task remains with the practitioner to assign the relevant flows to these categories, which in practice is often not explicitly or appropriately done.

Selection of impact categories and classification are mandatory elements of LCIA, which force the practitioner to make explicit choices on impact categories considered and not considered and on inventory results assigned and not assigned.

4 Selection of Impact Categories

According to ISO 14044 (2006b), the necessary 'selection of impact categories' components for each impact category includes:

- identification of the category endpoint(s),
- identification of appropriate LCI results that can be assigned to the impact category, taking into account the chosen category indicator and identified category endpoint(s)

ISO 14044 (2006b) states that this procedure "facilitates the collection, assignment, and modelling of appropriate LCI results" and "helps to highlight the scientific and technical validity, assumptions, value-choices and degree of accuracy in the model". It further states that "the category indicator can be chosen anywhere along the environmental mechanism between the LCI results and the category endpoint(s)".

With respect to the selection of impact categories, ISO defined the following requirements:[1]

- the selection of impact categories [...] shall be consistent with the goal and scope of the LCA study;
- the sources for impact categories [...] shall be referenced;

[1] Text referring to selection of category indicators and models has been left out of the requirements (indicated as [...]) as this Chapter does not deal with these topics.

- the selection of impact categories […] shall be justified;
- accurate and descriptive names shall be provided for the impact categories […];
- the selection of impact categories shall reflect a comprehensive set of environmental issues related to the product system being studied taking the goal and scope into consideration.

In addition a number of recommendations is given for the selection of impact categories, indicators and models:

- the impact categories […] should be internationally accepted i.e. based on an international agreement or approved by an international body;
- the impact categories should represent the aggregated emissions or resource use of the product system on the category endpoint(s) through the indicators;
- value-choices and assumptions made during the selection of impact categories […] should be minimised;
- the impact categories […] should avoid double counting unless required by the Goal and scope definition, for example when the study includes both human health and carcinogenicity.

Based on these ISO requirements and recommendations developers of LCA handbooks, guides and LCIA methods have come up with proposals for impact categories and have elaborated default lists of impact categories. There are hardly any institutes or practitioner(s) that develop impact categories on a case by case basis. Below, we will first discuss different ways of defining impact categories (e.g., midpoint and endpoint approaches) and then discuss different defaults lists proposed for midpoint and for endpoint approaches.

4.1 *Different Ways of Defining Impact Categories*

Based on ISO, a "category indicator can be chosen anywhere along the environmental mechanism between the LCI results and the category endpoint(s)". It is striking that ISO doesn't mention that the impact category can also be selected anywhere along the environmental mechanism between the LCI results and the category endpoint(s), as long as it is consistently chosen in relation to the corresponding category indicator(s). This has caused some confusion as from the ISO Standards one could read that names and definitions of impact categories remain the same, while category indicators can be different and also defined at different point along the environmental mechanism. A number of method-developers and practitioners have done otherwise and differentiated between midpoint impact categories (global warming, acidification, eutrophication, ozone layer depletion, etc.) and endpoint category indicators (for example, damage to human health, damage to ecosystem quality and loss of resources; endpoint impact categories are often formulated in terms of 'damage to' protection areas with all possible related confusion (Klöpffer and Grahl 2009, 2014)); the UNEP-SETAC group on LCIA even provided separate definitions for 'midpoint impact category' and 'damage impact category' (Jolliet et al. 2003a). Others have held on to the ISO-line of

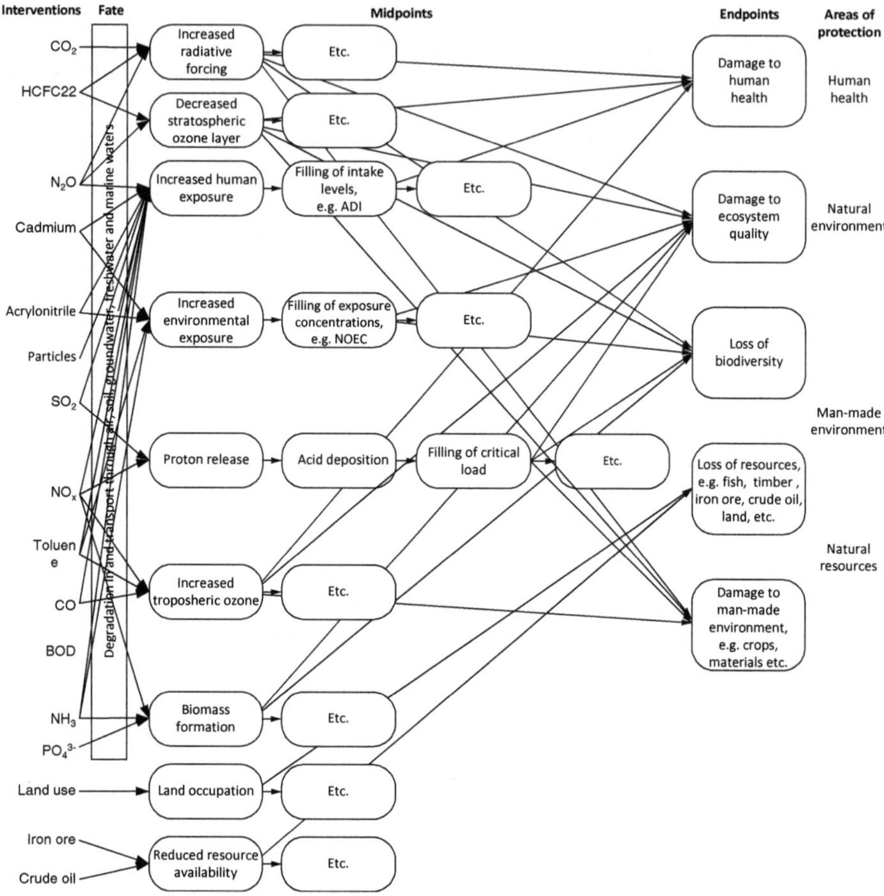

Fig. 2.2 Overview of the causal relationships between elementary flows (environmental interventions), midpoints and (category) endpoints (Freely adapted from figure presented by Udo de Haes et al. 1999)

thinking and adopted equal impact category names for both midpoint and endpoint indicator approaches (e.g., Itsubo and Inaba 2012).

One of the major characteristics of impact assessment methods is thus the point in the environmental mechanism at which the category indicators defined. As illustrated by Fig. 2.2, they may be defined close to the elementary flow or intervention (the midpoint, or problem-oriented approach, e.g. Heijungs et al. 1992; Udo de Haes 1996; Wenzel et al. 1997; Hauschild and Wenzel 1998; Udo de Haes et al. 1999; Jolliet et al. 2003b; Guinée et al. 2002; Hauschild and Potting 2005; Toffoletto et al. 2007; Frischknecht et al. 2009; Bare 2011; EC-JRC 2011; Goedkoop et al. 2012). The guiding principle in defining midpoint impact categories to that the midpoint should (ideally) be selected at the earliest point in the impact pathway beyond which the environmental processes are the same for all substances classified to that impact category (EC-JRC 2011; Goedkoop et al. 2012; Hauschild et al. 2013). Alternatively, they may be defined at the level of category endpoints (the endpoint, or damage approach, e.g. Steen 1999a, b; Goedkoop and Spriensma 2000; Itsubo and Inaba 2012). A cluster of category endpoints of recognisable value to society is referred to as an 'area of protection', for example human health, natural resources, the natural environment and the man-made environment. Different definitions of areas of protection exist (cf. Udo de Haes et al. 1999, Steen 1999a, b).

The result of the distinction between midpoint and damage or endpoint approaches is that there are default lists of midpoint impact categories and default lists of damage or endpoint impact categories. These different lists will be discussed below.

4.2 Default List of Impact Categories

In order to facilitate the LCA practitioner's work with easy-to-apply LCIA methods, several authors have developed default lists of impact categories, often supported by default classification lists (see Sect. 5.1). The definition of impact categories can – as also illustrated by Fig. 2.2 – be done in several ways either or not representing a specific midpoint or endpoint modelled. Several authors have developed different default lists of impact categories, of which a selection is presented in Table 2.2.

Table 2.2 shows that the diversity and overlap in impact categories among different methods is huge (Klöpffer and Grahl 2009, 2014). Sometimes it is merely a matter of semantics and the differences are likely to be minor, although that cannot be definitely concluded from the name of the impact category alone. For that, we need a more detailed description of, for example, the inventory items included and excluded. Sometimes differences are much more fundamental. For example, the way that EPS2000 (Steen 1999a, b) defines midpoint impact categories is fundamentally different from the way that Guinée et al. (2002), Hauschild and Potting (2005) and Bare (2011) have done. EPS2000 basically is an endpoint-approach and derives midpoint impact categories from endpoints. Guinée et al. (2002), Hauschild and Potting (2005) and Bare (2011) define midpoint impact categories basically from problem-oriented cross-media approach adopting environmental themes as defined in environmental policy (Anonymous 1992a). Despite the ISO-recommendation to minimise value-choices and assumptions during the selection of impact categories, the selection and definition of relevant problem fields is highly value-laden (Steen 1999a, b) as is also shown by Udo de Haes (1992): "A related question was which problem fields should be taken into account: which problems are to be regarded as environmental problems? [...] Thus, product safety was considered to be outside the scope of LCA. The opinions differed about occupational health [...]." Such normative discussion will continue and therefore 'default lists' may always change again, if it is only for the fact that new insight and problems may rise. Sometimes the differences are subtle but not futile, like in the case for resources where some approaches adopt just one category for abiotic

Table 2.2 Default lists of midpoint impact categories

Midpoint categories	1	2	3	4	5	6	7	8	9	10	11
1. Abiotic resource depletion								X			
2. Depletion of abiotic resources				X							X
3. Depletion of biotic resources				X							
4. Depletion of element reserves (element)						X					
5. Depletion of fossil reserves (coal)						X					
6. Depletion of fossil reserves (gas)						X					
7. Depletion of fossil reserves (oil)						X					
8. Depletion of mineral reserves (ore)						X					
9. Mineral extraction	X										
10. Mineral resources consumption							X				
11. Gravel									X		
12. Energy resources									X		
13. Non-renewable energy	X										
14. Fossil fuel consumption							X				
15. Fossil fuel depletion										X	
16. Resource depletion, water											X
17. Resources									X		
18. Fish and meat production capacity						X					
19. Crop production capacity						X					
20. Wood production capacity						X					
21. Forest resources consumption							X				
22. Freshwater									X		
23. Land competition				X							
24. Land occupation	X										
25. Agricultural land occupation		X									
26. Natural land transformation		X									
27. Rural land occupation		X									
28. Land use							X	X	X		X
29. Impacts of land use				X							
30. Share of species extinction [NEX]						X					
31. Loss of biodiversity				X							
32. Loss of life support functions				X							
33. Global warming/climate change	X	X		X	X		X	X	X	X	X
34. (Stratospheric) ozone (layer) depletion (destruction)	X	X		X	X		X	X	X	X	X
35. Photochemical oxidation (formation)/photochemical ozone formation/photochemical smog (formation)	X	X		X	X		X	X		X	X
36. Volatile organic compounds (NMVOCs)									X		
37. Acidification				X	X		X	X	X	X	X

(continued)

Table 2.2 (continued)

Midpoint categories	1	2	3	4	5	6	7	8	9	10	11
38. Base cation capacity [H⁺]						X					
39. Terrestrial acidification/ nutrification	X	X									
40. Eutrophication				X			X			X	
41. Eutrophication, aquatic											X
42. Aquatic eutrophication/freshwater eutrophication	X	X			X			X			
43. Terrestrial eutrophication					X			X			X
44. Marine eutrophication		X									
45. Nitrogen (nitrate)									X		
46. Freshwater aquatic ecotoxicity				X							X
47. Groundwater emissions									X		
48. Hazardous wastes in underground landfills									X		
49. Heavy metals									X		
50. Human health cancer										X	
51. Human health criteria pollutants										X	
52. Human health non-cancer										X	
53. Human toxicity	X	X		X	X		X	X			
54. Human toxicity cancer											X
55. Human toxicity non-cancer											X
56. Particulate matter formation		X									
57. Particulate matter/respiratory inorganics											X
58. PM10 and diesel soot									X		
59. Respiratory effects	X							X			
60. Life expectancy						X					
61. Endocrine disruptors									X		
62. Soil emissions									X		
63. Surface water emissions									X		
64. Urban area air pollution							X				
65. Indoor air contamination							X				
66. Impacts of ionising radiation				X							
67. Ionising radiation, ecosystems											X
68. Ionising radiation, human health											X
69. Ionizing radiation	X	X									
70. Aquatic ecotoxicity/freshwater ecotoxicity	X	X									
71. Ecotoxicity (aquatic and terrestrial)				X	X		X	X		X	
72. Freshwater sediment ecotoxicity				X							
73. Terrestrial ecotoxicity	X	X		X							
74. Marine ecotoxicity		X		X							

(continued)

Table 2.2 (continued)

Midpoint categories	1	2	3	4	5	6	7	8	9	10	11
75. Marine sediment ecotoxicity				X							
76. Malodourous air				X							
77. Malodourous water				X							
78. Carbon in bio-reactive landfills									X		
79. Casualties				X							
80. Desiccation				X							
81. Noise				X	X		X		X		
82. Plant protection products									X		
83. Production capacity for water (drinking water)						X					
84. Radioactive									X		
85. Radioactive wastes in final repositories									X		
86. Waste(s)							X		X		
87. Morbidity						X					
88. Severe morbidity and suffering						X					
89. Nuisance						X					
90. Severe nuisance						X					
91. Waste heat				X							

1 IMPACT2002+ (Jolliet et al. 2003b), *2* ReCiPe (Goedkoop et al. 2012), *3* EI99 (Goedkoop and Spriensma 2000), *4* CML 2002 (Guinée et al. 2002), *5* EDIP 2003 (Hauschild and Potting 2005), *6* EPS 2000 (Steen 1999a, b), *7* LIME2 (Itsubo and Inaba 2012), *8* LUCAS (Toffoletto et al. 2007), *9* Swiss Ecoscarcity 2006 (Frischknecht et al. 2009), *10* TRACI 2.0 (Bare 2011), *11* ILCD (EC-JRC 2011)

resources and other define categories for subclasses of abiotic resources including fossil fuels or different fossil fuels and minerals.

As a sort of reaction to and supplement of default lists of impact categories, several proposals for new impact categories have been published. These new proposals often address specific sectors of LCA studies. For example, in aquaculture and fisheries LCA studies, proposals for the following impact categories have been drafted over the past 10 years: biotic resource use (Papatryphon et al. 2004; Pelletier et al. 2007); water dependency (Aubin et al. 2009; d'Orbcastel et al. 2009); the area altered by farm waste, changes in nutrient concentration in the water column, the percentage of carrying capacity reached, the percentage of total anthropogenic nutrient release, release of wastes into freshwater, the number of escaped salmon, the number of reported disease outbreaks, parasite abundance on farms, and the percentage reduction in wild salmon survival (Ford et al. 2012). These proposals are often made by sector-experts who have very good and specific knowledge of their field and aim for better and more site-specific impact assessments requiring additional site-specific data. As a consequence, the methods proposed are generally not applicable or feasible for the average LCA case study.

Table 2.3 Default lists of damage or endpoint impact categories

Damage or endpoint categories	1	3	6	7
1. (Damage to) human health	X	X	X	
2. (Damage to) ecosystem quality	X	X		
3. Climate change (life support systems)	X			
4. Resources	X	X	X	
5. Ecosystem production			X	
6. Biodiversity			X	
7. Urban area air pollution				X
8. Global warming				X
9. Ozone layer destruction				X
10. Toxic chemicals (human toxicity)				X
11. Biological toxicity (ecotoxicity)				X
12. Acidification				X
13. Eutrophication				X
14. Photochemical oxidant				X
15. Land use				X
16. Mineral resources consumption				X
17. Fossil fuel consumption				X
18. Forest resources consumption				X
19. Indoor air contamination				X
20. Noise				X
21. Waste				X

1 IMPACT2002+ (Jolliet et al. 2003b), *3* EI99 (Goedkoop and Spriensma 2000), *6* EPS 2000 (Steen 1999a, b), *7* LIME2 (Itsubo and Inaba 2012)

Different default lists of impact categories have also been developed for damage or endpoint approaches and these are shown in Table 2.3.

Table 2.3 shows that the diversity in damage or endpoint impact categories among different methods is significant but not as huge as for midpoint impact categories. The differences are actually dominated by the approach taken in LIME. LIME chose to use the same names for midpoint and endpoint impact categories elaborating different indicators for each of them, which results in two similar lists of LIME impact categories in Tables 2.2 and 2.3.

Bare and Gloria (2008) observed that there is a need to discuss the range of impacts which could and should be included. In their paper, they present a meta-model to facilitate an expanded discussion of the taxonomy of impact, impact category, midpoint, endpoint, damage, etc. Their taxonomy meta-model includes the existing impacts found in LCIA literature, and then expands to be more comprehensive and includes a larger set of impacts than are normally included within LCIA. The taxonomy meta-model represents a first attempt to facilitate a standard vocabulary and structure in the field of LCA impact category discussions, and is to help ensure that the selection of impact categories is truly comprehensive.

5 Assignment of LCI Results to Impact Categories (Classification)

In the classification, the inventory results (elementary flows including resource uses, land uses, and all kind of chemical emissions) are assigned on a purely qualitative basis to the various selected impact categories. When working with a default list of elementary flows, for which characterisation factors have previously been derived, the actual work to be done by an LCA-practitioner as part of the classification step is significantly reduced. All inventory results have then – as far as scientific knowledge and data allows so – been pre-classified to pre-selected impact categories. Several LCA handbooks, guides and software tools listed above provide such default lists. For generally acknowledged and well-defined impact categories with a limited number of contributing flows, such as global warming/climate change, stratospheric ozone depletion, acidification and eutrophication, the default classification lists are highly similar between different LCIA methods. This may be different for other impact categories. Particularly for human and ecotoxicity related impact categories the coverage of chemicals not only differs significantly between methods, but is by definition far from complete as there are more than 100,000 different substances known on the so-called EINECS (**E**uropean **IN**ventory of **E**xisting **C**ommercial chemical **S**ubstances) list. Hauschild and Wenzel (1998) developed a screening tool supporting the classification of substances contributing to human or ecotoxicity. Based on some key characteristics of a substance, it is considered potentially toxic or not and is classified as such.

Despite default classification lists, there is always some work left for the LCA practitioner in the classification step. For example, for inventory results for which no pre-classification for any of the pre-selected impact categories is available, the practitioner will have to adopt an appropriate procedure of his own. Either the practitioner then develops additional characterisation factors for elementary flows for which characterisation factors are lacking and for which certain impacts are known, or inputs and outputs with lacking characterisation factors are reported separately from the characterisation results.

With respect to classification, ISO defined the following recommendations:

- Assignment of LCI results to impact categories should consider the following, unless otherwise required by the goal and scope:
  - assignment of LCI results that are exclusive to one impact category;
  - identification of LCI results that relate to more than one impact category, including
    - distinction between parallel mechanisms (e.g. SO_2 is apportioned between the impact categories of human health and acidification), and
    - assignment to serial mechanisms (e.g. NO_x can be classified to contribute to both ground-level ozone formation and acidification).

The second point is about how to handle LCI results that relate to more than one impact category and this topic will be discussed below separately first (Sect. 5.1). Subsequently, we will discuss the handling of elementary flows for which impact categories/characterisation factors are lacking (Sect. 5.2), for which ISO also provides guidance. Finally we will discuss how to handle missing information.

The ILCD handbook (EC-JRC 2010) adds to this by stating that "the practitioner is [...] responsible to ensure that the inventory elementary flows are correctly linked with the LCIA factors [...] and [...] to derive or develop missing impact factors if potentially relevant for the study". The ILCD handbook also explicitly mentions the fact that LCA practitioners are responsible for checking the completeness of default classification lists for their study, particularly for 'newly created or imported elementary flows' (i.e., elementary flows that were not on the default classification list and newly created by a practitioner or imported from another database, etc.). The ILCD handbook concludes that "it is one of the most widely found errors to not classify and characterise newly introduced flows despite of their environmental relevance".

5.1 Identification of LCI Results That Relate to More Than One Impact Category

Guinée (1995), Lindfors et al. (1995), Udo de Haes (1996) and Wenzel et al. (1997) discuss the topic of inventory results that relate to more than one impact category. They conclude that this topic mainly relates to multiple impacts of chemical releases and together they distinguish the following four categories of emissions (Guinée et al. 2002):

- Emissions with parallel impacts, i.e. emissions of substances that may theoretically contribute to more than one impact category but in practice only to one, e.g. an emission of SO_2 which may have either toxic or acidifying impacts.
- Emissions with serial impacts, i.e. emissions of substances that may in practice have successive impacts, e.g. emissions of heavy metals which may first have eco-toxicological impacts and subsequently, via food chains, impacts on human health.
- Emissions with indirect impacts, i.e. emissions of substances having a primary impact that in turn leads to one or more secondary impacts, e.g. aluminium toxicity induced by acidification, or methane contributing to photo-oxidant formation, with the produced ozone contributing in turn to climate change, which in turn may contribute to stratospheric ozone depletion.
- Emissions with combined impacts, i.e. emissions of substances having a mutual influence on each other's impacts, e.g. synergistic or antagonistic impacts of toxic substance mixes, or NO_x and VOC, both of which are required for photo-oxidant formation.

In order to avoid double counting, for emissions having parallel impacts, it is generally recommended in the literature that the respective contributions of such emissions to relevant impact categories be specified. However, no guidelines are available on how this task is to be performed. In general, such specification should be performed only in those cases where it really matters (where the contribution of the substance to one impact category substantially lessens its potential contribution to another, e.g. acidification or eutrophication by NH_3. Rough calculations show that SO_2, for example, is less relevant in this respect; see Heijungs et al. 1992). If it is unclear how such emissions are to be allocated, they are often assigned in their entirety to all relevant impact categories.

For emissions having serial and indirect impacts the literature generally recommends allocating such emissions in their entirety to all relevant (i.e. serial and indirect) impact categories unless characterisation factors for this purpose are lacking, as in the case of missing (indirect) GWP factors.

For emissions having combined impacts the literature generally recommends introducing assumptions regarding background concentrations of the other relevant substances. In practice this is currently only feasible for NO_x as a precursor in photo-oxidant formation, but not for synergistic or antagonistic impacts of toxic substance mixes, as knowledge on these issues is virtually entirely lacking.

Recently Ventura (2011) proposed two new approaches for classification and handling of inventory results that relate to more than one impact category: equiprobable classification and zone classification. Equiprobable classification is not entirely new since Guinée et al. (2002) already proposed it as one way of handling parallel impacts. Ventura (2011), however, also applies this principle to indirect impacts. The approach aims at avoiding double counting of impacts by simply equally dividing inventory results over all impact categories that they could potentially contribute to. Zone classification is a new approach and is based on two steps: (1) defining an impacted zone around the source, inside which the emitted chemicals are expected to majorly diffuse or spread, and (2) scoring the chemical to the occurrence of the chemical target inside the impacted zone (Ventura 2011). The zone classification approach has been applied to one process, which is not an LCA, and needs site-specific spatial data that are not available from general LCA databases. This approach is clearly not applicable for generic LCAs, and its feasibility in terms of data needs and if it works with functional unit based full LCAs will need to be tested further.

There has been notably little attention for this topic for the last decade or more (Reap et al. 2008) except for the proposals by Ventura (2011). This is due to the complexity of the problems involved including spatial and temporal dimensions of inventory results that are hard to cover in sufficient detail within LC(I)A. Depending on what is considered the prime purpose of LCA, this may be or may not be experienced as a problem. For example, if LCA is considered merely an encompassing systems analysis of environmental mass loadings (and not concentrations) based on the 'precautionary principle', the lack of progress – if possible at all – may not be considered that problematic.

5.2 Handling Missing Information

Finally, we discuss the issue of missing data or missing knowledge. In clause 4.4.2.5 ISO states that "After characterisation and before the optional elements described in 4.4.3, the inputs and outputs of the product system are represented, for example, by

- a discrete compilation of the LCIA category indicator results for the different impact categories referred to as an LCIA profile,
- a set of inventory results that are elementary flows but have not been assigned to impact categories e.g. due to lack of environmental relevance, and
- a set of data that does not represent elementary flows."

This implies that ISO requires inventory results that cannot be assigned to impact categories and data that do not represent elementary flows to be reported separately.

The main problem with missing information in the LCIA phase is that data are aggregated, reducing the number of data entries, but that data that are left out of this aggregation because of missing data or knowledge 'disappear' or 'get lost'.

The main problems in terms of missing information for the 'Selection of impact categories' and 'Classification of LCI results' steps concern:

- inventory results that cannot be assigned to an impact category;
- flows that are not specified in terms of environmental interventions, like energy, metals or solid waste.

The basic strategy is 'reporting'. Inventory results that cannot be assigned to an impact category but are assumed to be environmentally relevant, should be reported – in line with ISO clause 4.4.2.5 – separately (e.g. as 'missing important'; EC-JRC 2011), in addition to the LCIA category indicator results. Inventory results anticipated to be environmentally irrelevant may be excluded from the LCIA category indicator results, but this should be transparently justified in the LCA study report at stake, e.g. as 'missing unimportant' (EC-JRC 2010; Guinée et al. 2002).

Preferably, every effort should be made to avoid flows that are not specified in terms of environmental interventions. If that is not feasible, it again comes down to proper reporting. For example, all flows that cannot be specified in terms of elementary flows, should then be listed in a separate category including a qualitative (e.g. 'hazardous waste' and 'non-hazardous waste') and, wherever possible, quantitative (e.g. input of 10^{-12} truck) description (Guinée et al. 2002; EC-JRC 2011).

6 Conclusions and Research Recommendations

Selection of impact categories and classification are the first two mandatory steps of LCIA according to ISO. In practice these mandatory steps are implemented in default lists of impact categories and default lists of inventory items classified to the default impact categories, either as part of LCA handbooks, guides or software tools. These two steps have not always received proper attention in LCA case studies and also in LCA methodology studies or papers over the last decades.

For LCA case studies it is important to pay sufficient attention to both these steps. The selection of impact categories is often already made as part of the Goal and Scope definition. The selection of impact categories should basically be as comprehensive as needed for the specific goal of the study. As LCA is essentially a method looking at all possible impacts, LCA studies should preferably cover all relevant environmental issues related to the analysed (product or service) system, unless a limitation was set in the goal definition as e.g. in the case of Carbon footprint studies, where exclusively Climate change relevant interventions are considered (EC-JRC 2010). One may argue that such limitations violate the ISO requirement that "the selection of impact categories shall reflect a comprehensive set of environmental issues related to the product system being studied taking the goal and scope into consideration". However, since this ISO requirement exists of two parts that basically contradict each other, the argument may easily be refuted again. Most importantly, the initial exclusion of relevant impacts needs to be clearly documented and considered in the interpretation of the results, potentially limiting conclusions and recommendations of the study (EC-JRC 2010). Proper attention to classification in LCA case studies is also essential. LCA practitioners are responsible for checking the completeness of default classification lists for their study, and particularly the handling of missing information needs to be reported explicitly and transparently, and needs to be taken into consideration when defining conclusions and recommendations for the study at stake.

After the 1990s, the attention to selection of impact categories and classification in LCA methodology studies and papers has focused on several proposals for new impact categories and one proposal for classification of inventory results that relate to more than one impact category. Still, there are some issues that deserve attention from LCA method developers.

The first issue concerns the harmonisation in terms of naming impact categories. As shown in Table 2.2, there are various different names for seemingly similar impact categories, which is rather confusing. It could be very helpful to harmonise naming of impact categories that address the same endpoint(s) and inventory results (see, for example, the taxonomy proposals by Bare and Gloria (2008)). A minor issue of attention in this harmonisation process might be the way impact categories are named in general. Some are named quite 'negative' like ozone layer destruction, loss of biodiversity or respiratory diseases, while others are named more 'positive' like damage to human health, particulate matter formation or life expectancy.

The second issue is that considering the continuous flow of proposals for new impact categories, default lists of impact categories can potentially explode. As these new proposals often address specific sectors of LCA studies, it might be useful to distinguish between different types of default lists. For example, a default list of baseline midpoint impact categories could include categories as depletion of fossil resources, depletion of mineral resources, climate change, etc., while on top of this baseline default list a default list of aquaculture and fisheries midpoint impact categories could include biotic resource use, water dependency, the area altered by farm waste, etc., for specific aquaculture and fisheries LCA case studies.

The third issue concerns the classification of inventory results that relate to more than one impact category. As discussed above, it is arguable whether this is really a problem or not for LCA. However, if there were one, all-encompassing fate and exposure model available covering all impact categories, rather than the diversity of models used for the various impact categories today, parallel impacts would no longer constitute a problem. Current fate and exposure models specify the compartment (or target organism) in which the substance has its principal impact and which impact categories are thus potentially relevant (e.g. emissions to the air may end up in soil or water, with consequent terrestrial or aquatic ecotoxic impacts, respectively). Then, the 'Classification' step will be restricted solely to the assignment of elementary flows to defined impact categories (Guinée et al. 2002).

Finally, it could be sensible to create more clarity, for example in the next update of the ISO 14040 series of standards, on the issue of whether or not to differentiate between names for midpoint impact categories and damage or endpoint category indicators, including different names for different midpoint impact categories (there are many different midpoints possible between inventory results and endpoints) and possibly also between different damage or endpoint impact categories.

References

Anonymous (1992a) Life-cycle assessment. Proceedings of SETAC-Europe workshop on environmental life cycle assessment of products, 2–3 Dec 1991 in Leiden. SETAC-Europe, Brussels

Anonymous (1992b) Product life cycle assessment – principles and methodology. Nord 1992:9. Nordic Council of Ministers, Copenhagen

Aubin J, Papatryphon E, van der Werf HMG, Chatzifotis S (2009) Assessment of the environmental impact of carnivorous finfish production systems using life cycle assessment. J Clean Prod 17:354–361

Bare JC (2011) TRACI 2.0: the tool for the reduction and assessment of chemical and other environmental impacts 2.0. Clean Techn Environ Policy 13:687–696

Bare JC, Gloria TP (2008) Environmental impact assessment taxonomy providing comprehensive coverage of midpoints, endpoints, damages, and areas of protection. J Clean Prod 16 (10):1021–1035

Baumann H, Ekvall T, Svensson G, Rydberg T, Tillman A-M (1992) Aggregation and operative units. In: Life-cycle assessment; Proceedings of SETAC-Europe workshop on environmental life-cycle assessment of products, 2–3 Dec 1991 in Leiden. SETAC-Europe, Brussels

Consoli F, Allen D, Boustead I, de Oude N, Fava J, Franklin W, Quay B, Parrish R, Perriman R, Postlethwaite D, Seguin J, Vigon B (1993) Guidelines for life-cycle assessment: a 'code of practice', 1st edn. SETAC-Europe, Brussels

d'Orbcastel RE, Blancheton J-E, Aubin J (2009) Towards environmentally sustainable aquaculture: comparison between two trout farming systems using life cycle assessment. Aquac Eng 40:113–119

de Smet B (ed) (1990) Life-cycle analysis for packaging environmental assessment. Proceedings of the specialised workshop, 24–25 Sept 1990, Leuven. Procter & Gamble Technical Center, Strombeek-Bever

EC-JRC (2010) General guide for life cycle assessment—detailed guidance. ILCD handbook—International Reference Life Cycle Data System, European Union EUR24708 http://lct.jrc.ec.europa.eu/

EC-JRC (2011) Recommendations based on existing environmental impact assessment models and factors for life cycle assessment in European context. ILCD handbook—International Reference Life Cycle Data System, European Union EUR24571EN http://lct.jrc.ec.europa.eu/

Fava JA, Consoli F, Denison R, Dickson K, Mohin T, Vigon B (eds) (1993) A conceptual framework for life-cycle impact assessment. SETAC, Washington, DC

Ford JS, Pelletier NL, Ziegler F, Scholz AJ, Tyedmers PH, Sonesson U, Kruse SA, Silverman H (2012) Proposed local ecological impact categories and indicators for life cycle assessment of aquaculture: a salmon aquaculture case study. J Ind Ecol 16:254–265

Frischknecht R, Steiner R, Jungbluth N (2009) The ecological scarcity method – eco-factors 2006. A method for impact assessment in LCA. Environmental Studies No. 0906. Federal Office for the Environment, Bern. Available from: www.environment-switzerland.ch/uw-0906-e. Accessed 19 Dec 2012

Goedkoop M, Spriensma R (2000) The Eco-indicator 99: a damage oriented method for life cycle assessment, methodology report, 2nd edn. Pré Consultants, Amersfoort

Goedkoop M, Heijungs R, Huijbregts M, De Schryver A, Struijs J, van Zelm R (2012) ReCiPe 2008: a life cycle impact assessment method which comprises harmonised category indicators at the midpoint and the endpoint level. First edition (revised). Report I: Characterisation. 6 Jan 2009, http://www.lcia-recipe.net/. Accessed July 2012

Guinée JB (1992) Headings for classification. In: Life-cycle assessment; Proceedings of SETAC-Europe workshop on environmental life-cycle assessment of products, 2–3 Dec 1991 in Leiden. SETAC-Europe, Brussels

Guinée JB (1995) Development of a methodology for the environmental life-cycle assessment of products; with a case study on margarines. Thesis, Leiden University, Leiden

Guinée JB (ed), Gorrée M, Heijungs R, Huppes G, Kleijn R, de Koning A, van Oers L, Wegener Sleeswijk A, Suh S, Udo de Haes HA, de Bruijn JA, van Duin R, Huijbregts MAJ (2002) Handbook on life cycle assessment: operational guide to the ISO standards, vol 7, - Eco-efficiency in industry and science. Springer, Dordrecht

Hauschild M, Potting J (2005) Spatial differentiation in life cycle impact assessment – the EDIP2003 methodology. Environmental News No. 80. The Danish Ministry of the Environment, Environmental Protection Agency, Copenhagen

Hauschild M, Wenzel H (1998) Environmental assessment of products, vol 2, Scientific background. Chapman & Hall, London and Kluwer Academic Publishers, Hingham. ISBN 0-412-80810-2

Hauschild M, Goedkoop M, Guinée J, Heijungs R, Huijbregts M, Jolliet O, Margni M, De Schryver A, Humbert S, Laurent A, Sala S, Pant R (2013) Identifying best existing practice for characterization modelling in life cycle impact assessment. Int J Life Cycle Assess 18:683–697

Heijungs R, Guinée JB, Huppes G, Lankreijer RM, Udo de Haes HA, Wegener Sleeswijk A, Ansems AAM, Eggels PG, van Duin R, de Goede HP(1992) Environmental life cycle assessment of products. Guide & backgrounds, Oct 1992. Centre of Environmental Science, Leiden University, Leiden

ISO 14040 (1997) Environmental management – life cycle assessment – principles and framework. International Standards Organization, Geneva, Switzerland
ISO 14042 (2000) Environmental management – life cycle assessment – life cycle impact assessment. International Standards Organization, Geneva, Switzerland
ISO 14040 (2006a) Environmental management – life cycle assessment – principles and framework. International Standards Organization, Geneva, Switzerland
ISO 14044 (2006b) Environmental management – life cycle assessment – requirements and guidelines. International Standards Organization, Geneva, Switzerland
Itsubo N, Inaba A (2012) LIME2, Life-cycle impact assessment method based on endpoint modelling: summary. Available from: http://lca-forum.org/english/pdf/No12_Summary.pdf. Accessed 20 Dec 2012
Jolliet O, Brent A, Goedkoop M, Itsubo N, Mueller-Wenk R, Peña C, Schenk R, Stewart M, WeidemaB (2003a) Life cycle impact assessment programme of the life cycle initiative – final report of the LCIA definition study. Available from: http://sph.umich.edu/riskcenter/jolliet/Jolliet%202003%20LCIA_defStudy.pdf. Accessed 20 Dec 2012
Jolliet O, Margni M, Charles R, Humbert S, Payet J, Rebitzer G, Rosenbaum R (2003b) IMPACT 2002+: a new life cycle impact assessment methodology. Int J Life Cycle Assess 8:324–330
Klöpffer W, Grahl B (2009) Ökobilanz (LCA): Ein Leitfaden für Ausbildung und Beruf. Wiley-VCH, Weinheim
Klöpffer W, Grahl B (2014) Life cycle assessment (LCA) – a guide to best practice. Wiley-VCH, Weinheim
Lindfors LG, Christiansen K, Hoffman K, Virtanen Y, Juntilla V, Hansen OJ, Rønning A, Ekvall T, Finnveden G (1995) LCA-NORDIC technical reports no 10. Nordic Council of Ministers, Copenhagen
Papatryphon E, Petit J, Kaushik SJ, van der Werf HMG (2004) Environmental impact assessment of salmonid feeds using life cycle assessment (LCA). Ambio 33:316–323
Pelletier NL, Ayer NW, Tyedmers PH, Kruse SA, Flysjo A, Robillard G, Ziegler F, Scholz AJ, Sonesson U (2007) Impact categories for life cycle assessment research of seafood production systems: review and prospectus. Int J Life Cycle Assess 12:414–421
Reap J, Roman F, Duncan S, Bras B (2008) A survey of unresolved problems in life cycle assessment. Part II: impact assessment and interpretation. Int J Life Cycle Assess 13:374–388
Steen B (1999a) A systematic approach to environmental priority strategies in product development (EPS). Version 2000-general system characteristics; CPM report 1999:4, Chalmers University of Technology, Gothenburg
Steen B (1999b). A systematic approach to environmental priority strategies in product development (EPS). Version 2000-models and data of the default method; CPM report 1999:5, Chalmers University of Technology, Gothenburg
Toffoletto L, Bulle C, Godin J, Reid C, Deschênes L (2007) LUCAS – a new LCIA method used for a Canadian-specific context. Int J Life Cycle Assess 12:93–102
Udo de Haes HA (1992) Workshop conclusions on classification session. In: Life-cycle assessment; Proceedings of SETAC-Europe workshop on environmental life cycle assessment of products, 2–3 Dec 1991 in Leiden. SETAC-Europe, Brussels
Udo de Haes HA (1996) Discussion of general principles and guidelines for practical use. Part I. In: Udo de Haes HA (ed) Towards a methodology for life cycle impact assessment. SETAC-Europe, Brussels
Udo de Haes HA, Jolliet O, Finnveden G, Hauschild M, Krewitt W, Müller-Wenk R (1999) Best available practice regarding impact categories and category indicators in life cycle impact assessment. Background document for the second working group on life cycle impact assessment of SETAC-Europe (WIA–2). Int J Life Cycle Assess 4:66–74, and 4:167–174
Ventura A (2011) Classification of chemicals into emission-based impact categories: a first approach for equiprobable and site-specific conceptual frames. Int J Life Cycle Assess 16:148–158
Wenzel H, Hauschild M, Alting L (1997) Environmental assessment of products, vol 1, Methodology, tools and case studies in product development. Chapman & Hall, London and Kluwer Academic Publishers, Hingham. ISBN 0 412 80800 5

Chapter 3
Climate Change

Annie Levasseur

Abstract Climate change is defined as the warming of the climate system due to human activities. Emission of greenhouse gases (GHGs), which cause an increase in radiative forcing, is the main contributor, and the only climate forcing agent currently considered in life cycle impact assessment (LCIA) methodologies. The direct consequence is an increase in the temperature of atmosphere and oceans, which leads to several types of higher-level impacts such as sea level rise, extreme meteorological events and perturbations in rainfalls, which in turn cause damages to human health and ecosystem quality. All the LCIA methodologies use GWPs (Global Warming Potentials), developed by the Intergovernmental Panel on Climate Change (IPCC), as midpoint characterisation factors since they are based on state-of-the art and peer-reviewed publications and have a relatively low associated uncertainty. Some LCIA methodologies also propose endpoint characterisation factors. However, these factors are considered highly uncertain because of the complexity of the impact pathway so that further research is still needed to improve robustness of the models. Recent new developments are addressing the accounting of biogenic CO_2 emissions, the timing of GHG emissions, and the development of characterisation factors for terrestrial albedo changes induced by human activities.

Keywords Albedo • Climate change • Global warming • Greenhouse gas • GWP • LCA • Life cycle assessment • LCIA • Life cycle impact assessment • Radiative forcing

1 Impact Pathway

The climate change impact category refers to the warming of the climate system due to human activities, which is also called anthropogenic global warming. The emission of greenhouse gases (GHG) to the atmosphere is the leading cause of global warming. However, other climate forcing agents such as changes in

A. Levasseur (✉)
CIRAIG, Department of Chemical Engineering, École Polytechnique de Montréal, stn Centre-ville, P.O. Box 6079, Montréal, QC H3C 3A7, Canada
e-mail: annie.levasseur@polymtl.ca

M.Z. Hauschild, M.A.J. Huijbregts (eds.), *Life Cycle Impact Assessment*, LCA Compendium – The Complete World of Life Cycle Assessment, DOI 10.1007/978-94-017-9744-3_3

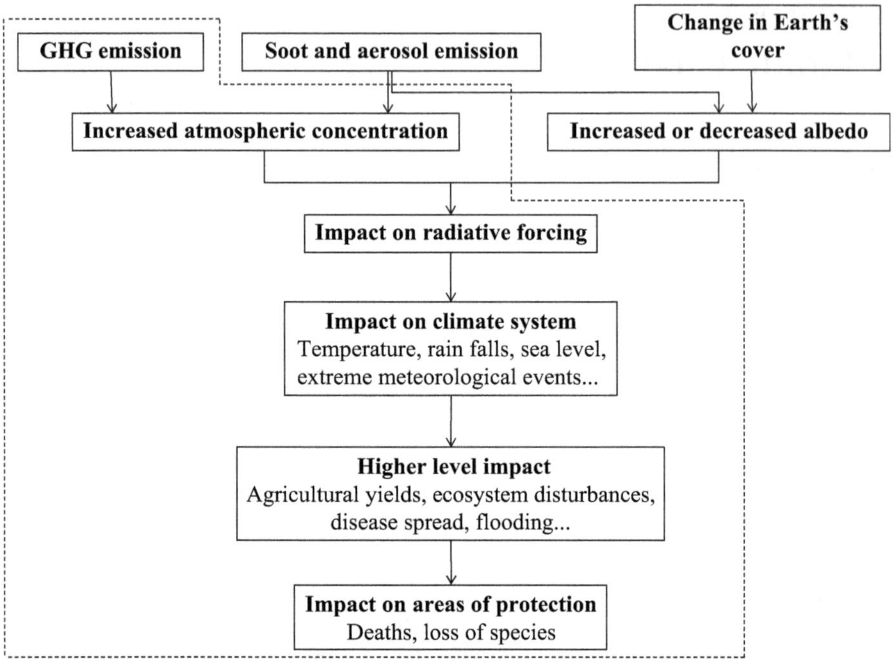

Fig. 3.1 Impact pathway of climate change (mechanisms outside the dotted lines are not considered in current LCIA methodologies)

terrestrial albedo and soot or aerosol emissions also have an impact on climate. Figure 3.1 illustrates the impact pathway for the climate change impact category which is further explained below.

IPCC (Intergovernmental Panel on Climate Change) Assessment Reports provide deeper information on the climate change impact pathway (IPCC 2013, 2014).

GHGs have the property of absorbing radiations at specific infrared wavelengths emitted by the Earth's surface and by clouds, resulting in a net warming effect called the greenhouse effect. Terrestrial albedo is defined as the fraction of solar radiation that is reflected by the Earth's surface. Human activities such as deforestation may induce changes in terrestrial albedo, resulting in a net warming or cooling effect. Emissions of soot and aerosols also have an impact on the Earth's temperature following different mechanisms such as direct (reflecting solar radiation) or indirect (modifying cloud properties which reflect solar radiation) changes in terrestrial albedo or by direct absorption of solar radiation.

All these climate forcing agents have an impact on radiative forcing, which is defined as the perturbation of the Earth's energy balance. Positive radiative forcing causes an increase in the atmosphere and ocean average surface temperature, as well as other associated climate impacts such as changes in rainfalls, extreme meteorological events, and raising sea level. These climate perturbations cause secondary effects such as perturbation of agricultural yields, important changes in physical characteristics of Earth (e.g. desertification, reduction in ice cover, perturbation of ocean currents), flooding, etc.

Finally, these secondary effects lead to damages on two areas of protection: human health and ecosystem quality, increasing the number of deaths and morbidities as well as the number of threatened species. Consequences of climate change on human health are diverse and may affect people all around the world, both in developed and developing countries (Patz et al. 2005; McMichael et al. 2006). For instance, an increased frequency of heat waves may cause higher daily death counts in cities (Li et al. 2013), prevalence of some vector-borne and infectious diseases may increase (Baker-Austin et al. 2013), malnutrition due to crop failures may affect large populations, and floods and extreme meteorological events may cause deaths as well as several other consequences such as population displacements.

2 Contributing Substances

As shown in Fig. 3.1, GHG emissions are the only elementary flows considered in current life cycle impact assessment (LCIA) methodologies. The four principal GHGs resulting from human activities are carbon dioxide (CO_2), nitrous oxide (N_2O), methane (CH_4) and halocarbons. The main anthropogenic sources of carbon dioxide are fossil fuel combustion and the effects of land use change on plant and soil carbon (IPCC 2007). Agricultural activities such as fertilization and land use change are responsible for most human-induced nitrous oxide emissions (IPCC 2013). Methane is mainly released by agriculture (e.g. ruminant animals, rice culture biomass burning), with smaller contributions from industrial sources (e.g. landfill, natural gas processing) (IPCC 2013). Halocarbons are used as refrigeration agents and in other industrial processes. However, as they also are ozone depleting substances, their atmospheric load is decreasing following international regulations to protect the ozone layer (IPCC 2013). The Intergovernmental Panel on Climate Change (IPCC) publishes a list of GHGs in each of its assessment reports. The most recent list has been published in 2013 and can be found in Table 8.A.1 of the Fifth Assessment Report, Working Group I (IPCC 2013).

3 Scale, Spatial Variability, and Temporal Variability

The lifetime of the vast majority of greenhouse gases is higher than the tropospheric mixing time, which is considered to be around 1 year. Thus, climate change is considered a global impact category, and impacts do not depend on where emissions occur. This means that the increase in radiative forcing caused by a given amount of GHG will be the same wherever it is released. However, impacts on climate system from an increasing radiative forcing are not the same all around the world, and endpoint LCIA methodologies have to consider the spatial variability of these effects.

Since CO_2 has a very long atmospheric residence time compared to other GHGs, impacts are not assessed over an infinite time frame. Indeed, an infinite time horizon would lead to an infinite impact for CO_2 and a zero impact for other GHGs. The use of shorter time horizons (e.g. 20, 100 and 500 years) thus better addresses the urgency issue of avoiding passing a climate tipping point beyond which adverse and irreversible changes may occur, while neglecting some long-term effects (Jørgensen and Hauschild 2013). The selection of a time horizon in impact assessment is a value judgement rather than a scientific decision (Fearnside 2002; Shine 2009), and life cycle assessment (LCA) results may be very sensitive to this choice (e.g. Howarth et al. 2011). It is best practice in LCA to integrate impacts over an infinite (or very long) time frame in order to include impacts on future generations. However, a shorter time horizon may also be chosen because future impacts are less certain and may be avoided by adaptation or remediation (Udo de Haes et al. 1999).

There is an increasing concern for the temporal variability of GHG emissions in recent literature. It has been shown that the use of a fixed time horizon for characterisation factors in LCA can lead to an inconsistency in time frames when comparing the impact of product systems over long time frames (Levasseur et al. 2010; O'Hare et al. 2009). The solutions proposed to overcome this limitation take into account the timing of GHG emissions when assessing their impact. Other time-related aspects of life cycle GHG assessments, such as valuing temporary carbon storage or amortizing land use change-related emissions, also require consideration for the timing of emissions (Brandão et al. 2013; Cherubini et al. 2011; Clift and Brandão 2008; Kendall et al. 2009; Levasseur et al. 2012a, b; Manomet Center for Conservation Sciences 2010; Pingoud et al. 2011). Temporal aspects of climate change are still debated and undergoing research. New developments on this topic are expected in a near future and are discussed further in Sect. 6.

4 Midpoint Methodologies

All LCIA methodologies offer midpoint characterisation factors for the climate change impact category, using Global Warming Potentials (GWP) developed by the IPCC. The GWP concept represents the cumulative radiative forcing of a given GHG over a fixed time horizon, relative to the same value calculated for CO_2, the reference substance. GWP values are calculated for all GHGs and for three time horizons (i.e. 20, 100 and 500 years). The time-dependent atmospheric mass loading for CO_2 needed to calculate GWPs is given by the Bern carbon cycle climate model using a constant background atmospheric CO_2 concentration (Joos et al. 2001, 2013). For other GHGs, a first-order atmospheric decay equation is used. GWPs are updated in every IPCC assessment report, considering the latest developments in the field. These reports are based on state-of-the art and peer-reviewed publications and are themselves peer-reviewed by several experts.

The uncertainty of GWP values is estimated to ±35 % for the 5–95 % confidence range (Forster et al. 2007).

There are two major differences between the LCIA methodologies regarding midpoint characterisation factors for climate change: (1) the version of the assessment report from which GWPs are taken, and (2) the time horizon chosen for the calculation (see Table 3.1).

The European Commission's ILCD Handbook for LCIA recommends that midpoint characterisation factors be based on the latest IPCC assessment report in order to use the most up-to-date and scientifically-robust consensus-based model available, but not all methodologies do (European Commission 2011). Some of the LCIA methodologies are using a 100-year time horizon for GWP values since this is the time frame chosen for the application of the Kyoto Protocol. Others are using a 500-year time horizon, which is closest to infinity, in order to include more long-term effects. Finally, some of the LCIA methodologies are offering characterisation factors for different time horizons, which allow the user to test the sensitivity of the results to this subjective decision.

Even though there is a large-scale consensus on the use of GWP as a climate change metric, some authors have discussed limitations and suggested improvements or opened the door to alternative metrics. Indeed, some researchers have shown that GWPs could be underestimated for some GHGs since indirect effects such as gas-aerosol interactions are not modelled (Shindell et al. 2009). They also remind that the use of an integrated metric for a 100-year time horizon (GWP100) reduces the impact of short-lived GHGs compared to long-lived ones, and that it does not allow assessing climate impacts such as rate of change or surface temperature response to regionally distributed forcings. Other researchers have shown that using the alternative metric GTP (Global Temperature Change Potential) developed by Shine et al. (2005) could be more appropriate to assess the relative impact of GHGs in some cases (Fuglestvedt et al. 2010; Peters et al. 2011). Contrary to GWP which compares GHGs based on the cumulative radiative forcing over a given time horizon, GTP expresses the potential temperature increase caused by GHGs at a given time following the emission. The use of an instantaneous metric in LCA (GTP) compared to a cumulative one (GWP) may lead to different conclusions, especially if short-lived GHGs are important. GTPs result in increased uncertainties since temperature is further than radiative forcing in the impact pathway. These recent publications uncover the need for further discussions in the LCA community on the choice of metrics and time horizons to assess climate change impacts (Peters et al. 2011).

5 Endpoint Methodologies

Climate change affects humans and ecosystems in numerous ways. Higher temperatures may increase health problems caused by heat waves or favour the proliferation of infectious diseases. Extreme weather events or floods may cause more

Table 3.1 Climate change in LCIA methods

Method	Midpoint	Unit	Endpoint human health	Unit	Endpoint ecosystems	Unit	Reference
EPS 2000	–	–	Human mortality (caused by heat stress, starvation, flooding, and malaria) human morbidity (caused by starvation and malaria)	pers. year	Crop production, wood production, and extinction of species	Respectively kg, kg and NEX (Normalised extinction of species)	Steen (1999)
IMPACT 2002+	GWP500	$kgCO_2$–eq	–	–	–	–	Humbert et al. (2005)
EDIP 2003	GWP100	$kgCO_2$–eq	–	–	–	–	Hauschild and Potting (2005)
TRACI	GWP100	$kgCO_2$–eq	–	–	–	–	Bare et al. (2003)
ReCiPe	GWP20 (Individualist) GWP100 (Hierarchical) GWP500 (Egalitarian)	$kgCO_2$–eq	Malnutrition, diarrhoea, cardiovascular diseases, coastal and inland flooding, and malaria	DALY	Extinction of species	species. year	Goedkoop et al. (2009)

accidental deaths. Perturbations in rainfalls and warmer oceans may have important effects on crop yields, water availability and marine productivity, leading to malnutrition and impoverishment. A warming climate would also have repercussions on all living species, leading to displacements or extinctions.

Because of the broad range of impacts caused by climate change and of uncertain cause-effect pathways, endpoint modelling can be quite complex. Nonetheless, some LCIA methodologies model climate change impacts to the endpoint level: Ecoindicator 99, ReCiPe, EPS 2000, LIME, and the coming method IMPACT World$^+$. The EPS 2000 and ReCiPe methods are further discussed since they are the most recent endpoint methods with English documentation publicly available. De Schryver et al. (2009) and Hanafiah et al. (2011) also developed endpoint characterisation factors for climate change impacts on human health and ecosystems for the first, and on freshwater fish species extinction for the second. The ReCiPe approach is partly based on the work done by De Schryver et al. (2009).

EPS 2000 (Steen 1999) is the older of these two methodologies. It models the impacts of a CO_2 emission on human mortality (caused by heat stress, starvation, flooding, and malaria), human morbidity (caused by starvation and malaria), crop production, wood production, and extinction of species. Studies coming from different sources are used in combination with several assumptions to estimate the total impact caused by expected CO_2 emissions over the same 100-year timeframe for each type of modelled effect. In a next step, EPS considers that 1 kg CO_2 contributes to 1.26×10^{-16} of the total global warming effect occurring over a 100-year period (from 1990 to 2090) according to the IPCC scenario IS92A; i.e., the total impact is then multiplied by 1.26×10^{-16} to get the impact per kg CO_2 emitted. The impact for other GHGs is then calculated by multiplying the obtained value by the respective GWP from the IPCC First Assessment Report published in 1990 for the 100-year time horizon.

ReCiPe (Goedkoop et al. 2009) models impacts of a CO_2 emission on human health for five different effects (i.e. malnutrition, diarrhoea, cardiovascular diseases, coastal and inland flooding, and malaria), as well as on ecosystem diversity. The increase in atmospheric temperature caused by a unit mass CO_2 emission is first determined using a study correlating equilibrium temperature changes with avoided cumulative fossil CO_2 emissions over a given period as calculated by different climate models (Meinshausen 2005). Then, damage factors for human health and ecosystem diversity are developed per unit temperature increase. For human health, the relative risk of dying from one of the five studied effects due to a given temperature increase are calculated from a World Health Organization report (WHO 2003). The DALYs (Disability-adjusted life year) associated to each type of health effects are coming from the report 'The Global Burden of Disease' for the 1990 period (Murray and Lopez 1996). Damage factors for ecosystem diversity are developed using a summary of different studies linking the risk of extinction with an increase in temperature published in Nature (Thomas et al. 2004). The impact of other GHGs on human health and ecosystem diversity are calculated by multiplying the impact of CO_2 by the respective GWP from the IPCC Fourth Assessment Report published in 2007.

Similar to Ecoindicator, ReCiPe proposes characterisation factors for three cultural perspectives (individualist, hierarchist, and egalitarian) for which some choices and assumptions differ depending on different subjective views. For the climate change category, the aspects that vary with cultural perspectives are the time horizon used for GWPs, the assumed degree of future adaptation which influences the relative risk value used for the different health effects, and the assumed possibility for species to migrate when climate conditions are no longer viable.

Endpoint characterisation factors have a much greater level of uncertainty than midpoint characterisation factors due to the large number of mechanisms involved, of which many are not included in the present models, and to the difficulty in attempting to predict how humans and living species may adapt in the future. Indeed, socioeconomic conditions significantly affect the capacity of a population to adapt or prevent reverse effects, such as malnutrition, diseases or floods. Magnitude of impacts also depends on several region-specific natural factors, making them even more difficult to model.

The EPS 2000 methodology roughly estimates uncertainty ranges for endpoint characterisation factors, pointing out that the models used are speculative and that it is very difficult to predict future health care and adaptation. Damages on human health and ecosystem quality are considered uncertain by a factor of ten, and damages on wood and crop production are considered uncertain by a factor of two to three. The ReCiPe methodology handles uncertainty using the cultural perspectives. Indeed, since the individualist perspective assumes that humans have a high capacity of adaptation and that short time horizons should be used, and that the egalitarian perspective assumes the opposite (worst case scenario and long time-horizons), the hierarchist perspective being in the middle, the characterisation factors for these three perspectives illustrate the variability of impacts with some given aspects of uncertainty.

6 New Developments and Research Needs

Further research is needed regarding endpoint characterisation in order to improve robustness of the models. Indeed, several types of impacts on human health are not modelled yet such as several infectious diseases (dengue, cholera, tick-borne diseases, etc.) or increase in air pollution, for instance (Confalonieri et al. 2007), and the increase of mortality and morbidity due to an increased number of conflicts in a more unstable future world; due to a changing climate, studies are difficult to model in a meaningful way. The damage models used for both human health and ecosystem quality are thus considered highly uncertain. For midpoint characterisation factors, cumulative radiative forcing calculations have a relatively low uncertainty, and the IPCC is updating GWPs on a regular basis using latest research developments.

As mentioned in Sect. 3, several developments are currently occurring regarding the consideration of the timing of GHG emissions in LCA. The dynamic LCA approach enables the user to calculate the cumulative radiative forcing of a temporally detailed GHG inventory over a given time horizon using characterisation factors that depend on the time elapsed between each emission and the selected time horizon. The general dynamic LCA framework, which can be applied to any product system to integrate temporal variability in LCA, could also be used to calculate the time-dependent impact of a temporally detailed inventory for any other impact category if characterisation factors are developed.

Similar approaches have been proposed by the ILCD Handbook (European Commission 2010) and the British specification PAS 2050 for carbon footprint (BSI 2008) for CO_2 emissions only and for a fixed 100-year time horizon. These developments aimed at addressing the issue of inconsistency in time frames when assessing products or projects over long time periods, while enabling for the consideration of temporary carbon storage in biomass-based products. However, no consensus exists yet on the subject, as concluded by a group of experts at a workshop held at the European Commission's Joint Research Centre in 2010 (Brandão and Levasseur 2011).

Other specific timing issues related to GHG emissions have also been raised by researchers in the field of bioenergy and forestry. Indeed, land use change emissions associated to biofuel production must be amortized over a given number of years following the land use change process. Different approaches have been proposed to treat this particular question, all of them being based on the concept of cumulative radiative forcing (Kendall et al. 2009; Kløverpris and Mueller 2013; O'Hare et al. 2009). Some concerns have also been raised about the impact of biogenic CO_2 emissions, usually considered neutral. Since a forest takes a long time to grow up, CO_2 released by burning wood will induce some global warming before an equivalent amount of carbon is sequestered from the atmosphere by growing trees. The GWP concept was adapted into a GWP_{bio}, calculated for different biomass species and different time horizons, to better characterise biogenic carbon emissions (Cherubini et al. 2011).

Another new development in climate change impact assessment is on the consideration of terrestrial albedo. Indeed, human-induced land use changes such as deforestation, reforestation or urbanisation cause changes in surface albedo that may have an important impact on global warming. Researchers have shown that in some locations, the change in albedo caused by deforestation can compensate for the loss of carbon sequestration (Bernier et al. 2011; Betts 2000; Schwaiger and Bird 2010). This climate forcing agent is not yet considered by LCIA methodologies. However, a methodology has recently been proposed for the development of characterisation factors for albedo changes (Muñoz et al. 2010).

This ongoing research will significantly improve robustness of LCA results in terms of climate change assessment by including impact pathways and climate forcing agents that are currently not modelled, reducing uncertainties, and consistently considering short-term as well as long-term effects. As shown by several authors, these improvements may affect the conclusions drawn from LCA studies enabling better decision-making.

References

Baker-Austin C, Trinanes JA, Taylor NGH, Siitonen A, Martinez-Urtaza J (2013) Emerging *Vibrio* risk at high latitudes in response to ocean warming. Nat Clim Chang 3:73–77

Bare JC, Norris GA, Pennington DW, McKone TE (2003) TRACI: the tool for the reduction and assessment of chemical and other environmental impacts. J Ind Ecol 6:49–78

Bernier PY, Desjardins RL, Karimi-Zindashty Y, Worth D, Beaudoin A, Luo Y et al (2011) Boreal lichen woodlands: a possible negative feedback to climate change in eastern North America. Agric Meteorol 151:521–528

Betts RA (2000) Offset of the potential carbon sink from boreal forestation by decreases in surface albedo. Nature 408:187–190

Brandão M, Levasseur A (2011) Assessing temporary carbon storage in life cycle assessment and carbon footprinting: outcomes of an expert workshop. Publications Office of the European Union, Luxembourg

Brandão M, Levasseur A, Kirschbaum MUF, Weidema BP, Cowie AL, Jørgensen SV et al (2013) Key issues and options in accounting for carbon sequestration and temporary storage in life cycle assessment and carbon footprinting. Int J Life Cycle Assess 18:230–240

BSI (2008) PAS 2050 (2008) Specification for the assessment of the life cycle greenhouse gas emissions of goods and services. British Standards Institution, London

Cherubini F, Peters GP, Berntsen T, Stromman AH, Hertwich E (2011) CO_2 emissions from biomass combustion for bioenergy: atmospheric decay and contribution to global warming. GCB Bioenergy 3:413–426

Clift R, Brandão M (2008) Carbon storage and timing of emissions. CES working papers, Centre for Environmental Strategy, University of Surrey, Surrey

Confalonieri U, Menne B, Akhtar R, Ebi KL, Hauengue M, Kovats RS et al (2007) Human health. In: Parry ML, Canziani OF, Palutikof JP, van der Linden PJ, Hanson CE (eds) Climate change 2007: impacts, adaptation and vulnerability. Contribution of working group II to the fourth assessment report of the Intergovernmental Panel on Climate Change. Cambridge University Press, Cambridge/New York

De Schryver AM, Brakkee KW, Goedkoop MJ, Huijbregts MAJ (2009) Characterization factors for global warming in life cycle assessment based on damages to humans and ecosystems. Environ Sci Technol 43:1689–1695

European Commission – Joint Research Centre – Institute for Environment and Sustainability (2010) International Reference Life Cycle Data System (ILCD) handbook – general guide for life cycle assessment – detailed guidance. Publications Office of the European Union, Luxembourg

European Commission – Joint Research Centre – Institute for Environment and Sustainability (2011) International Reference Life Cycle Data System (ILCD) handbook – recommandations for life cycle impact assessment in the European context. Publications Office of the European Union, Luxembourg

Fearnside PM (2002) Why a 100-year time horizon should be used for global warming mitigation calculations. Mitig Adapt Strateg Glob Chang 7:19–30

Forster P, Ramaswamy V, Artaxo P, Berntsen T, Betts R, Fahey DW et al (2007) Changes in atmospheric constituents and in radiative forcing. In: Solomon S, Quin D, Manning M, Chen Z, Marquis M, Averyt KB et al (eds) Climate change 2007: the physical science basic. Contribution of working group I to the fourth assessment report of the Intergovernmental Panel on Climate Change. Cambridge University Press, Cambridge/New York

Fuglestvedt JS, Shine KP, Berntsen T, Cook J, Lee DS, Stenke A et al (2010) Transport impacts on atmosphere and climate: metrics. Atmos Environ 44:4648–4677

Goedkoop M, Heijungs R, Huijbregts M, De Schryver A, Struijs J, van Zelm R (2009) ReCiPe 2008. A life cycle impact assessment method which comprises harmonised category indicators at the midpoint and the endpoint level. Ministry of Housing, Spatial Planning and Environment, The Hague

Hanafiah MM, Xenopoulos MA, Pfister S, Leuven RSEW, Huijbregts MAJ (2011) Characterization factors for water consumption and greenhouse gas emissions based on freshwater fish species extinction. Environ Sci Technol 45:5272–5278

Hauschild M, Potting J (2005) Spatial differentiation in life cycle impact assessment – the EDIP 2003 methodology. Danish Ministry of the Environment, Copenhagen

Howarth RW, Santoro R, Ingraffea A (2011) Methane and the greenhouse-gas footprint of natural gas from shale formations. Clim Chang 106:679–690

Humbert S, Margni M, Jolliet O (2005) Impact 2002+: user guide. École Polytechnique Fédérale de Lausanne, Lausanne

IPCC (2007) Changes in atmospheric constituents and in radiative forcing. In: Climate change 2007: the physical science basis. Contribution of working group I to the fourth assessment report of the Intergovernmental Panel on Climate Change. Cambridge University Press, Cambridge/New York

IPCC (2013) In: Stocker TF, Qin D, Plattner GK, Tignor M, Allen SK, Boschung J et al (eds) Climate change 2013 – the physical science basis. Contribution of working group I to the fifth assessment report of the Intergovernmental Panel on Climate Change. Cambridge University Press, Cambridge/New York

IPCC (2014) In: Field CB, Barros VR, Dokken DJ, Mach KJ, Mastrandea MD, Bilir TE et al (eds) Climate change 2014 – impacts, adaptation and vulnerability. Part A: global and sectoral aspects. Contribution of working group II to the fifth assessment report of the Intergovernmental Panel on Climate Change. Cambridge University Press, Cambridge/New York

Joos F, Prentice IC, Sitch S, Meyer R, Hooss G, Plattner GK et al (2001) Global warming feedbacks on terrestrial carbon uptake under the Intergovernmental Panel on Climate Change (IPCC) emission scenarios. Glob Biogeochem Cycles 15:891–907

Joos F, Roth R, Fuglestvedt JS, Peters GP, Enting IG, von Bloh W et al (2013) Carbon dioxide and climate impulse response functions for the computation of greenhouse gas metrics: a multimodel analysis. Atmos Chem Phys 13:2793–2825

Jørgensen SV, Hauschild MZ (2013) Need for relevant timescales when crediting temporary carbon storage. Int J Life Cycle Assess 18:747–754

Kendall A, Chang B, Sharpe B (2009) Accounting for time-dependent effects in biofuel life cycle greenhouse gas emissions calculations. Environ Sci Technol 43:7142–7147

Kløverpris J, Mueller S (2013) Baseline time accounting: considering global land use dynamics when estimating the climate impact of indirect land use change caused by biofuels. Int J Life Cycle Assess 18:319–330

Levasseur A, Lesage P, Margni M, Deschênes L, Samson R (2010) Considering time in LCA: dynamic LCA and its application to global warming impact assessments. Environ Sci Technol 44:3169–3174

Levasseur A, Lesage P, Margni M, Brandão M, Samson R (2012a) Assessing temporary carbon sequestration and storage projects through land use, land-use change and forestry: comparison of dynamic life cycle assessment with ton-year approaches. Clim Chang 115:759–776

Levasseur A, Lesage P, Margni M, Samson R (2012b) Biogenic carbon and temporary storage addressed with dynamic life cycle assessment. J Ind Ecol 17:117–128

Li T, Horton RM, Kinney PL (2013) Projections of seasonal patterns in temperature-related deaths for Manhattan, New York. Nat Clim Chang 3:717–721. doi:10.1038/nclimate1902

Manomet Center for Conservation Sciences (2010) Massachusetts biomass sustainability and carbon policy study: report to the Commonwealth of Massachusetts Department of Energy Resources. In: Walter T (ed). Manomet Center for Conservation Sciences, Brunswick

McMichael AJ, Woodruff RE, Hales S (2006) Climate change and human health: present and future risks. Lancet 367:859–869

Meinshausen M (2005) Emission and concentration implications of long-term climate targets. Swiss Federal Institute of Technology, Zurich

Muñoz I, Campra P, Fernandez-Alba AR (2010) Including CO_2-emission equivalence of changes in land surface albedo in life cycle assessment. Methodology and case study on greenhouse agriculture. Int J Life Cycle Assess 15:672–681

Murray CJL, Lopez AD (1996) The global burden of disease: a comprehensive assessment of mortality and disability from diseases, injuries, and risk factors in 1990 and projected to 2020, vol 1, Global burden of disease and injury series. Harvard School of Public Health, World Bank, World Health Organisation

O'Hare M, Plevin RJ, Martin JI, Jones AD, Kendall A, Hopson E (2009) Proper accounting for time increases crop-based biofuels' greenhouse gas deficit versus petroleum. Environ Res Lett 4:024001

Patz JA, Campbell-Lendrum D, Holloway T, Foley JA (2005) Impact of regional climate change on human health. Nature 438:310–317

Peters GP, Aamaas B, Lund MT, Solli C, Fuglestvedt JS (2011) Alternative "global warming" metrics in life cycle assessment: a case study with existing transportation data. Environ Sci Technol 45:8633–8641

Pingoud K, Ekholm T, Savolainen I (2011) Global warming potential factors and warming payback time as climate indicators of forest biomass use. Mitig Adapt Strateg Glob Chang 17:369–386

Schwaiger HP, Bird DN (2010) Integration of albedo effects caused by land use change into the climate balance: should we still account in greenhouse gas units? For Ecol Manag 260:278–286

Shindell DT, Faluvegi G, Koch DM, Schmidt GA, Unger N, Bauer SE (2009) Improved attribution of climate forcing to emissions. Science 326:716–718

Shine KP (2009) The global warming potential – the need of an interdisciplinary retrial. Clim Chang 96:467–472

Shine KP, Fuglestvedt JS, Hailemariam K, Stuber N (2005) Alternatives to the global warming potential for comparing climate impacts of emissions of greenhouse gases. Clim Chang 68:281–302

Steen B (1999) A systematic approach to environmental priority strategies in product development (EPS). Version 2000 – models and data of the default method. Centre for Environmental Assessment of Products and Material Systems, Chalmers University of Technology, Göteborg

Thomas CD, Cameron A, Green RE, Blakkenes M, Beaumont LJ, Collingham YC et al (2004) Extinction risk from climate change. Nature 427:145–148

Udo de Haes HA, Jolliet O, Finnveden G, Hauschild MZ, Krewitt W, Müller-Wenk R (1999) Best available practice regarding impact categories and category indicators in life cycle impact assessment. Int J Life Cycle Assess 4:66–74

WHO (2003) In: McMichael AJ, Campbell-Lendrum DH, Corvalán CF, Ebi KL, Githeko AK, Scheraga JD et al (eds) Climate change and human health: risks and responses. World Health Organization, Geneva

Chapter 4
Stratospheric Ozone Depletion

Joe L. Lane

Abstract The stratospheric ozone layer plays a critical role in regulating conditions on Earth, but has been substantially depleted by CFC (chlorofluorocarbon) and other halocarbon emissions. This has increased transmission of UVB radiation to the surface, and been implicated in a range of negative human and ecosystem health impacts.

Midpoint-level LCA has traditionally utilised the steady-state Ozone Depletion Potential factors that are prominent in policy making. Current ozone-depletion endpoint models incorporate skin cancer, cataract damages, and certain changes in ecosystem productivity caused by excess UVB exposure. Other health, ecosystem and agri-production impacts are still to be incorporated into the LCA framework.

As the ozone layer recovers following regulated halocarbon emission reductions, scientific attention turns to the question of longer term ozone layer management. While growing anthropogenic emissions of N_2O (nitrous oxide) might pose a threat to ozone layer recovery, the mitigating effects of CH_4 (methane) and CO_2 (carbon dioxide) emissions will more than compensate for this. Global stratospheric ozone is expected to exceed pre-industrial levels sometime this century, albeit with a very different spatial distribution. Predictions are that UVB levels will remain elevated in the tropics, but become depressed in other regions. That latter situation might increase the incidence of diseases associated with insufficient UVB exposure.

Whatever the policy response to these new challenges, it seems the interface of ozone layer science and management will become increasingly complex. It may be that the metrics used for ozone layer analysis will also need to evolve, if LCA is to remain relevant to this new management paradigm.

Keywords N_2O • CFC • Cancer • Carbon dioxide • Greenhouse gases • Chlorofluorocarbons • LCA • Life cycle assessment • LCIA • Life cycle impact assessment • Methane • Nitrous oxide • Ozone layer

---

J.L. Lane (✉)
School of Chemical Engineering, The University of Queensland, Brisbane, QLD 4072, Australia
e-mail: j.lane1@uq.edu.au

M.Z. Hauschild, M.A.J. Huijbregts (eds.), *Life Cycle Impact Assessment*, LCA Compendium – The Complete World of Life Cycle Assessment, DOI 10.1007/978-94-017-9744-3_4

1 Stratospheric Ozone Chemistry

Ozone (O_3) is a natural constituent of the Earth's atmosphere, most heavily concentrated in the lower part of the stratosphere (Fig. 4.1). While varying in thickness and altitude, this band of elevated ozone concentration extends around the entire globe, and is commonly referred to as the *ozone layer*.

Ozone is an extremely reactive substance, and its presence in the stratosphere is the result of a continual cycle of formation and breakdown processes. Ozone breakdown occurs both chemically and by photo dissociation. This latter pathway is the only part of the ozone cycle that absorbs wavelengths spanning the UVB spectrum (Rowland 2006), intercepting the vast majority of the UVB radiation reaching the outer atmosphere. Since ~90 % of atmospheric ozone resides in the stratosphere, the amount of UVB radiation reaching the Earth's surface is very sensitive to stratospheric ozone concentrations (Fahey and Hegglin 2011).

Net stratospheric ozone concentrations are strongly influenced by a small group of reaction pathways, predominantly associated with halogen, NO_x, and HO_x free radicals. These reactions are catalytic in nature, whereby the reaction step that destroys the ozone molecule is followed by another that regenerates the original free radical molecule. The stable nature of the stratosphere provides relatively little opportunity for these catalytic reaction chains to be interrupted.

The extent of degradation via these catalytic pathways is buffered by the presence of stratospheric 'reservoirs' of compounds that are unreactive with

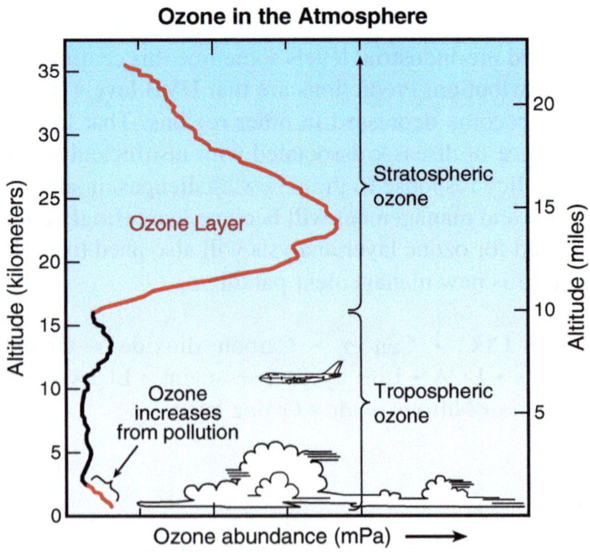

Fig. 4.1 Vertical profile of global mean ozone abundance across the troposphere and stratosphere, taken from Fahey and Hegglin (2011). The troposphere extends from the Earth's surface to an altitude of 10–15 km, and contains the turbulent atmospheric activity that defines surface weather patterns. The stratosphere is a much more stable environment, extending to approximately 50 km above the Earth's surface. The majority of atmospheric ozone is concentrated between the altitudes of 15 and 35 km – this band is known as the ozone layer

ozone ($ClONO_2$, HCl and HNO_3). The reservoir compounds bind up molecules that would otherwise be present in the form of reactive radicals (Cl^-, NO_x, HO_x), although they do break down slowly over time. The greater the relative abundance of the reservoir compounds, the lesser opportunity there is for ozone breakdown to occur (Rowland 2006).

Another controlling factor for the rate of ozone degradation is stratospheric temperature, which has a strong influence on the abundance of ozone-destroying NO radicals; and on the chemical breakdown of ozone molecules to form stable O_2. Outside of the polar regions, the influence of lower temperatures will be a net increase in ozone abundance (Bekki et al. 2011; Portmann et al. 2012).

Somewhat different processes occur in the polar regions, during the period from late winter to early spring. Conditions at this time encourage a different set of catalytic reaction chains, which are far more destructive than occur elsewhere in the stratosphere. At the same time, low UV levels limit the generation of ozone by photo dissociation of O_2, and local ozone levels become severely depleted. These transient 'ozone holes' have occurred regularly over the Antarctic since the early 1980s (Rowland 2006; Fahey and Hegglin 2011), and more recently over the Arctic region (Manney et al. 2011).

2 Anthropogenic Emissions

A strong scientific consensus exists that anthropogenic emissions caused substantial levels of stratospheric ozone depletion over the latter parts of the twentieth century (WMO 2007; Douglass et al. 2011; Fahey and Hegglin 2011; Montzka et al. 2011) (see Fig. 4.2).

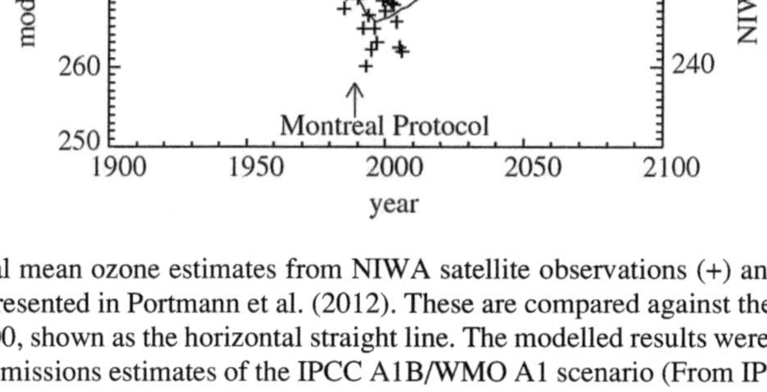

Fig. 4.2 Global mean ozone estimates from NIWA satellite observations (+) and modelled time series (—) as presented in Portmann et al. (2012). These are compared against the modelled value for the year 1900, shown as the horizontal straight line. The modelled results were generated using the mid-range emissions estimates of the IPCC A1B/WMO A1 scenario (From IPCC 2000; WMO 2007)

Historically, both the scientific and policy communities have reserved the term 'ozone depleting substances (ODS)' exclusively for describing the halocarbon substances that are controlled under the Montreal Protocol (see below). So as to avoid confusion with other literature, this chapter uses the more generic term of 'ozone affecting substances (OAS)' which encompasses all (halogenated and non-halogenated) emissions that can cause substantial decreases or increases in stratospheric ozone abundance. OAS emission types are discussed here in three groups.

2.1 Halocarbon Emissions

While the different halogenated emission types vary substantially in their characteristics, the general pathways to ozone destruction are the same. Source gases containing chlorine or bromine are sufficiently stable to reach the stratosphere in their original form, where they break down when exposed to the high intensity of UV radiation present in the upper stratosphere. This releases the chlorine and bromine atoms, providing a source of reactive radicals that can initiate catalytic ozone destruction pathways. Halogenated substances containing fluorine or iodine molecules, but no chlorine or bromine, pose little threat to the integrity of the stratospheric ozone layer (Rowland 2006; Fahey and Hegglin 2011).

By 1987, concerns over risk to the ozone layer led to the initiation of the Montreal Protocol. This regulated the phase-out of harmful halocarbon emissions from most anthropogenic sources, notably the use of CFCs, HCFCs and Halons as refrigerants, solvents and fire extinguisher agents. These international actions halted the rapid increases in stratospheric chlorine and bromine, and global ozone levels started to recover again from the early 1990s (Fig. 4.2).

A collection of recent studies demonstrate how important those changes were to preserving the ozone layer function. Their findings indicate the ozone layer would have collapsed if the Montreal Protocol had not been implemented (Newman et al. 2009; Garcia et al. 2012), leading to substantial increases in surface levels of erythemal-UV radiation (Newman and McKenzie 2011), and substantial increases in incidence of skin cancer (Newman and McKenzie 2011; van Dijk et al. 2013).

Even with these successes, there remains a legacy of halocarbon emissions that will continue for many years (Table 4.1). Large 'banks' of CFC and Halon substances remain in the economy, expected to leak slowly into the atmosphere over the coming decades. While less potent, the production of HCFCs is still growing in developing countries, and substantial 'banks' are being accumulated. Carbon Tetrachloride and Methyl Bromide present a somewhat different management challenge, given concerns that there are substantial sources falling outside the current regulatory regime (Yvon-Lewis et al. 2009; Xiao et al. 2010; Montzka et al. 2011).

Table 4.1 Forecast anthropogenic emissions of halocarbons and N_2O over the period 2011–2050; and key atmospheric characteristics of each substance type

Emission group	Time-integrated, ODP-weighted, emission forecasts (Mt CFC-11-eq.)[a], and the main anthropogenic emission sources		Atmospheric lifetime (y)[b]	Steady-state ODP (kg-CFC11e/kg)[b]
CFC banks	1.27 Mt	Refrigerants and other compounds stored in goods that were manufactured prior to the CFC phase out	45–1,020	0.57–1.0
Halon banks	1.09 Mt	Fire extinguishers manufactured prior to the Halon phase out.	16–65	3–10
HCFC production and use	0.66 Mt	Production and use as industrial solvents; Production of refrigerants and other compounds.	1.3–17.2	0.01–0.12
HCFC banks		Refrigerants and other compounds stored in goods that were manufactured prior to the CFC phase out		
Carbon Tetra-chloride (CCl_4)	0.54 Mt	Use as industrial solvent and feedstock for chemicals manufacture[c]	26	0.82
Methyl Bromide (CH_3Br)	0.26 Mt	Use for Quarantine and pre-shipment services (QPS)[d]	0.8	0.66
Methyl Chloroform (CH_3CCl_3)	0.004 Mt	Production and use as industrial solvents.	5	0.16
Nitrous Oxide (N_2O)	6 Mt	Agricultural fertiliser use; Biomass burning; Fuel combustion; Industrial processes	114	0.017–0.019

[a]Emission forecasts adapted from Table 5-4 of Daniel et al. (2011)
[b]Lifetime estimates and semi-empirical ODPs for the halogenated compounds were taken from Table 5-1 of Daniel et al. (2011). For N_2O, the atmospheric lifetime value is taken from Table 2-14 of IPCC (2007). Three steady-state ODP values for N_2O were found in the literature – all for year 2000 atmospheric conditions, albeit calculated using different atmospheric modelling packages (Ravishankara et al. 2009; Daniel et al. 2010; Fleming et al. 2011)
[c]The CCL_4 estimate makes an allowance for unattributed emissions that are not currently being accounted for using (bottom up) emission inventories consistent with the Montreal Protocol (Montzka et al. 2011)
[d]QPS uses are exempt from the Montreal Protocol. Other notable anthropogenic sources of CH_3Br are not included in this estimate – e.g. emissions from the growth of certain crops, and from the combustion of biofuels and leaded gasoline

2.2 *Nitrous Oxide Emissions*

Similar to many halocarbon OAS, nitrous oxide (N_2O) is a long-lived substance that is very stable in the troposphere. The vast majority of surface emissions will reach the stratosphere, where 90 % of that N_2O will convert to stable N_2. A portion of the remainder breaks down into NO radicals that can initiate catalytic ozone destruction cycles (Rowland 2006; Portmann et al. 2012; Revell et al. 2012b).

This potential for N_2O emissions to damage the ozone layer has been recognised since the early 1970s (Crutzen 1970; Johnston 1971). Despite that, N_2O was never included in the list of controlled substances under international regulations. However, now that halocarbon emission controls have been implemented so successfully, scientific attention has returned to the role that anthropogenic N_2O emissions might play in determining the future status of the ozone layer (e.g. Ravishankara et al. 2009). Anthropogenic sources contribute approximately 40 % of the current global N_2O emissions budget (IPCC 2007; Syakila and Kroeze 2011), and there is considerable uncertainty surrounding expected future growth in emission rates. Based on mid-range projections, anthropogenic N_2O emissions will exert far greater potential to deplete the ozone layer than will the remaining halocarbon emission sources (Table 4.1).

In recognition that anthropogenic N_2O emissions could impede future ozone layer recovery, a recent synthesis from the WMO Scientific Assessment Panel for the first time included substantive coverage of the effect that N_2O can have on the ozone layer (WMO 2011).

2.3 Carbon Dioxide and Methane Emissions

All greenhouse gases absorb and re-emit radiation towards the Earth's surface, increasing the heat retained in the lower atmosphere (see Chap. 3). The particular radiative properties of CO_2 and water vapour mean that they will also reduce temperatures in the stratosphere (Forster et al. 2011). As noted above, reduced temperatures will slow the rate of ozone breakdown in the stratosphere, leading to increased overall ozone abundance.

Atmospheric modelling studies consistently indicate that ongoing growth in CO_2 emissions will expedite ozone layer recovery; then deliver a 'super recovery' (see Fig. 4.2) whereby global ozone abundance goes over and above pre-industrial levels (Austin et al. 2010; Eyring et al. 2010; Oman et al. 2010; Plummer et al. 2010; Fleming et al. 2011). By the year 2100, atmospheric CO_2 is expected to be the strongest anthropogenic influence on ozone layer status (Fleming et al. 2011).

The net effect of future anthropogenic CH_4 emissions will also expedite ozone layer recovery, albeit through a complex mix of counteracting influences (Plummer et al. 2010; Fleming et al. 2011; Revell et al. 2012b). Some portion of surface CH_4 emissions will react to form water vapour, and therefore contribute to lowering stratospheric temperatures. CH_4 emissions will also have direct chemical effects (both generating and depleting stratospheric ozone by different pathways), and indirect chemical effects (e.g. impeding ozone destruction by binding reactive chlorine into HCl reservoirs) (Rowland 2006; Revell et al. 2012b). Over the longer term, ozone increases induced by anthropogenic CH_4 might approximately offset the depletion induced by anthropogenic N_2O (Fleming et al. 2011).

3 Effects on Human Health and Ecosystems

Depletion of the ozone layer will increase the transmission of UVB to the Earth's surface, which can then cause a variety of human health and ecosystem effects (Fig. 4.3).

Excessive exposure to UVB is strongly linked to risk of skin cancer and certain types of eye diseases (Norval et al. 2011; UNEP EEAP 2012), accounting for two thirds of the UV-related health burden quantified in a previous WHO study (Lucas et al. 2008). Skin cancers are the most common form of cancer in high-risk countries such as Australia, and the incidence of melanoma and non-melanoma cancers has grown in many countries over the last four decades (Norval et al. 2011).

A number of other health outcomes are also linked to excess UVB exposure. The incidence of Merkel Cell Carcinoma, a particularly aggressive form of skin cancer, is growing rapidly in a number of populations (Agelli et al. 2010; Girschik et al. 2011; Kuwamoto 2011; UNEP EEAP 2012). At the other extreme, relatively mild but prevalent sunburn cases could even be considered to have a substantial overall health burden (Lucas et al. 2008). Excessive UVB exposure likely contributes to other diseases of the eye, in particular pterygium and (to a lesser extent) ocular melanoma. A third category of UVB related health effects concerns the suppression of immune response to viral and bacterial infections. While the mechanisms for this effect pathway seem well understood, there is little data available to quantify the scale of this health burden (Norval et al. 2011).

Insufficient UVB exposure is also considered a public health concern in many regions of the world, particularly in winter months in the mid-high latitude zones of the globe. UVB exposure is the main source of vitamin D for many population groups, hence low exposure to UVB can induce a Vitamin D deficiency. This has been associated with a range of negative health effects, including multiple sclerosis, diabetes, and some infectious diseases (Norval et al. 2011; UNEP EEAP 2012). While quantitative cause-effect relationships have not yet been established across this spectrum of issues, preliminary analysis illustrates that UVB deficiency could be as substantial as the human health burden that is currently associated with UVB excess (Lucas et al. 2008).

Negative implications of exposure to UVB radiation have been established for many different aquatic species at different trophic levels. However, as for terrestrial environments, there is little information available on broader ecosystem responses to UV change (Hader et al. 2011; UNEP EEAP 2012).

There have been strong causalities established between changes in UVB levels and changes in terrestrial ecosystems. Some field studies show large decreases in plant productivity, as a result of extreme changes in surface UVB radiation in high latitude regions. Others suggest that an increase in UVB levels can have positive productivity effects, by impeding herbivorous insect activity. In terms of more general ecosystem response, there is insufficient data available to establish the net scale or direction of ecological impacts associated with changing levels of UVB radiation (Ballare et al. 2011; UNEP EEAP 2012).

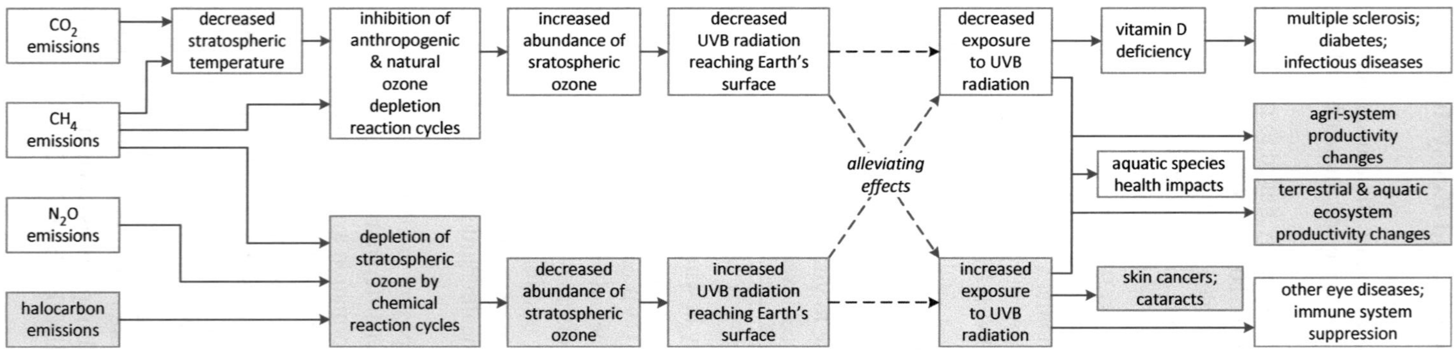

Fig. 4.3 A simplified selection of impact pathways associated with anthropogenic emissions and their effect on the stratospheric ozone layer. Those pathways included in conventional LCA impact assessment models are highlighted in grey. Pathways not shown include the potential for decreased stratospheric temperature to enhance the formation of polar ozone layer 'holes'; and the climate-change mediated effect on atmospheric circulation patterns that will influence the spatial distribution of stratospheric ozone abundance

The effect that changes in UVB radiation will have on agri-production systems is also somewhat unclear. While there are many studies showing crop yields being impeded by increased exposure, there are also concerns that this generalised conclusion has been biased by studies of extreme and unrealistic changes (Kakani et al. 2003; Wargent and Jordan 2013). Recent analysis suggests a complex mix of positive and negative effects (Kataria et al. 2013; Martinez-Luscher et al. 2013; Mazza et al. 2013), with hopes growing that there may be opportunities to use high UVB radiation levels to enhance agri-production outputs (Wargent and Jordan 2013).

4 Spatial and Temporal Variability

4.1 Spatial Variation

Due to the nature of circulation patterns in the atmosphere, the source location of most OAS emissions has little influence on the scale or distribution of ozone layer effects. The exception to this is a group of very-short lived halocarbons, whose chances of reaching the stratosphere will depend on the emission location and tropospheric conditions. It is thought that large quantities of brominated very-short lived substances do reach the stratosphere, although these are predominantly of natural origin (Montzka et al. 2011).

The extent of ozone layer depletion varies substantially at different latitudes (Fig. 4.4), being strongly influenced by stratospheric conditions (e.g. solar radiation levels; temperature) and overall atmospheric circulation patterns. Net ozone generation rates are highest in the tropics, because of the greater temperatures and levels of solar radiation that reach the stratosphere. Ozone abundance in tropical regions has changed little over the past decades. Net depletion rates are most severe in the low temperature polar regions, particularly associated with the transient

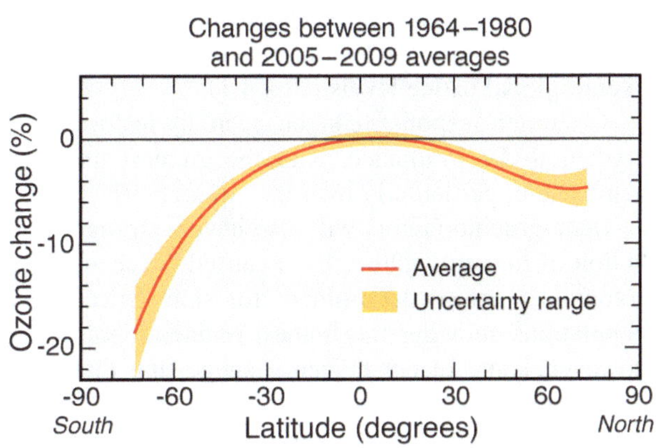

Fig. 4.4 The distribution (by latitude) of ozone layer depletion, from Fahey and Hegglin (2011). These changes are based on measured, rather than modelled, datasets for annual-averaged, total column ozone levels

Box 4.1 Commonly Used Metrics for Analysis of Ozone-Affecting Substances

Ozone Depletion Potentials (ODP) factors are ratios of the ozone change induced by a unit emission of a particular substance, benchmarked against the ozone change induced by a unit emission of CFC-11 (Solomon et al. 1992). ODP values are most commonly generated using steady state calculation approaches. This accounts for the full extent of ozone destruction caused by an emission, regardless of how long it might take for that to eventuate, and regardless of where it happens in the stratosphere. Steady state ODPs can be calculated from modelled predictions of ozone loss; or on a semi-empirical basis directly from observed data on the atmospheric behaviour of chlorine and bromine.

EESC (Equivalent Effective Stratospheric Chlorine) provides a measure of the stratospheric abundance of reactive halogens at any point in time (Newman et al. 2007). N_2O has also been successfully integrated into EESC calculations (Daniel et al. 2010). EESC is frequently used in atmospheric modelling studies for trending the stratospheric response to emission changes, and integrating these trends over time. EESC results are frequently expressed as global-mean values, and in this regard, are analogous to steady-state ODP calculations. For both metrics, global-mean values are weighted by surface area, and therefore more strongly influenced by changes in mid-latitude (and tropical) regions than changes in polar regions.

Erythemal-UV is a term commonly used to express the magnitude of biologically damaging UV radiation. For this, the amount of solar radiation in different wavelengths is weighted by the potency of the UV in those wavelengths to cause sunburn (erythema). The overall intensity of erythemal-UV irradiance is most strongly influenced by radiation levels in the UVB spectrum.

Antarctic ozone layer 'hole' (Fahey and Hegglin 2011). However, it is the smaller decreases at mid-latitudes that make the greatest contribution to reductions in the overall global ozone levels (Box 4.1).

A similar response can be seen in historical changes to surface levels of erythemal-UV irradiance, with the greatest increases occurring in the southern hemisphere, particularly over the Antarctic (Fig. 4.5).

Demographic factors will also have a strong influence on the extent and distribution of human health effects caused by changes in UVB radiation levels. Skin colour is a major determinant for skin cancer development, with fair skinned populations showing the highest incidence rates (Norval et al. 2011). There is, however, some recent evidence suggesting that non-melanoma cancer types are more common in dark skinned populations than previously thought (UNEP EEAP 2012). Incidence of skin cancer and UV-related eye disease also increases with

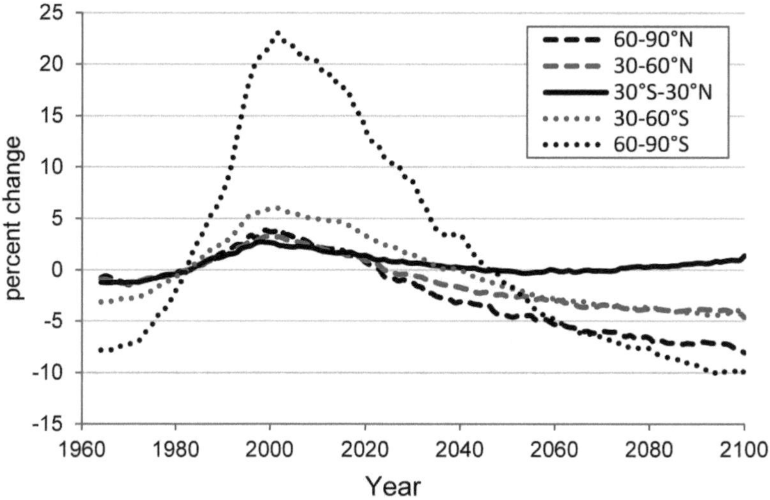

Fig. 4.5 Modelled trends of surface UV for the erythemal spectrum (wavelengths relevant to skin cancer), excluding the potentially important influence of changes in cloud cover, surface reflectivity or tropospheric aerosols (Adapted from Fig. 4.4 of McKenzie et al. (2011), with the permission of The Royal Society of Chemistry (RSC) on behalf of the Centre National de la Recherche Scientifique (CNRS) and the RSC)

population age (Norval et al. 2011), suggesting that the future influence of aging populations is likely to be non-trivial (van Dijk et al. 2013). Differing cultural attitudes to sun protection might also be expected to influence the distribution of human health effects (e.g. Callister et al. 2011; Cancer Council Australia 2007; Nowson et al. 2012).

4.2 Evolving Conditions Over Time

Most ozone affecting substances have long atmospheric lifetimes, and their effect on the ozone layer is manifested over many years. Furthermore, the extents to which halocarbon and non-halocarbon emissions will affect the ozone layer are strongly interlinked, with these interactions also playing out over long timeframes.[1] The future evolution of emission rates will therefore have important implications for estimating the scale of ozone layer depletion that occurs over time.

The significance of these changes can be seen in the steady-state ODP factors provided in Table 4.2. As described in previous sections, the changes in background conditions over this timespan will affect the marginal ozone destructiveness of both

[1] See Portmann et al. (2012) for a concise summary of some important interdependencies that lead to non-linear ozone responses in atmospheric modelling.

Table 4.2 Steady state ODP factors (kg-CFC11e/kg) compiled from Tables 1 and 2 of Fleming et al. (2011)

Year	1850	1950	2000	2100
CFC11	1	1	1	1
N_2O	0.025	0.024	0.019	0.018

These factors are calculated from Ozone changes were modelled for an emission pulse in the year shown, assuming the background conditions for that year remain constant. Each ODP factor was calculated relative to the modelled ozone change induced by an emission of CFC-11 in that same year. The year 2100 atmospheric conditions are based on the WMO-A1 (WMO 2007) and IPCC-A1B (IPCC 2000) scenarios

CFC-11 and N_2O emissions. However, the varying ODP values calculated for N_2O indicate that the two substances respond very differently to the evolving atmospheric conditions.

The changes shown for N_2O also illustrate the complex nature of the interdependencies involved. The ODP for N_2O calculated with year 2000 conditions is much lower than for earlier times, largely because the increased year 2000 levels of atmospheric chlorine, CH_4 and CO_2 all have the effect of inhibiting NO_X-driven degradation of ozone. However, the situation changes looking forward to the year 2100, as these strong influences largely cancel each other out. On the one hand, further increases in atmospheric CO_2 (via stratospheric cooling) and CH_4 (via chemical effects and stratospheric cooling) will dampen the ozone potency of N_2O. On the other hand, lesser atmospheric chlorine means an increase in N_2O potency, as less reactive NO_x gets bound into $ClONO_2$ reservoirs (see Ravishankara et al. 2009; Plummer et al. 2010; Fleming et al. 2011; Portmann et al. 2012; Revell et al. 2012a).

This evolving chemistry is also expected to change the spatial distribution of ozone abundance in the stratosphere. Each of the non-halocarbon species (N_2O, CH_4, CO_2) influence the ozone layer via different pathways, and each has differing spatial implications for ozone status. Also important is the effect that anthropogenic climate change will have on atmospheric circulation patterns, increasing the rate of bulk air and ozone transfer from the tropics (where ozone generation rates are highest) towards the mid-latitudes and poles (see Portmann and Solomon 2007; Plummer et al. 2010; Bekki et al. 2011; Fleming et al. 2011; Revell et al. 2012b; Garny et al. 2013).

This complex mix of influences will change the spatial distribution of global stratospheric ozone in two important ways. Firstly, tropical ozone levels are expected to decrease below those of pre-industrial times, despite overall global ozone levels recovering strongly. Secondly, ozone layer recovery will be much more rapid in the mid-latitudes than in polar regions; and occur sooner in the northern hemisphere than in the south (Austin et al. 2010; Oman et al. 2010; Plummer et al. 2010; Fleming et al. 2011; Garny et al. 2013).

Trends in surface UVB levels are expected to follow a similar (but inverse) pattern (Fig. 4.5), with future UVB levels being higher in regions where they are already high (i.e. the tropics), and lower in the higher latitude regions where winter

UVB levels are already low (Bais et al. 2011; McKenzie et al. 2011). This suggests the future might involve a more complex mix of human health risks related to ozone layer status, which are increased problems associated with UVB excess in the tropics, along with increased problems caused by UVB deficiency in other regions.

5 Midpoint Assessment Methodologies for LCA

Ozone Depletion Potential (ODP) factors (see Box 4.1) for halocarbon emissions have been a cornerstone of midpoint-level impact assessment since the early days of the LCA methodology. The majority of LCIA methods have favoured the use of steady state Ozone Depletion Potential (ODP) values, adopting the halocarbon factors that are updated periodically by the WMO.

It has been rare for non-steady state ODP values to be actively promoted for consideration in midpoint analysis, although the EDIP97 method (Hauschild and Wenzel 1998) and the CML-IA method (Guinée et al. 2002) do include a set of timeframe-specific factors published in the early 1990s (Solomon and Albritton 1992). Unfortunately, there has been little further development of non-steady state approaches in the published literature since that time. As a consequence, LCA practitioners do not have ready access to timeframe-dependent ODPs that also incorporate the latest knowledge on atmospheric behaviour of the different ozone-affecting substances.

There is a generally high degree of confidence ascribed to the ozone layer predictions that are synthesised through the WMO Scientific Assessment Panel review process (Daniel et al. 2011), even though certain fundamental uncertainties in the available atmospheric models are acknowledged (e.g. Bekki et al. 2011; Thompson et al. 2012). Much of these uncertainties are thought to affect the modelling of different halocarbons in a similar manner, and therefore cancel out when the ODP factors are calculated relative to a reference halocarbon substance.

The use of steady-state ODP factors for midpoint LCA modelling continues to be a highly policy-relevant approach. ODP values are formally adopted under the Montreal Protocol, and regularly used for national and international reporting on aspects to do with ozone layer management. In a recent review from the WMO Scientific Assessment Panel, the value of the ODP metric was again promoted because of its simplicity and transparency (Daniel et al. 2011). That report contains an updated set of steady state ODP factors for halocarbon emissions, reflecting developments in atmospheric modelling capability.

However, it remains to be seen whether the conventional ODP metric will retain its relevance into the future. The focus on halocarbon emissions in the WMO assessment reports is in part a legacy of the Montreal Protocol mandate, and does not necessarily imply that halocarbon emissions should still be the highest priority concern. Attention in the science community is now shifting more towards the longer term management of ozone layer recovery, and it is less clear how useful the ODP metric will be in this context.

6 Endpoint Assessment Methodologies for LCA

Early attempts by the LCA research community to model the damages related to stratospheric ozone depletion (e.g. Steen 1999; Goedkoop and Spriensma 2000) were hampered by a lack of quantitative science in formats that could readily support the development of LCA impact models.

More recently, two new approaches to LCA damage assessment have incorporated substantial improvements in basic understanding of the process and effects of ozone layer depletion.

The first of these was the LIME method (Hayashi et al. 2006), which provides characterisation factors for halocarbon compounds. The LIME model uses regressions to link emissions to changes in global ozone, then to changes in surface levels of UVB radiation. Human health damages (in DALYs) are calculated by combining distributions for this change in UVB radiation, with global distributions of skin colour (for skin cancer) and age (for cataracts), with literature based dose–response functions.

A strength of the LIME method is that it provides great breadth in the range of effects that are covered. Characterisation factors are also included for the impacts of UVB changes on net primary productivity for terrestrial (lowland conifer forest) and high-latitude aquatic (phytoplankton) ecosystems. The LIME method also characterises impacts on 'social assets', providing results for UVB-induced changes to yields of global food crops and managed timber production.

The second endpoint model of note is from Struijs et al. (2010).[2] That study also focused exclusively on halocarbon substances, but differs from the LIME method in that it provides characterisation factors only for the human health effects of skin cancer and cataracts.

The Struijs et al. (2010) modelling of skin cancer and cataract effects is based on a much higher spatial resolution for estimating the degree of human exposure to increased UVB radiation, and a more sophisticated treatment of demographics. They also make use of global scale models (van Dijk et al. 2008) to predict the evolution of surface erythemal-UV distributions, accounting for the latest available empirical data as well as climate-sensitive predictions of future cloud cover.

The Struijs et al. (2010) model also has the advantage of being well-aligned with best practice approaches to estimating changes in ozone abundance. Fate factors are calculated from time-integrated estimates of change in global-mean EESC (see Box 4.1). This allows a more robust consideration of the ozone response for longer lived substances, whose destructiveness might change over time as background stratospheric conditions change. The EESC modelling approach adopted standard conventions to minimise uncertainty, although there have been changes to the

[2] The Struijs et al. (2010) model evolved from the version provided with the ReCiPe method (Struijs et al. 2009). Only the more recent version is discussed here, as it contains a number of notable improvements, and the two models are otherwise structurally very similar.

recommended best-practice modelling parameterisation (see Daniel et al. 2011) since those EESC simulations were undertaken.

More fundamental challenges are introduced by the need to calculate absolute, rather than relative, measures of ozone change. In this regard, the strong influence of stratospheric temperature and atmospheric circulation patterns represent notable sources of uncertainty. The available atmospheric models do provide fairly consistent conclusions in regard to both these issues. However, calibration of the models against empirical temperature and circulation rate data has been problematic, and this remains a substantial concern (Bekki et al. 2011; Thompson et al. 2012).

Uncertainties are also introduced when translating ozone change into quantitative predictions for disease-relevant levels of UVB radiation. This step is made difficult by limitations in the ability to model the influence of other confounding factors (e.g. cloud cover), and uncertainty over the spectrum of wavelengths most relevant to each of the different endpoint effects (McKenzie et al. 2011). Additional complications exist for estimating exposure to the eyes, where quantified surface levels of UVB radiation are not necessarily a good proxy for exposure (McKenzie et al. 2011; Norval et al. 2011). Furthermore, the influence of changing social behavioural patterns is likely to be strong (Norval et al. 2011).

For the damage pathways that are included in the two 'best-practice' endpoint LCA models (skin cancer; cataracts; terrestrial productivity response; aquatic productivity response), there has been a considerable increase in the body of published data since they were developed. Notwithstanding that, many important data gaps remain. Even for skin cancer, the most well studied of this group, the lack of data from developing countries and dark-skinned populations is seen as a concern (Lucas et al. 2006; Wright et al. 2012).

Other uncertainties in the Struijs et al. (2010) calculation of ozone, and changes in UVB exposure, are introduced with choices made in their modelling design. These are discussed further in Sect. 7.

7 New Developments and Research Needs

7.1 *Including Non-halocarbons in LCA Analysis of Stratospheric Ozone Depletion*

Conventional LCA already has a strong focus on emissions of N_2O, CH_4 and CO_2, due to their links to climate change. Also including these substances in life-cycle ozone depletion assessment could potentially change the conclusions that might be drawn in certain studies (e.g. Lane and Lant 2012). Doing so would be methodologically consistent with the marginal-assessment approach favoured in Life-cycle Impact Assessment (see Huijbregts et al. 2011), given these substances will be the key drivers of ozone layer status into the future.

While there is not yet any policy imperative for such a change, it remains to be seen how the international community will respond to the growing scientific focus on the importance of non-halocarbon emissions. For N_2O, there have already been calls for the Montreal Protocol to be used to regulate anthropogenic emissions (Kanter et al. 2013).

If N_2O, CH_4 and CO_2 are to be included in conventional LCIA of ozone layer effects, then further critical review is required into the appropriate means for doing so. The atmospheric modelling community has been cautious in their recommendation of the available ODP factors for non-halocarbons, because the ozone-layer effects of N_2O, CH_4 and CO_2 are manifested in very different ways to those of halocarbons. This disparity increases the chance that substance comparisons could be unevenly affected by inherent bias in the underpinning atmospheric models (Daniel et al. 2011; Fleming et al. 2011).

Additional complications exist when considering CO_2 and CH_4 emissions. Negative ODP factors have been calculated for these two substances, reflecting their contribution to increasing overall global ozone levels (Fleming et al. 2011). While not a common practice, it is perfectly feasible for negative characterisation factors to be used in LCA (e.g. De Schryver et al. 2009).

Conventionally calculated ODP values might, however, have limited relevance as a proxy for the effects of CO_2 and CH_4emissions. These substances will be the primary driver of any future ozone layer 'super-recovery', which in turn will have very different human health implications (diseases associated with insufficient exposure to UVB radiation) than those associated with no-recovery (diseases associated with excessive exposure to UVB radiation). Combining both these impact pathways might require a holistic reconsideration of the design of LCA metrics (midpoint and endpoint) for assessing ozone-affecting substances.

7.2 *Sensitivity to Changing Atmospheric Conditions*

Forecast atmospheric changes over the course of the twenty-first century will substantially change the conditions that regulate the stratospheric ozone cycle. For all halocarbon and non-halocarbon OAS, this will mean changes in the extent to which current and future anthropogenic emissions impact on the ozone layer. Furthermore, the different substances will respond differently to these changing stratospheric conditions. Because the substance interdependencies are so strong, predicting the ozone depletion caused by a marginal OAS emission will also be very dependent on the assumptions made about future emission trends for halocarbons, N_2O, CH_4 and CO_2. These emission forecasts are themselves a source of great uncertainty.

Steady state ODP factors are calculated on the basis that atmospheric conditions remain constant over time, and are therefore unable to capture such complexity. LCA practitioners may therefore benefit from having access to a number of possible characterisation factor sets, each reflecting a different set of underpinning atmospheric conditions.

This would not, however, address the fundamental discrepancy between the steady-state calculation approach and the fate factors used in the Struijs et al. (2010) endpoint model. The latter were modelled for an evolving set of atmospheric conditions in response to a temporally varying profile of halocarbon and other greenhouse gas emissions. These midpoint and endpoint approaches could potentially reach very different conclusions when comparing substances that respond differently to changing background conditions, whether that is because of different atmospheric lifespans (e.g. CFC-12 vs. HCFC-123), or because they are affected by different chemical pathways (e.g. CFC-12 vs. N_2O).

7.3 *Including 'Post-recovery' Effects*

Another source of inconsistency between midpoint and endpoint metrics, once again related to differing approaches to time-integration, would be the choice to truncate the life cycle impact assessment at some pre-conceived point of ozone layer 'recovery'.

The defining feature of steady-state ODP factors is that they account for the full amount of ozone destruction caused by an emission, regardless of whether that is expressed over 1 year (e.g. HCFC-123) or over 100 years (e.g. CFC-12). In contrast, the Struijs et al. (2010) endpoint model truncates damage assessment at the year 2049, on the premise that health effects beyond that date would be at 'normal' background levels. Using that approach, endpoint analysis would only account for some portion of the ozone depletion impacts caused by longer lived emissions, but all of those caused by shorter-lived emissions. Once again, it seems likely that these midpoint and endpoint approaches could reach different conclusions when applied in certain contexts.

A further complication with the truncation approach is that the choice of 'recovery' benchmark is not straight forward. Three issues warrant consideration in this regard. Firstly, the policy convention of benchmarking recovery against conditions in the year 1980 (as followed by Struijs et al. 2010) represents a somewhat arbitrary choice for LCA. Many recent studies indicate that ozone levels in 1980 were already impacted by anthropogenic activity (Austin et al. 2010; Eyring et al. 2010; Oman et al. 2010; Plummer et al. 2010; Fleming et al. 2011; Portmann et al. 2012). Secondly, the timing for a return to pre-industrial levels of global ozone will be strongly influenced by the choice of greenhouse gas emission scenarios (Eyring et al. 2010). Thirdly, this benchmarking of global ozone abundance disguises the expectation that 'recovery' dates will vary substantially at different latitudes. Recovery over the northern hemisphere mid-latitudes, where much of the world's fair skinned population resides, will happen much sooner than elsewhere on the globe. At the other extreme, ozone abundance over the tropics might still be lower than pre-industrial levels, well into the twenty-second century.

This complexity suggests that LCA practice would benefit from having a choice of characterisation factors, reflecting (a) different time frames of concern, and

(b) different forecasts for the evolution of greenhouse gas emission rates. It also raises the question of what to do about the likelihood of 'post-recovery' health effects associated with low (insufficient) exposure to UVB radiation.

7.4 Using Time-Integrated Midpoint Characterisation Factors

As an alternative to using steady-state calculations, midpoint level characterisation factors could instead be generated with time-integrated modelling applied over defined time frames. The latter approach is already the norm for midpoint-level analysis of climate change impacts. Time frame-dependent ODP calculations are methodologically possible (see Solomon and Albritton 1992), and could potentially provide some flexibility in how the issues of evolving background atmospheric conditions (Sect. 7.2) and recovery date truncation (Sect. 7.3) are considered.

Given the likely sensitivity of this approach to the choice of greenhouse gas emission scenarios, consideration should be given to maintaining consistency with any assumptions used for developing climate change characterisation factors.

7.5 Spatially Resolved Modelling of Changes in UVB Levels

The two most current endpoint-level LCA models both estimate the change in harmful UV levels based on historical relationships between global ozone abundance (or its proxies), and the spatial distribution of UVB radiation levels at the surface. Both assume that the spatial distribution of marginal changes in these UVB radiation levels will be the same for all emission types.

This approach does not reflect the expectation that future surface UVB distributions will evolve in a very different manner to the way they changed in the past. Furthermore, the assumption that all substances have a similar spatial influence may be less valid if non-halocarbons are to be considered in the analysis of ozone depletion effects.

Further investigation would be required to identify whether or not these differences would have a material effect on calculated endpoint characterisation factors; and how sensitive this might be to changing greenhouse gas emission scenarios.

7.6 Estimating the Damages

Opportunities exist to improve on the dose–response information used to convert exposure rates into estimates of health damage, and to extend the range of disease types that are considered in LCA. Recent years have delivered substantial new insight into trends of melanoma and non-melanoma skin cancer incidence, and also for the growing occurrence of Merkel Cell Carcinoma. Previous quantitative studies (e.g. Lucas et al. 2008) might provide a template for LCA endpoint analysis to include less severe effects such as sunburn, pterygium, and recurring viral infections.

Endpoint damage modelling will also need to adequately consider the substantial uncertainty involved in predicting future behavioural patterns. Behavioural change could potentially have a very significant influence on the human health burdens associated with UVB exposure.

For the consideration of ecosystem and agri-system effects, existing endpoint LCA models should be updated to better reflect contemporary thinking. Many new studies exist to inform the quantification of plant response to changes in UVB radiation levels, although the weight of conflicting evidence may make it difficult to produce generalised dose–response functions.

Finally, for LCA endpoint analysis to embrace a new phase of ozone layer management, a more fundamental reconsideration of endpoint modelling priorities might be required. First of all, greater attention to human and ecosystem effects in tropical zones might be warranted, given the expectation that UVB levels in that region will remain elevated for the foreseeable future. Secondly, growing concerns about the health implications of under-exposure to UVB radiation suggest that this impact pathway should be considered for inclusion in the LCA framework.

Acknowledgements Prof. Robyn Lucas, of the Australian National University, provided guidance on the topic of human health damages, and useful feedback on a draft of this chapter. Her contributions were greatly appreciated. Thanks also to Prof. Alkis Bais (Aristotle University of Thessaloniki) for assistance with producing Fig. 4.5.

References

Agelli M, Clegg LX, Becker JC, Rollison DE (2010) The etiology and epidemiology of Merkel cell carcinoma. Curr Probl Cancer 34(1):14–37. doi:10.1016/j.currproblcancer.2010.01.001

Austin J, Scinocca J, Plummer D et al (2010) Decline and recovery of total column ozone using a multimodel time series analysis. J Geophys Res Atmos 115:D00M10. doi:10.1029/2010JD013857

Bais AF, Tourpali K, Kazantzidis A et al (2011) Projections of UV radiation changes in the 21st century: impact of ozone recovery and cloud effects. Atmos Chem Phys 11:7533–7545. doi:10.5194/acp-11-7533-2011

Ballare CL, Caldwell MM, Flint SD et al (2011) Effects of solar ultraviolet radiation on terrestrial ecosystems. Patterns, mechanisms, and interactions with climate change. Photochem Photobiol Sci 10:226–241. doi:10.1039/c0pp90035d

Bekki S, Bodeker GE, Bais AF, Butchart N, Eyring V, Fahey DW et al (2011) Future ozone and its impact on surface UV. In: Scientific assessment of ozone depletion: 2010, vol 52, Global Ozone Research and Monitoring Project—Report. World Meteorological Organization, Geneva, p 516

Callister P, Galtry J, Didham R (2011) The risks and benefits of sun exposure: should skin colour or ethnicity be the main variable for communicating health promotion messages in New Zealand? Ethn Health 16:57–71. doi:10.1080/13557858.2010.527925

Cancer Council Australia (2007) Risks and benefits of sun exposure – position statement. Website: http://www.cancer.org.au/policy-and-advocacy/position-statements/sun-smart/#jump_1. Accessed 2 May 2014

Crutzen PJ (1970) The influence of nitrogen oxides on the atmospheric ozone content. QJR Meteorol Soc 96:320–325. doi:10.1002/qj.49709640815

Daniel JS, Fleming EL, Portmann RW et al (2010) Options to accelerate ozone recovery: ozone and climate benefits. Atmos Chem Phys 10:7697–7707. doi:10.5194/acp-10-7697-2010

Daniel JS, Velders GJM, Morgenstern O, Toohey DW, Wallington TJ, Wuebbles D et al (2011) A focus on information and options for policymakers. In: Scientific assessment of ozone depletion: 2010, vol 52, Global Ozone Research and Monitoring Project—Report. World Meteorological Organization, Geneva, p 516

De Schryver AM, Brakkee KW, Goedkoop MJ, Huijbregts MAJ (2009) Characterization factors for global warming in life cycle assessment based on damages to humans and ecosystems. Environ Sci Technol 43:1689–1695. doi:10.1021/es800456m

Douglass A, Fioletev V, Godin-Beekmann S, Muller R, Stolarski RS, Webb A et al (2011) Stratospheric ozone and surface ultraviolet radiation. In: Scientific assessment of ozone depletion: 2010, vol 52, Global Ozone Research and Monitoring Project—Report. World Meteorological Organization, Geneva, p 516

Eyring V, Cionni I, Lamarque JF et al (2010) Sensitivity of 21st century stratospheric ozone to greenhouse gas scenarios. Geophys Res Lett 37:L16807. doi:10.1029/2010GL044443

Fahey DW, Hegglin MI (2011) Twenty questions and answers about the ozone layer: 2010 update. In: Scientific assessment of ozone depletion: 2010, vol 52, Global Ozone Research and Monitoring Project—Report. World Meteorological Organization, Geneva, p 516

Fleming EL, Jackman CH, Stolarski RS, Douglass AR (2011) A model study of the impact of source gas changes on the stratosphere for 1850–2100. Atmos Chem Phys 11:8515–8541. doi:10.5194/acp-11-8515-2011

Forster PM, Thompson DWJ, Baldwin MP, Chipperfield MP, Dameris M, Haigh JD et al (2011) Stratospheric changes and climate. In: Scientific assessment of ozone depletion: 2010, vol 52, Global Ozone Research and Monitoring Project—Report. World Meteorological Organization, Geneva, p 516

Garcia RR, Kinnison DE, Marsh DR (2012) "World avoided" simulations with the Whole Atmosphere Community Climate Model. J Geophys Res Atmos 117. doi:10.1029/2012JD018430

Garny H, Bodeker GE, Smale D, Dameris M, Grewe V (2013) Drivers of hemispheric differences in return dates of mid-latitude stratospheric ozone to historical levels. Atmos Chem Phys 13 (15):7279–7300. doi:10.5194/acp-13-7279-2013

Girschik J, Thorn K, Beer TW et al (2011) Merkel cell carcinoma in Western Australia: a population-based study of incidence and survival. Br J Dermatol 165:1051–1057. doi:10.1111/j.1365-2133.2011.10493.x

Goedkoop M, Spriensma R (2000) The Eco-indicator 99: a damage oriented method for life cycle impact assessment. Pre Consultants, The Hague

Guinée JB, Gorée M, Heijungs R, Huppes G, Kleijn R, Koning A, Oers L, Wegener Sleeswijk A, Suh S, Udo de Haes HA, Bruijn H, Duin R, Huijbregts MAJ (2002) Handbook on life-cycle assessment: operational guide to the ISO standards. Kluwer Academic Publishers, Dordrecht

Hader DP, Helbling EW, Williamson CE, Worrest RC (2011) Effects of UV radiation on aquatic ecosystems and interactions with climate change. Photochem Photobiol Sci 10(2):242–260. doi:10.1039/c0pp90036b

Hauschild MZ, Wenzel H (1998) Environmental assessment of products, vol 2, Scientific background. Chapman & Hall, London and Kluwer Academic Publishers, Hingham. ISBN 0412 80810 2

Hayashi K, Nakagawa A, Itsubo N, Inaba A (2006) Expanded damage function of stratospheric ozone depletion to cover major endpoints regarding life cycle impact assessment. Int J Life Cycle Assess 11(3):150–161. doi:10.1065/lca2004.11.189

Huijbregts MAJ, Hellweg S, Hertwich E (2011) Do we need a paradigm shift in life cycle impact assessment? Environ Sci Technol 45(9):3833–3834. doi:10.1021/es200918b

IPCC (2000) Special report on emissions scenarios: a special report of Working Group III of the Intergovernmental Panel on Climate Change. Cambridge, UK, p 599

IPCC (2007) In: Solomon S, Qin S, Manning M, Chen Z, Marquis M, Averyt K, Tignor M, Miller H (eds) Climate Change 2007: the physical science basis. Contribution of working group I to the fourth assessment report of the Intergovernmental Panel on Climate Change. Cambridge, UK, p 996

Johnston H (1971) Reduction of stratospheric ozone by nitrogen oxide catalysts from supersonic transport exhaust. Science 173:517–522. doi:10.1126/science.173.3996.517

Kakani VG, Reddy KR, Zhao D, Sailaja K (2003) Field crop responses to ultraviolet-B radiation: a review. Agric For Meteorol 120(1–4):191–218. doi:10.1016/j.agrformet.2003.08.015

Kanter D, Mauzerall DL, Ravishankara AR et al (2013) A post-Kyoto partner: considering the stratospheric ozone regime as a tool to manage nitrous oxide. Proc Natl Acad Sci U S A 110:4451–4457. doi:10.1073/pnas.1222231110

Kataria S, Guruprasad KN, Ahuja S, Singh B (2013) Enhancement of growth, photosynthetic performance and yield by exclusion of ambient UV components in C3 and C4 plants. J Photochem Photobiol B Biol 127:140–152. doi:10.1016/j.jphotobiol.2013.08.013

Kuwamoto S (2011) Recent advances in the biology of Merkel cell carcinoma. Hum Pathol 42:1063–1077. doi:10.1016/j.humpath.2011.01.020

Lane J, Lant P (2012) Including N_2O in ozone depletion models for LCA. Int J Life Cycle Assess 17(2):252–257. doi:10.1007/s11367-011-0362-y

Lucas R, McMichael T, Smith W, Armstrong B (2006) In: Pruss-Ustun A, Zeeb H, Mathers C, Repacholi MH (eds) Solar ultraviolet radiation: global burden of disease from solar ultraviolet radiation. World Health Organization, Geneva, p 250

Lucas RM, McMichael AJ, Armstrong BK, Smith WT (2008) Estimating the global disease burden due to ultraviolet radiation exposure. Int J Epidemiol 37:654–667. doi:10.1093/ije/dyn017

Manney GL, Santee ML, Rex M et al (2011) Unprecedented Arctic ozone loss in 2011. Nature 478:469–475. doi:10.1038/nature10556

Martinez-Luscher J, Morales F, Delrot S, Sanchez-Diaz M, Gomes E, Aguirreolea J, Pascual I (2013) Short- and long-term physiological responses of grapevine leaves to UV-B radiation. Plant Sci 213:114–122. doi:10.1016/j.plantsci.2013.08.010

Mazza CA, Giménez PI, Kantolic AG, Ballaré CL (2013) Beneficial effects of solar UV-B radiation on soybean yield mediated by reduced insect herbivory under field conditions. Physiol Plant 147:307–315. doi:10.1111/j.1399-3054.2012.01661.x

McKenzie RL, Aucamp PJ, Bais AF, Bjorn LO, Ilyas M, Madronich S (2011) Ozone depletion and climate change: impacts on UV radiation. Photochem Photobiol Sci 10(2):182–198. doi:10.1039/c0pp90034f

Montzka SA, Reimann S, Engel A, Kruger K, O'Doherty S, Sturgess WT et al (2011) Ozone-depleting substances (ODSs) and related chemicals. In: Scientific assessment of ozone depletion: 2010, vol 52, Global Ozone Research and Monitoring Project—Report. World Meteorological Organization, Geneva, p 516

Newman PA, McKenzie R (2011) UV impacts avoided by the Montreal Protocol. Photochem Photobiol Sci 10(7):1152–1160. doi:10.1039/c0pp00387e

Newman PA, Daniel JS, Waugh DW, Nash ER (2007) A new formulation of equivalent effective stratospheric chlorine (EESC). Atmos Chem Phys 7(17):4537–4552. doi:10.5194/acp-7-4537-2007

Newman PA, Oman LD, Douglass AR et al (2009) What would have happened to the ozone layer if chlorofluorocarbons (CFCs) had not been regulated? Atmos Chem Phys 9(6):2113–2128. doi:10.5194/acp-9-2113-2009

Norval M, Lucas RM, Cullen AP, de Gruijl FR, Longstreth J, Takizawa Y, van der Leun JC (2011) The human health effects of ozone depletion and interactions with climate change. Photochem Photobiol Sci 10(2):199–225. doi:10.1039/c0pp90044c

Nowson CA, McGrath JJ, Ebeling PR et al (2012) Vitamin D and health in adults in Australia and New Zealand: a position statement. Med J Aust 196:686–687. doi:10.5694/mja11.10301

Oman LD, Plummer DA, Waugh DW et al (2010) Multimodel assessment of the factors driving stratospheric ozone evolution over the 21st century. J Geophys Res. doi:10.1029/2010JD014362

Plummer DA, Scinocca JF, Shepherd TG, Reader MC, Jonsson AI (2010) Quantifying the contributions to stratospheric ozone changes from ozone depleting substances and greenhouse gases. Atmos Chem Phys 10(18):8803–8820. doi:10.5194/acp-10-8803-2010

Portmann RW, Solomon S (2007) Indirect radiative forcing of the ozone layer during the 21st century. Geophys Res Lett. doi:10.1029/2006GL028252

Portmann RW, Daniel JS, Ravishankara AR (2012) Stratospheric ozone depletion due to nitrous oxide: influences of other gases. Philos Trans R Soc Lond B Biol Sci 367:1256–1264. doi:10.1098/rstb.2011.0377

Ravishankara AR, Daniel JS, Portmann RW (2009) Nitrous oxide (N_2O): the dominant ozone-depleting substance emitted in the 21st century. Science 326(5949):123–125. doi:10.1126/science.1176985

Revell LE, Bodeker GE, Smale D et al (2012a) The effectiveness of N_2O in depleting stratospheric ozone. Geophys Res Lett 39(15). doi:10.1029/2012GL052143

Revell LE, Bodeker GE, Huck PE, Williamson BE, Rozanov E (2012b) The sensitivity of stratospheric ozone changes through the 21st century to N_2O and CH_4. Atmos Chem Phys 12 (23):11309–11317. doi:10.5194/acp-12-11309-2012

Rowland FS (2006) Stratospheric ozone depletion. Philos Trans R Soc Lond B Biol Sci 361 (1469):769–790. doi:10.1098/rstb.2005.1783

Solomon S, Albritton DL (1992) Time-dependent ozone depletion potentials for short-term and long-term forecasts. Nature 357(6373):33–37. doi:10.1038/357033a0

Solomon S, Mills M, Heidt LE, Pollock WH, Tuck AF (1992) On the evaluation of ozone depletion potentials. J Geophys Res 97(D1):825–842

Steen B (1999) A systematic approach to environmental priority strategies in product development (EPS) Version 2000 – models and data of the default method. Chalmers University of Technology, Göteborg

Struijs J, Van Wijnen HJ, van Dijk A, Huijbregts MAJ (2009) Ozone depletion. In: Goedkoop M, Heijungs R, Huijbregts MAJ, De Schryver A, Struijs J, Van Zelm R (eds) ReCiPe 2008: a life cycle impact assessment method which comprises harmonised category indicators at the midpoint and the endpoint level, 1st edn. Pre Consultants, CML University of Leiden, Radboud University, RIVM, Bilthoven, pp 37–53

Struijs J, van Dijk A, Slaper H, van Wijnen HJ, Velders GJM, Chaplin G, Huijbregts MAJ (2010) Spatial- and time-explicit human damage modeling of ozone depleting substances in life cycle impact assessment. Environ Sci Technol 44(1):204–209. doi:10.1021/es9017865

Syakila A, Kroeze C (2011) The global nitrous oxide budget revisited. Greenh Gas Meas Manag 1:17–26. doi:10.3763/ghgmm.2010.0007

Thompson DWJ, Seidel DJ, Randel WJ et al (2012) The mystery of recent stratospheric temperature trends. Nature 491:692–697. doi:10.1038/nature11579

UNEP EEAP (2012) Environmental effects of ozone depletion and its interactions with climate change: progress report 2011. Photochem Photobiol Sci 11:13–27. doi:10.1039/C1PP90033A

van Dijk A, den Outer PN, Slaper H (2008) Climate and ozone change effects on ultraviolet radiation and risks (COEUR) – using and validating earth observation. RIVM, Bilthoven, p 70

Van Dijk A, Slaper H, den Outer PN et al (2013) Skin cancer risks avoided by the Montreal Protocol – worldwide modeling integrating coupled climate-chemistry models with a risk model for UV. Photochem Photobiol 89:234–246. doi:10.1111/j.1751-1097.2012.01223.x

Wargent JJ, Jordan BR (2013) From ozone depletion to agriculture: understanding the role of UV radiation in sustainable crop production. New Phytol 197:1058–1076. doi:10.1111/nph.12132
WMO (2007) Scientific assessment of ozone depletion: 2006, Global Ozone Research and Monitoring Project – Executive Summary. World Meteorological Organization, Geneva, p 572
WMO (2011) Scientific assessment of ozone depletion: 2010, vol 52, Global Ozone Research and Monitoring Project-Report. World Meteorological Organization, Geneva, p 516
Wright CY, Norval M, Summers B et al (2012) The impact of solar ultraviolet radiation on human health in sub-Saharan Africa. S Afr J Sci 108:23–28
Xiao X, Prinn RG, Fraser PJ et al (2010) Atmospheric three-dimensional inverse modeling of regional industrial emissions and global oceanic uptake of carbon tetrachloride. Atmos Chem Phys 10(21):10421–10434. doi:10.5194/acp-10-10421-2010
Yvon-Lewis SA, Saltzman ES, Montzka SA (2009) Recent trends in atmospheric methyl bromide: analysis of post-Montreal Protocol variability. Atmos Chem Phys 9(16):5963–5974. doi:10.5194/acp-9-5963-2009

Chapter 5
Human Toxicity

Olivier Jolliet and Peter Fantke

Abstract This chapter reviews the human toxicological impacts of chemicals and how to assess these impacts in life cycle impact assessment (LCIA), in order to identify key processes and pollutants. The complete cause-effect pathway – from emissions of toxic substances up to damages on human health – demonstrates the importance to account for both outdoor and indoor exposure, including consumer products. Analysing the variations in intake fraction (the fraction of the emitted or applied chemical that is taken in by the consumer and the general population), effect factor and characterisation factor across all chemicals and impact pathways characterizes the contribution of each factor to the total variation of 10–12 orders of magnitude in impacts per kg across all chemicals. This large variation between characterisation factors for different chemicals as well as the 3 orders of magnitude uncertainty on characterisation factors means that results should by default be reported and interpreted in log scales when comparing scenarios or substance contribution! We conclude by outlining future trends in human toxicity modelling for LCIA, with promising developments for (a) better estimates of degradation half-lives, (b) the inclusion of ionization of chemicals in human exposure including bioaccumulation, (c) metal speciation, (d) spatialised models to differentiate the variability associated with spatialisation from the uncertainty, and (e) the assessment of chemical exposure via consumer products and occupational settings. As a whole, the assessment of toxicity in LCA has progressed on a very sharp learning curve during the past 20 years. This rapid progression is expected to continue in the coming years, focusing more on direct exposure of workers to chemicals during manufacturing and of consumers during product use.

The first section of this chapter outlines the complete cause-effect pathway, from emissions of toxic substances to intake by the population up to damages in terms of

O. Jolliet (✉)
School of Public Health, Department of Environmental Health Sciences, University of Michigan, 109 S. Observatory, Ann Arbor, MI 48109-2029, USA
e-mail: ojolliet@umich.edu

P. Fantke
Division for Quantitative Sustainability Assessment, Department of Management Engineering, Technical University of Denmark (DTU), Produktionstorvet 424, 2800 Kgs., Lyngby, Denmark

M.Z. Hauschild, M.A.J. Huijbregts (eds.), *Life Cycle Impact Assessment*, LCA Compendium – The Complete World of Life Cycle Assessment, DOI 10.1007/978-94-017-9744-3_5

human health effects. Section 2 outlines the framework for assessing human toxicity in LCIA. Section 3 discusses the contributing substances and their coverage in LCIA methods. Section 4 provides an overview of the main LCIA methods available to address human toxicological impacts. Section 5 presents the range of variation of factor across chemicals, the main sources of uncertainty and good interpretation practice of results from human toxicity assessments. Section 6 finally discusses new developments and research needs.

Keywords Cause-effect-pathway • Consumer products • Dose–response relationship • Endpoint • Environmental fate • Exposure • Health effects • Human toxicity • Intake fraction • LCA • LCIA • Life cycle assessment • Life cycle impact assessment • Midpoint • Occupational health • USEtox

1 Introduction

The scope and methodology of a life cycle assessment (LCA) differs from that of many approaches adopted for toxicological assessments in a regulatory context. Regulatory assessments of chemical emissions usually have the objective to avoid unacceptable risks of a toxicological effect to a sensitive subpopulation or an individual person. Thus, very conservative (worst case) or conservative, but realistic (reasonable worst case) assumptions are applied in assessment frameworks and extrapolation factors to ensure safety (Covello and Merkhoher 1993; IGHRC Interdepartmental Group on Health Risks from Chemicals 2003; van Leeuwen and Vermeire 2007). LCA, in contrast, primarily aims at comparing products (goods and services) and their relative impacts associated with the release of quantities of chemical into the environment. LCA therefore aims at comparing best estimates of risks based on the likelihood of a toxicological impact integrated over time and space (Udo de Haes et al. 1999).

Historically, Life Cycle toxicity assessment has thus borrowed most of its data and methods from risk assessment practices, adapting them for comparative assessment and contributing to define new population-based concepts better suited for such comparisons. The learning curve has been very sharp, from very basic methods in the early nineties based on residence time of substances in the environment, to the development of multi-media models such as USEtox that are now also used in risk assessment for chemical screening (Wambaugh et al. 2013).

Traditionally, LCA has mainly focused on environmentally-mediated toxic emissions and impacts. Recent discoveries show that larger exposures can occur near-field e.g. during the use stage of consumer products (Wambaugh et al. 2013). As a product-oriented approach with well-defined usage, LCA is a priori well suited to assess potential impacts of consumer products, but is paradoxically still underdeveloped in this area. It is therefore necessary to extend LCA practices with an improved focus on product use and to develop appropriate metrics for assessing these direct impacts during the use stage.

2 Impact Pathway and Framework

2.1 Impact Pathway

Figure 5.1 presents the environmental mechanism for human toxicity impacts, where the product life cycle generates releases of chemicals into the environment and thus an increase in the chemical mass in indoor and outdoor air, in water, soil and biota. Humans are exposed to these chemical releases directly by inhaling the air and drinking the water or indirectly by ingesting chemicals that bioaccumulate in the human food chain (Huijbregts et al. 2005b; Rosenbaum et al. 2011). Consumers and workers may also be exposed by direct dermal contact to chemicals in products applied to the skin or by contact with chemicals embedded in different products used in various ways in daily private and professional life (Jolliet et al. 2012a). Once taken into humans, chemicals distribute inside the body, may damage target organs, induce the onset of various diseases, causing periods with disability or loss of life in case of death. This framework is essentially common to both LCA and RA (Risk Assessment), the difference between these approaches being primarily related to the kind of default data (best estimate versus conservative estimates) and assumptions (average consumption versus high end consumption) that are made to characterise intakes and resulting health responses.

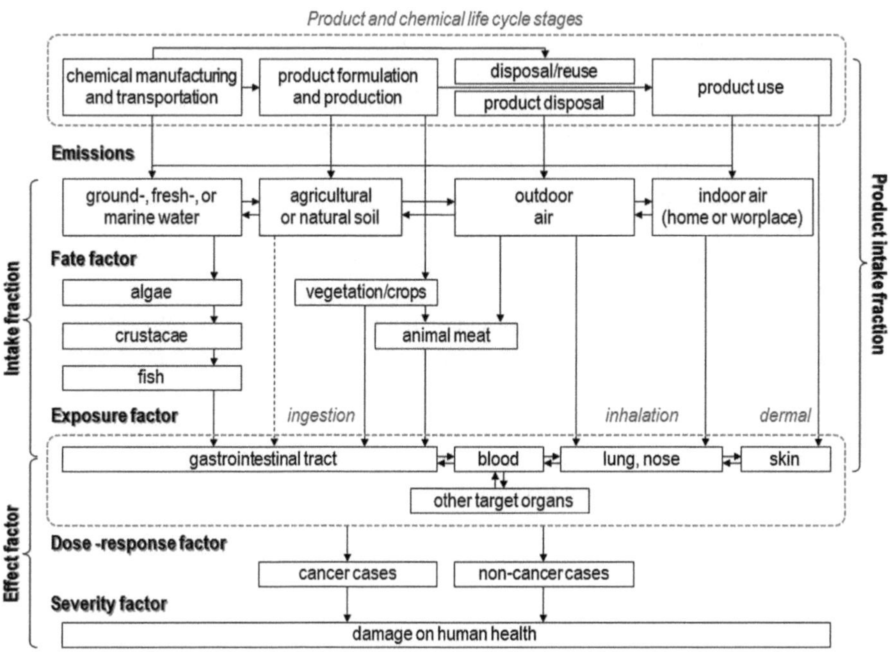

Fig. 5.1 Cause-effect pathway for human toxicity impacts

Box 5.1 Criteria for Evaluation of Characterisation Models for Human Toxicity

Criteria for sound modelling of human toxicity in LCIA

- Emissions from indoor sources and into urban areas are considered separately and advection out of a region or beyond the boundaries of a continent is not considered a final loss.
- Influential fate processes and characteristics are considered including typically volatilisation from soil, water, and plant surfaces, deposition onto soil, water, and plants, degradation in all media, sedimentation, adsorption to colloid matter, phase partitioning, and the intermittent character of rain.
- Main exposure pathways are covered, i.e. inhalation of air, ingestion of drinking water, ingestion of food (crops, meat, dairy products, fish, and eggs), and dermal uptake, accounting for biomagnification, and ensuring that carry over rates do comply with mass balance principles even in case of high bioaccumulation.
- Occupational exposure during manufacturing and consumer exposure during product use are considered.
- Human data are preferably used and dose–response slopes are derived from benchmark measures, i.e. an effective dose inducing a 10 % response over background, ED_{10} (Crettaz et al. 2002; Pennington et al. 2002), or a 50 % response over background, ED_{50} (Huijbregts et al. 2005a), in the exposed population for a lifetime exposure as a point of departure, thereby avoiding the use of arbitrary safety factors. If for non-cancer effects ED_{10} or ED_{50} data are not available, extrapolations are based on the no-observed adverse effect level (NOAEL) and lowest observed adverse effect level (LOAEL) (Huijbregts et al. 2005a).
- Chemicals that are negative in all cancer bioassays are accounted for differently than chemicals with non-available data and exposure route to route extrapolation is included (Rosenbaum et al. 2011).
- Severity and aggregation are transparent and intermediary results are kept separate. Severity factors are based on the latest data from the World Health Organization (WHO) and the Global Burden of Disease (Lim et al. 2012).

Building on European Commission (2010b, 2011), Box 5.1 defines criteria for sound modelling of the human toxicity cause-effect chain in LCIA that can be used.

2.2 *Framework*

The links of the cause-effect chain for human toxicity can be modelled according to the successive steps of environmental fate, exposure, dose–response, and severity of response. The resulting human health-related impact score *IS* often expressed in [disease case] or in [disability-adjusted life years, DALY] can be determined from the chemical mass m [$kg_{emitted}$] emitted to the environment, adapting the framework proposed by Rosenbaum et al. (2007):

$$IS = m \cdot \underbrace{\underbrace{FF \cdot XF}_{iF} \cdot \underbrace{DRF \cdot SF}_{EF}}_{CF} = m \cdot \underbrace{iF \cdot EF}_{CF} = m \cdot CF \tag{5.1}$$

where (left side in Fig. 5.1):

(a) The **fate factor** FF [day] links the quantity of a chemical released into the environment to the chemical masses (or concentrations) in a given environmental compartment. The fate factor can be interpreted as the increase of chemical mass in a certain compartment [kg] due to an emission of the chemical in the same or another compartment [$kg_{emitted}$/day]. FF is proportional to the fraction transferred from emitting to receiving compartment multiplied by the residence time in this receiving compartment.

(b) The **exposure factor** XF [1/day] relates the amount of a chemical found in a given environmental compartment to the amount of the chemical taken in by a human population. It can be distinguished between direct intake (e.g. breathing air, drinking water), and indirect intake through bioaccumulation processes in animal tissues (e.g. meat, milk and fish) and uptake through dermal contact. The exposure factor is the equivalent rate of ingestion of the environmental medium (air, water) by humans.

(a&b) The human **intake fraction** $iF = FF \cdot XF$ [$kg_{intake}/kg_{emitted}$] is the combination of the fate and exposure factors. The intake fraction can be interpreted as the fraction of an emitted chemical mass that is taken in by a human population (Bennett et al. 2002a, b).

(c) The **dose–response slope factor** DRF [cases/kg_{intake}] relates the quantity of chemical taken in by a human population to the likelihood (or potential risk) of developing an adverse effect, often expressed in cases of cancer or non-cancer per unit mass of the chemical taken in.

(d) The optional **severity factor** SF [DALY/case] distinguishes between differences in the severity of disabilities caused by a disease in terms of affected life years, e.g. discriminating between a lethal cancer and a skin irritation. It is often expressed as disability adjusted life years (DALY) accounting for both years of life lost (YLL) and years of life lived (YLD) with a disability-specific weighting factor. It is also possible to only consider the likelihood of an adverse effect without any severity weighting by simply omitting the severity factor for the calculations as e.g. documented in the first version of USEtox (Rosenbaum et al. 2008), but this implies that all effect types should be kept separate unless an equal severity weighting is implicitly assumed.

(c&d) The human toxicological **effect factor** $EF = DRF \cdot SF$ [DALY/kg_{intake}] is the combination of the dose–response slope factor and the severity factor. It can be interpreted as the increase in the number of DALY in the exposed population per unit mass of chemical taken in by that population.

(a&d) The human toxicological **characterisation factor** $CF = FF \cdot XF \cdot DRF \cdot SF = iF \cdot EF$ [DALY/$kg_{emitted}$] combines all these steps and expresses the increase in the number of DALY in the exposed population per unit mass of a chemical emitted into the environment. Excluding severity leads to a CF at midpoint level for human toxicological impacts, referred to as human toxicity potential, *HTP* [comparative toxic units, CTU/$kg_{emitted}$], with $HTP = FF \cdot XF \cdot DRF = iF \cdot DRF$. However, summing up the effects of different substances may implicitly assume equal severity within or across the subclasses of cancer and non-cancer effects. For the specific case of USEtox, the multiplication of the intake fraction by the dose–response factor yields a result (per kg emitted) expressed in *case of cancer* or *cases of non-cancer* called 'Comparative Toxic Units for human (CTU_h)' in USEtox, to emphasise the comparative nature of this results and its high absolute uncertainty.

Various adaptations of this framework have recently been proposed to account for more direct worker occupational health exposure and consumer exposure to chemicals contained in products. In these cases, the exposure is modified, while the same effect factor can be applied to then determine the cancer and non-cancer cases and eventually DALY.

- Impacts of releases to indoor air can be accounted for using air renewal rate and building occupancy to calculate an indoor intake fraction, usually higher than the intake fraction for outdoor emissions (Hellweg et al. 2009; Wenger et al. 2012).
- For occupational exposure, chemical emissions at the workplace are often unknown (Hellweg et al. 2005; Kikuchi and Hirao 2008). Chemical intake in each sector can be assessed by multiplying the number of hours worked by measured concentrations (Kijko et al. 2013). For accidents at workplace, Scanlon et al. (2013, 2014) directly relate accident and health statistics in a sector to corresponding numbers of DALY, yielding DALYs per unit physical output of each sector.
- Pesticide residues in food crops can be assessed using a dynamic model to determine the intake fraction, since dynamics of pesticides in the crop-environment system and resulting residues in harvested crop parts depend on the time lag between application of the pesticide and harvest of the crop (Fantke et al. 2011a, b, 2012a). Since dynamic assessment models are difficult to combine with existing LCIA approaches for assessing human toxicity of chemical emissions, a parameterised version to account for pesticide residues in food crops was recently proposed (Fantke et al. 2012b, 2013).
- For direct exposure to products (e.g. cosmetics, household products), Jolliet et al. (2012a) propose, instead of relating intakes to chemical emissions that are often unknown in terms of time and location (Finnveden et al. 2009), to relate intakes directly to the chemical mass in products, defining the **product intake fraction** *PiF* [kg_{intake}/$kg_{in\ product}$]. The product intake fraction is defined as the fraction of the chemical in product that is taken in by the consumer and the rest of the population during the use and disposal stages of the product life cycle (right side in Fig. 5.1).

3 Contributing Substances

As discussed by Rosenbaum in the ecotoxicity chapter (Chap. 8, see this volume), the human toxicity and ecotoxicity categories are "facing the challenge of having to characterize several tens of thousands of chemicals. The CAS registry currently contains more than 70 million unique organic and inorganic substances (www.cas.org/about-cas/cas-fact-sheets) of which roughly 100,000 may play an important industrial role".

As described in Table 5.1, current LCIA models provide characterisation factors for between 180 and 1,250 substances for human toxicity, with the USEtox model presently offering the largest substance coverage. Calculation methods and model validity strongly depend on the considered class of substances:

- For organic substances, environmental multi-media models are best designed to assess a large number of non-polar, non- ionisable compounds, for which fate and exposure can potentially be extrapolated from basic chemical properties for several tens of thousands of chemicals using databases such as the EPI Suite database (United States – Environmental Protection Agency 2012). Further adaptations are needed for polar and ionisable chemicals or for specific classes of chemicals such as perfluorinated compounds that show specific environmental behaviours.
- For inorganics and metals, partition coefficients are based on measured values rather than predicted from substance properties, but these values may highly depend on local conditions such as pH, redox potential or dissolved organic carbon levels. Metal speciation and specific removal processes need to be better considered and currently limit the validity of most human toxicity characterisation factors for metals, especially for the potentially high long-term emissions from landfills to groundwater that are found in some of the mostly used LCA inventory databases.
- For all substances, dose–response toxicity data are most of the time the main limiting factor in terms of substance coverage, since (a) chronic data and more generally animal-based experiment have only been performed for a limited number of substances and may be very expensive to carry out for additional substances, and (b) QSAR approaches have so far very limited validity to predict human toxicity. In the specific case of USEtox cancer effect factors, data are available for 1,024 substances among which 59 % show a positive response in rats and mice and 39 % are true zero (i.e. no carcinogenicity). Non-cancer effect factors via inhalation and/or ingestion are available for 437 chemicals.

Finally, in LCA human toxicity and health effects of fine particulates, photochemical oxidants, or ionising radiations are in most LCIA methods addressed in specific impact categories that are partly detailed in other chapters of this volume (particulates – Chap. 6, photochemical oxidants – Chap. 7, ionising radiations – not included). Impact results for these categories may nevertheless be compared with cancer and non-cancer scores at endpoint levels for LCIA method that enable calculation in common damage units such as DALYs.

Table 5.1 Overview of human toxicity assessment methods and their integration into LCIA methods

LCIA method	Multimedia model	Substance coverage	Effect classes	Severity	Spatial resolution	Population density	Near-field exposure	References method [model]
IMPACT World+	USEtox 1.1 beta	>1,250 organics and metals interim	Cancer and non-cancer cases	DALY based on Huijbregts et al. (2005a)	Continent-specific	Urban archetype	Indoor air (Wenger et al. 2012)	Bulle et al. (2012) [Rosenbaum et al. 2008, 2011]
ILCD LCIA	USEtox 1.0	>1,250 organics and metals interim	Cancer and non-cancer cases	DALY based on Huijbregts et al. (2005a)	Global default	Urban archetype	None	EC European Commission (2010a, b, 2011), Hauschild et al. (2013) [Rosenbaum et al. 2008, 2011]
ReCiPe	USES-LCA 2.0	>1,000 organics and metals	Non-linear cancer and non-cancer cases	DALY based on Huijbregts et al. (2005a)	Europe	Urban archetype	None	Goedkoop et al. (2009) [van Zelm et al. 2009]
IMPACT 2002+/ LIME 2	IMPACT 2002	>800 organics and metals	Cancer and non-cancer cases	DALY based on Crettaz et al. (2002), Pennington et al. (2002)/DALY based on various sources	Spatial Europe, North America, single box conti-nents/spatial Japan	None/ urban archetype	None	Jolliet et al. (2003)/Itsubo and Inaba (2012) [Humbert et al. 2009; Pennington et al. 2005]
TRACI 2/ TRACI 1	USEtox 1.0/ CalTOX 2.2	>1,250 organics and metals interim/ >380 substances	Cancer and non-cancer cases	None	Global default, US at state level	Urban archetype	None	Bare (2011)/Bare et al. (2002) [Rosenbaum et al. 2008, 2011/McKone 1993]
EDIP 2003	EDIP 1997	>180 organics and metals	Cancer and non-cancer cases	None	Site-dependent	None	None	Hauschild and Potting (2005), Potting and Hauschild (2005) [Hauschild and Wenzel 1998; Wenzel et al. 1997]
CML 2002	USES-LCA 1.0	>850 organics and metals	Cancer and non-cancer cases	None	None	Urban archetype	None	Guineé et al. (2001) [Huijbregts et al. 2000]

4 Midpoint and Endpoint Methodologies

LCIA methods that address human toxicity are based on different underlying multimedia models and corresponding factors covering fate, exposure, and effects.

Table 5.1 provides an overview of the main LCIA methods in use with their main underlying models and characteristics.

As stated in European Commission (2011), LCIA characterisation models and factors for human toxicity, effects must be based on models that account for a chemical's fate in the environment, human exposure, and differences in toxicological response. Analysing the main human toxicity models against the criteria described in Box 5.1, Hauschild et al. (2013) evaluated the scientific consensus model USEtox best as between fully compliant and compliant in all essential aspects, and USES-LCA implemented in ReCiPe and IMPACT 2002 implemented in IMACT 2002+ as compliant in all essential aspects. CalTOX scored as compliant for several aspects, while the older work on USES-LCA implemented in CML 2002 and EDIP 2003 was evaluated as compliant in some aspects only. USEtox was the preferred choice, since it reflects the latest consensus amongst different modellers and their associated models, being parsimonious while falling in the range of previous models (Hauschild et al. 2008). USEtox also offers the largest substance coverage with more than 1,250 human toxicological characterisation factors (Rosenbaum et al. 2008), reflects more up to date knowledge and data on cancer effect factors than other approaches and aims for global representativeness by parameterising its unit world according to globally representative conditions. USEtox has now been implemented in TRACI 2.0 – the US-EPA LCIA method, and also in the ILCD LCIA method. IMPACT World + has implemented the beta version of USEtox 1.1 (http://usetox.org/1.1_beta) with continent-specific landscapes, an indoor air emission compartment, updated data and exposure pathways for pesticides and refined processes to account for ionisable chemicals. ReCiPe/ USES LCA and IMPACT 2002+ are useful alternative methods and models for sensitivity analysis to test result robustness.

Since cancer and non-cancer have different severity but are not differentiated in the USEtox so-called 'Comparative Toxic Units', the ILCD and IMPACT World + methods apply severity factors taken from Huijbregts et al. (2005a) as 11.5 Disability Adjusted Life Years (DALY) per case of cancer and 2.7 DALY per case of non-cancer.

5 Chemical Spatial and Temporal Variability and Uncertainty

5.1 Variability Inhuman Intake Fraction (iF)

For inhalation, the population density is the key factor driving the intake, except for very persistent and mobile chemicals that are taken in by the global population independently from their place of emission. For most chemicals in USEtox, inhalation, above-ground produce, and fish are the important exposure pathways with key driving factors being the (a) compartment and place of emission, (b) chemical partitioning, (c) degradation processes, (d) bioaccumulation and bioconcentration, and (e) dietary habits of the population (in no particular order and to various degrees depending on the respective dominating exposure pathway).

Building on Rosenbaum et al. (2011):

- *Inhalation – urban vs. rural air*: The urban *iF* from inhalation exposure is usually at least an order of magnitude higher than the continental rural *iF* ($iF^{inh}_{urban} \approx 10^{-4} > iF^{inh}_{rural} \approx 10^{-5}$ to 10^{-7} for most chemicals). Therefore, it is important to differentiate between emissions to urban and rural areas. The difference in *iF* is due to the higher population density and lower dilution volumes in urban areas, leading to a higher population intake via inhalation. The inhalation-related urban *iF* is relatively constant for all substances, since it mainly depends on the residence time of air in the urban area.
- *Inhalation – continental vs. global scale*: Substances with long half-lives and high air-water partition coefficients K_{aw} may also achieve high inhalation intake fractions of $> 10^{-4}$, e.g. chlorodifluoromethane (HCFC-22). This high *iF* is a result of long persistence and eventual long range transport to other parts of the globe.
- *Indoor air*: Inhalation exposure of indoor air releases is a direct function of the air renewal rate of the building, leading to high intake fractions typically between 10^{-3} and 10^{-2} due to a restricted volume of dilution and higher population density than outside. Inhalation of compounds with low vapor pressure may be reduced due to removal by absorption and eventual degradation on the room surfaces, whereas Weschler and Nazaroff (2012) suggest that gaseous dermal intake may dominate for such compounds and also lead to comparatively high intake fractions.
- *Drinking water ingestion*: Drinking water is the dominant exposure pathway for substances with low octanol-air partition coefficient K_{oa}. However, *iF* values associated with this pathway tend to be generally low (10^{-10} to 10^{-5}).
- *Food ingestion – above-ground produce, meat and dairy produce*: Intake fraction by ingestion of above-ground produce is the product of bioaccumulation in this food compartment, the mass in the air compartment, and the population intake of this food produce. The bioaccumulation factor *BAF* [L/kg] is determined by diffusion and deposition from the air to the food crop surface and is primarily dependent on the plant-air partition coefficient K_{pa}[L/kg]. Lipophilic compounds with octanol-water partition coefficient K_{ow} higher than 10^4 may also bioaccumulate in the food chain and are mostly found in meat, eggs and dairy products with intake fractions up to 10^{-3}.
- *Food ingestion – fish*: In the range of high octanol-water partition coefficients K_{ow} between 10^3 and 10^9, the fish ingestion exposure pathway dominates for emission to water, e.g. for polychlorinated biphenyls (PCB) or dioxin-like compounds. In this range, the ingestion *iF* increases with increasing K_{ow}, peaking at $K_{ow} \approx 10^7$ and decreasing thereafter. $iF > 10^{-2}$ should be interpreted with care (Bennett et al. 2002a) and may be linked to overestimated bioaccumulation factors.

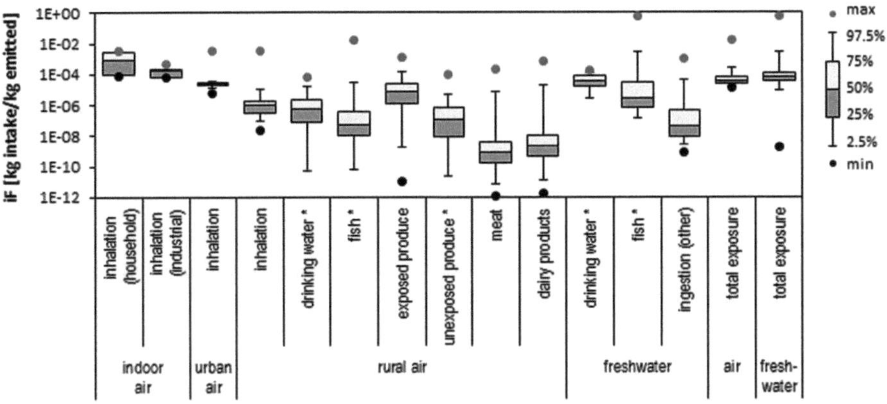

Fig. 5.2 Variations in human intake fractions for different compartments of emission *iF* [kg_{intake}/$kg_{emitted}$] for organic and inorganic substances in the USEtox 1.01 results database. Total exposure from emissions to air assumes 50 % emissions to rural air and 50 % to urban air. For categories marked with *, minimum values are low, between $4 \cdot 10^{-20}$ and $2 \cdot 10^{-17}$

- *Dermal uptake*: For dermal applications of e.g. cosmetics, the product intake fraction, *PiF*, via dermal uptake tends to increase with lipophilic chemicals and lower molecular weights, yielding *PiF* values between 10^{-3} and 1 from the application of products, such as body lotions that are left during hours on the body leading to the highest uptakes (Mitragotri et al. 2011).

Figure 5.2 presents the overall variation in intake fractions for various compartments of emission and exposure pathways. As expected, household and industrial indoor intake fractions are high, with limited variations between chemicals. The outdoor urban air *iF* is close to 10^{-4} for most chemicals, whereas the rural intake fraction is typically to 2 orders of magnitude lower. For water related pathways, drinking water *iF* are below $2 \cdot 10^{-4}$ with a median at $4 \cdot 10^{-5}$. The median fish ingestion-related *iF* is lower than for drinking water at $3 \cdot 10^{-6}$, but the highly bioaccumulating chemicals may lead to intake fractions higher than 10^{-2}. When combining all pathways for ambient emissions together, variation of *iF* across chemicals is restricted to max 3 orders of magnitude, since at least one exposure pathway maintains a moderate or high intake fraction for most chemicals. For directly dermally applied chemicals in e.g. cosmetics, the direct dermal uptake by the consumer is in general orders of magnitude higher than indirect intake by the overall population via leaching to wastewater (Jolliet and McKone 2012).

5.2 *Variability in Dose–Response and Severity*

In the case of USEtox, a linear relationship is assumed to calculate the dose–response as a response level of 0.5 divided by the lifetime ED_{50}

$$ED_{50,n\text{-}cancer} = \left| \frac{NOEL \cdot NOEL\_to\_ED50 \cdot body\ weight \cdot lifetime \cdot day\_to\_year}{dog\_to\_human \cdot subchronic\_to\_chronic \cdot mg\_to\_kg} \right|^{-1}$$

$$= \left| \frac{0.15\ [mg/kg_{bw}/day] \cdot 9 \cdot 70\ [kg_{bw}] \cdot 70\ [years] \cdot 365\ [days/year]}{1.5 \cdot 2 \cdot 10^6\ [mg/kg]} \right|^{-1}$$

$$= 0.61\ [kg/person/lifetime]$$

Fig. 5.3 Calculation steps for deriving ED50 values for human non-cancer effects from no observed effect level for the herbicide MCPA

For *cancer*, the lifetime ED_{50} is calculated in priority from human based data, but this information is only available for few substances. For the majority of the chemicals, the lifetime ED_{50} is thus derived from animal cancer tests from the chronic TD_{50} (tumorigenicdose-rate in milligrams per kilogram per day for 50 % of the animals over background in a standard lifetime) in the carcinogenic Potency database, CPDB (Gold 2011).

For *non-cancer*, most ED_{50} values have been estimated from chronic no observed effect level NOEL by a NOEL-to-ED_{50} conversion factor of 9 or in case only LOEL was available, by a LOEL-to-ED_{50} conversion factor of 2.25 (Huijbregts et al. 2010). In case no data were available for a specific exposure route, an analysis of route-to-route extrapolation supports the assumption of equal potency or slope factor between inhalation and ingestion route for most chemicals. Figure 5.3 presents how the ED_{50} is calculated from the NOAEL, taking the phenoxy herbicide 2-methyl-4-chlorophenoxyacetic acid (MCPA, CAS-RN: 94-74-6) as an example. In case no chronic data are available, Rosenbaum et al. (2011) determined an acute to chronic extrapolation, directly relating the animal acute LD_{50} (kg/day) to the human lifetime ED_{50} [kg/person/lifetime].

For effect factors, variation across chemicals is large with the 95th percentile of the effect factors spanning 5 orders of magnitude and extreme variations of up to 9 orders of magnitude, the highest impacts being related to dioxins (Fig. 5.4). Median effect factors for metals tend to be higher than for organics, whereas highest impacts per kg taken in are found for dioxin-like organics.

5.3 *Variability in Human Toxicological Characterisation Factors*

In most methods, the characterisation factor for human toxicity is calculated based on the lifetime ED50 – the estimated lifetime dose for humans that causes an increase in disease probability of 50 % (in kilograms per person per lifetime).

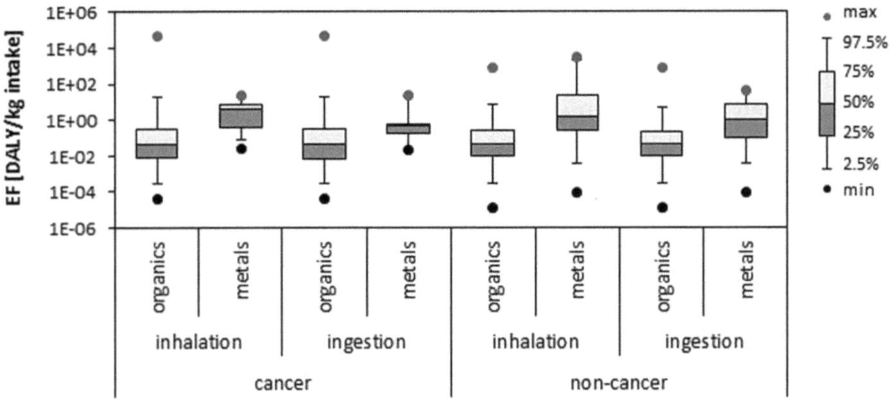

Fig. 5.4 Variations in human cancer and non-cancer effect factors inhalation and ingestion exposure *EF* [DALY/$kg_{emitted}$] for organic and inorganic substances in the USEtox 1.01 results database

Keeping inhalation and ingestion route separate and differentiating between the contributions of cancer and non-cancer impacts, Eq. 5.1 can be rewritten as:

$$CF = \left[iF^{inh} \cdot \underbrace{\frac{0.5}{ED^{inh}_{50h\ cancer}}}_{DRF^{inh}_{cancer}} + iF^{ing} \cdot \underbrace{\frac{0.5}{ED^{ing}_{50h\ cancer}}}_{DRF^{ing}_{cancer}} \right] \cdot SF_{cancer} + \left[iF^{inh} \cdot \underbrace{\frac{0.5}{ED^{inh}_{50h\ non\text{-}cancer}}}_{DRF^{inh}_{non\text{-}cancer}} + iF^{ing} \cdot \underbrace{\frac{0.5}{ED^{ing}_{50h\ non\text{-}cancer}}}_{DRF^{ing}_{non\text{-}cancer}} \right] \cdot SF_{non\text{-}cancer} \quad (5.2)$$

Characterisation factors typically vary by 10–12 orders of magnitude depending on the compartment of emission with the 95th percentile of the value spanning a 6 orders of magnitude interval. As expected impact per kg is the highest for near field indoor exposure which tends to be 2 to 4 orders of magnitude higher than for outdoor emissions, followed by emissions to air and freshwater for outdoor emissions.

5.4 Uncertainty

Three different types of uncertainty are particularly relevant in LCIA modelling of human toxicological impacts: (1) uncertainty due to lack of knowledge of the 'true' value of a model input parameter, i.e. parameter uncertainty, (2) uncertainty caused by arbitrary choices in a model, i.e. decision rule uncertainty, and (3) uncertainty caused by the loss of information resulting from the simplification of reality by the use of models, i.e. model structure uncertainty (Huijbregts 1998).

Degradation half-lives are in most cases the model input parameters driving uncertainty for the environmental fate part of the assessment (Fantke et al. 2012a; Hertwich et al. 1999), on which all existing human toxicity characterisation models rely. In the subsequent exposure assessment part, the factors for the bioconcentration BCF, bioaccumulation BAF, and biotransformation BTF account for the highest uncertainties (Arnot et al. 2010; McKone and Maddalena 2007). Finally, in the effect assessment part, the low dose extrapolation and dose–response modelling are responsible for the highest uncertainties (Crettaz et al. 2002; Pennington et al. 2002).

Spatial variation may be addressed using spatially explicit models at continental (variations of 2 orders of magnitude, Kounina et al. 2013; Rochat et al. 2006), regional (Pennington et al. 2005) or even local level (Jolliet et al. 2012b). This is especially important to capture the influence of population density in large urban area, for which a high resolution of around 10 km to max 50 km grid is needed. An efficient alternative is to use an urban archetype as offered by several LCIA models (Table 5.1) and to define the fraction of emission in densely populated area as already available e.g. in the Ecoinvent database. When emission locations are unknown, it is better to use as a proxy a 50 % emission to air in urban areas and 50 % in rural areas (or according to the proportion of population in urban and rural areas), than assuming the entire emission to rural air.

The time horizon plays an important role for persistent substances such as metals. Rather than totally discounting impacts after a given cut-off, it is interesting to present both the impacts occurring within the first 100 years and the long-term impacts occurring afterwards.

In term of substances, the multimedia models are best adapted for non-polar non-ionic organic substances. The model validity is limited and further research is needed to implement metal speciation and specific removal processes such as precipitation which play an important role for e.g. Aluminium removal in groundwater. Another limitation in terms of validity in existing LCIA models for human toxicity is related to ionising substances. An approach based on work by Franco and Trapp (2008, 2010) has been implemented in USEtox 1.1 beta (http://usetox.org/1.1_beta) and USES-LCA (van Zelm et al. 2012) to account for ionising chemicals in the environmental fate assessment. However, in the exposure assessment, ionising chemicals remain unaddressed in terms of corrected bioconcentration, bioaccumulation and biotransformation.

For the quantification of the uncertainty of human toxicity characterisation factors, e.g. Hofstetter (1998) provided expert-based estimates yielding a 95 % confidence interval limit of a factor 2–80 assuming a log normal distribution. Based on comparisons among different models, e.g. Rosenbaum et al. (2008) suggested an additional model uncertainty of a factor 10. This generally results in a factor 100 for the uncertainty of recommended characterisation factors and a factor 1,000 for factors that are characterised as 'interim' in the USEtox context, where interim *CF* often show large differences when applying different LCIA models and are therefore to be used with caution. Huijbregts et al. (2005a) and Fantke et al. (2012a) provide more detailed uncertainty factors for each factor of Equation 6–1 in order to enable the user to perform uncertainty analyses. Furthermore, spatial differentiation may influence results, especially for chemicals with short lifetimes: the population density around the point of emission in case of inhalation being the dominant route (taken into account through the introduction of an urban box e.g. in USEtox), the agriculture production intensity in case of food dominant pathways, the vicinity of the emission relative to a drinking water source, etc., Jolliet et al. (2012b), have created a multi-scale multimedia global model (Pangea) – de facto, a spatialised version of USEtox, that can assess intake fraction on a scale of 10 km up to global scale, the resolution being adaptive as a function of e.g. population density and proximity to large emission sources. The influence of spatial variation may be partially cancelled out by other factors, such as having multiple sources of emissions or may be negligible relative to other sources of uncertainty or variation for many contaminants.

5.5 Good Interpretation Practices for Human Toxicity

One has to first acknowledge that life cycle toxic emissions reported in many inventories and processes from a database like ecoinvent are often mostly correlated to energy (Frischknecht et al. 2007; Frischknecht and Rebitzer 2005). This could be expected for emissions that are related to combustion processes that are well covered by databases such as ecoinvent. However, for other toxic releases, the correlation to energy may also be due the poor coverage of non-energy related emissions and to the limitation and lack of data characterising toxic emissions during the manufacturing, use and disposal stages. This is therefore crucial, when making decisions on toxic impacts to first obtain good estimates of the main toxic emissions in the foreground processes and system.

As with all LCA results, best-estimates and related uncertainty ranges must be used for decision support, reflecting the current state of scientific knowledge (Udo de Haes et al. 1999). As shown in Sect. 4, uncertainty of human toxicity characterisation factors are high, of typically 2–3 orders of magnitude. This is nevertheless much lower than the roughly 12 orders of magnitude variation between the characterisation factors of different chemicals (Fig. 5.5). As discussed in Rosenbaum et al. (2008), characterisation factors for human toxicity are preferably used in a way that reflects this large variation of 12 orders of magnitude between characterisation factors for different chemicals as well as the 3 orders of magnitude uncertainty on the individual factors. First this means that results should be by default reported and interpreted in log scales when comparing scenarios or substance contribution. This also means that two chemicals contributing with 1 and 90 % (0.01 and 0.9) to the total human toxicity indicator can be considered as having

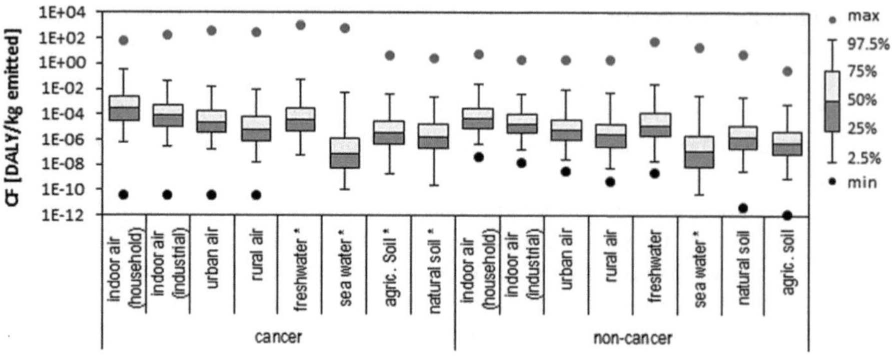

Fig. 5.5 Variations in human toxicity characterisation factors for different compartments of emission *CF* [DALY/$kg_{emitted}$] for organic and inorganic substances in the USEtox 1.01 results database. For categories marked with *, minimum values are low, between $1 \cdot 10^{-24}$ and $1 \cdot 10^{-16}$

similar impacts. The impacts are, however, significantly larger than those of a chemical contributing to less than 1 per million of the total category indicator (0.00001 %).

In practice, this means that for the LCA practitioner these human toxicity-related characterisation factors can be useful to identify the priority contaminants in product life cycles. The factors for human-toxicological impacts thus enable the identification of chemicals contributing more than e.g. one 1000th to the total indicator result. This will allow the practitioner to identify the chemicals that contribute the most to the indicator and, perhaps more importantly, to disregard the other substances, whose impact is not significant for the considered application. This is important in the interpretation phase as well as where refinement of the study may be needed.

6 New Developments and Research Needs

Several issues are essential in the development of the next generation of life cycle toxicity methods and are presently the subject of ongoing research:

Better predictors for estimating degradation half-lives are required in priority. These degradation rates should also include photolysis and hydrolysis, processes that are not accounted for in most of the presently used databases, such as the U.S. EPI Suite (United States – Environmental Protection Agency 2012), that are used as input into current LCIA models. Such predictors will largely be developed from reviewing existing experimental data, such as for pesticides (Fantke and Juraske 2013), and analysing chemical as well as microbiological properties and conditions.

Ionisation of chemicals needs to be fully considered in environmental fate, human exposure and effect assessment in LCIA models based on work from Franco and

Trapp (2008, 2010) and van Zelm et al. (2012) regarding environmental fate processes, and from Armitage et al. (2013) and Fu et al. (2009) for exposure assessment-related bioconcentration.

For metals, speciation and specific removal processes need to be better considered in a way consistent with what has been developed for ecotoxicity (Gandhi et al. 2010, 2011; Dong et al. 2014), for example the precipitation of aluminium in groundwater.

Region-specific models can be used to reduce uncertainty associated with spatialisation. To analyse the spatial variations in toxic impacts, an adaptive multimedia model named PANGEA is currently being developed, determining intake fractions at local (10 km grid around the emission source), regional (200 km grid within the continent of emission), and continental (world divided in continents) levels (Jolliet et al. 2012b). Special emphasis will be thereby given to further develop modelling of exposure in the food chain linked to highest intake fractions and high level of uncertainties.

Typically, additivity is assumed in LCIA with respect to human toxicological effects of substance mixtures, to which we are naturally exposed. This is facilitated by assessing substances individually in available LCIA models. However, in some cases, the assumption of additivity is not fully valid, and synergistic, antagonistic or even more complex interactions between chemicals may occur. An example is the interaction between different active ingredients or between the active ingredient and adjuvants in pesticide formulations (Benachour et al. 2007; Benachour and Sralini 2009; Sanborn et al. 2004). When such interactions are known, methods could be developed to adapt characterisation factors as a function of the level of other substances in the mixture. A more realistically approach in the LCA context is to provide specific factors for frequently observed mixture, but this is to be kept for cases where very large differences occur depending on the mixture composition, e.g. for impacts of PAHs (Li et al. 2014).

An important need is to develop characterisation factors for chemicals used in consumer products, building on the product intake fraction concept (Jolliet et al. 2012a). Products of special interest are cosmetics, cleaning products, flame retardants, plasticizers, antibacterial agents, that may in particular contain substances having endocrine disrupting effects. For such indoor emission, the inhalation intake fraction is well defined and relatively easy to estimate. Additional insights and data are especially needed to improve estimates of dermal intake from direct contact and from gaseous phase, that have been suggested to dominate exposure for certain classes of semi-volatile organic compounds (Weschler and Nazaroff 2012).

Finally, worker exposure to organic chemicals per hour of blue collar work is now available for hundreds of industrial sectors per hour worked (Hellweg et al. 2005; Kijko et al. 2013) using measured concentration by worker inspectors. A similar approach could be built for accident and acute disease upon the statistics collected by (Scanlon et al. 2013, 2014).

As a whole, the assessment of toxicity in LCA has progressed on a very sharp learning curve during the past 20 years. This rapid progression is expected to continue in the coming years, focusing more on direct exposure of workers to chemicals during manufacturing and of consumers during product use.

References

Armitage JM, Arnot JA, Wania F, Mackay D (2013) Development and evaluation of a mechanistic bioconcentration model for ionogenic organic chemicals in fish. Environ Toxicol Chem 32 (1):115–128

Arnot JA, Mackay D, Parkerton TF, Zaleski RT, Warren CS (2010) Multimedia modeling of human exposure to chemical substances: the roles of food web biomagnification and biotransformation. Environ Toxicol Chem 29(1):45–55

Bare J (2011) TRACI 2.0: The tool for the reduction and assessment of chemical and other environmental impacts 2.0. Clean Technol Environ Policy 13(5):687–696

Bare JC, Norris GA, Pennington DW, McKone T (2002) Traci: The tool for the reduction and assessment of chemical and other environmental impacts. J Ind Ecol 6:49–78

Benachour N, Sralini G-E (2009) Glyphosate formulations induce apoptosis and necrosis in human umbilical, embryonic, and placental cells. Chem Res Toxicol 22(1):97–105

Benachour N, Moslemi S, Sipahutar H, Seralini G-E (2007) Cytotoxic effects and aromatase inhibition by xenobiotic endocrine disrupters alone and in combination. Toxicol Appl Pharmacol 222(2):129–140

Bennett DH, Margni MD, McKone TE, Jolliet O (2002a) Intake fraction for multimedia pollutants: a tool for life cycle analysis and comparative risk assessment. Risk Anal 22(5):905–918

Bennett DH, McKone TE, Evans JS, Nazaroff WW, Margni MD, Jolliet O, Smith KR (2002b) Defining intake fraction. Environ Sci Technol 36:207A–211A

Bulle C, Jolliet O, Humbert S, Rosenbaum R, Margni M (2012) IMPACT World+: a new global regionalized life cycle impact assessment method. Society of Environmental Toxicology and Chemistry 6th World Congress/Europe 22nd annual meeting, Berlin, 20–24 May 2012

Covello VT, Merkhoher MW (1993) Risk assessment methods: approaches for assessing health and environmental risks. Plenum Press, New York

Crettaz P, Pennington DW, Rhomberg L, Brand K, Jolliet O (2002) Assessing human health response in life cycle assessment using ED10s and DALYs: part 1, Cancer effects. Risk Anal 22(5):931–946

Dong Y, Gandhi N, Hauschild MZ (2014) Development of ecotoxicity characterization factors for 14 metals (Al, Ba, Be, Cd, Co, Cr(III), Cs, Cu, Fe(II), Fe(III), Mn, Ni, Pb, Sr and Zn) in freshwater. Chemosphere 112:26–33

European Commission (2010a) International Reference Life Cycle Data System (ILCD) Handbook: analysis of existing environmental impact assessment methodologies for use in life cycle assessment, 1st edn. Brussels

European Commission (2010b) International Reference Life Cycle Data System (ILCD) Handbook: framework and requirements for LCIA models and indicators, 1st edn. Brussels

European Commission (2011) International Reference Life Cycle Data System (ILCD) Handbook: recommendations for life cycle impact assessment in the European context, based on existing environmental impact assessment models and factors, 1st edn. Brussels

Fantke P, Juraske R (2013) Variability of pesticide dissipation half-lives in plants. Environ Sci Technol 47(8):3548–3562

Fantke P, Charles R, de Alencastro LF, Friedrich R, Jolliet O (2011a) Plant uptake of pesticides and human health: dynamic modeling of residues in wheat and ingestion intake. Chemosphere 85(10):1639–1647

Fantke P, Juraske R, Antón A, Friedrich R, Jolliet O (2011b) Dynamic multicrop model to characterize impacts of pesticides in food. Environ Sci Technol 45:8842–8849

Fantke P, Friedrich R, Jolliet O (2012a) Health impact and damage cost assessment of pesticides in Europe. Environ Int 49:9–17

Fantke P, Wieland P, Juraske R, Shaddick G, Sevigné E, Friedrich R, Jolliet O (2012b) Parameterization models for pesticide exposure via crop consumption. Environ Sci Technol 46 (23):12864–12872

Fantke P, Wieland P, Wannaz C, Friedrich R, Jolliet O (2013) Dynamics of pesticide uptake into plants: from system functioning to parsimonious modeling. Environ Model Software 40:316–324

Finnveden G, Hauschild MZ, Ekvall T, Guinée J, Heijungs R, Hellweg S, Koehler A, Pennington D, Suh S (2009) Recent developments in life cycle assessment. J Environ Manage 91(1):1–21

Franco A, Trapp S (2008) Estimation of the soil-water partition coefficient normalized to organic carbon for ionizable organic chemicals. Environ Toxicol Chem 27(10):1995–2004

Franco A, Trapp S (2010) A multimedia activity model for ionizable compounds: validation study with 2,4-dichlorophenoxyacetic acid, aniline, and trimethoprim. Environ Toxicol Chem 29 (4):789–799

Frischknecht R, Rebitzer G (2005) The ecoinvent database system: a comprehensive web-based LCA database. J Clean Prod 13(13–14):1337–1343

Frischknecht R, Jungbluth N, Althaus H-J, Doka G, Dones R, Heck T, Hellweg S, Hischier R, Nemecek T, Rebitzer G, Spielmann M, Wernet G (2007) Overview and methodology. Final report ecoinvent v2.0 No. 1, Swiss Centre for Life Cycle Inventories, Duebendorf

Fu W, Franco A, Trapp S (2009) Methods for estimating the bioconcentration factor of ionizable organic chemicals. Environ Toxicol Chem 28(7):1372–1379

Gandhi N, Diamond ML, van de Meent D, Huijbregts MAJ, Peijnenburg WJGM, Guinée J (2010) New method for calculating comparative toxicity potential of cationic metals in freshwater: application to copper, nickel, and zinc. Environ Sci Technol 44(13):5195–5201

Gandhi N, Diamond ML, Huijbregts MAJ, Guinée JB, Peijnenburg WJGM, van de Meent D (2011) Implications of considering metal bioavailability in estimates of freshwater ecotoxicity: examination of two case studies. Int J Life Cycle Assess 16(8):774–787

Goedkoop M, Heijungs R, Huijbregts M, De Schryver A, Struijs J, van Zelm R (2009) ReCiPe 2008: a life cycle impact assessment method which comprises harmonised category indicators at the midpoint and the endpoint level. First edition report I: characterisation. Ministry of Housing, Spatial Planning and Environment (VROM), Den Haag

Gold LS (2011) The Carcinogenic Potency Database (CPDB). University of California, Berkeley, Lawrence Berkeley National Laboratory, National Library of Medicine. http://potency.berkeley.edu

Guineé JB, Gorrée M, Heijungs R, Huppes G, Kleijn R, de Koning A, van Oers L, Sleeswijk AW, Suh S, Udo de Haes HA, de Bruijn H, van Duin R, Huijbregts MAJ (2001) Life cycle assessment: an operational guide to the ISO standards. Leiden University, Leiden

Hauschild MZ, Potting J (2005) Spatial differentiation in life cycle impact assessment: the EDIP2003 methodology. Danish Ministry of the Environment, Copenhagen

Hauschild MZ, Wenzel H (1998) Environmental assessment of products, vol 2, Scientific background. Chapman and Hall, London

Hauschild MZ, Huijbregts MAJ, Jolliet O, Macleod M, Margni MD, van de Meent D, Rosenbaum RK, McKone TE (2008) Building a model based on scientific consensus for life cycle impact assessment of chemicals: the search for harmony and parsimony. Environ Sci Technol 42 (19):7032–7037

Hauschild MZ, Goedkoop M, Guinée J, Heijungs R, Huijbregts M, Jolliet O, Margni M, De Schryver A, Humbert S, Laurent A, Sala S, Pant R (2013) Identifying best existing practice for characterization modeling in life cycle impact assessment. Int J Life Cycle Assess 18 (3):683–697

Hellweg S, Demou E, Scheringer M, McKone TE, Hungerbühler K (2005) Confronting workplace exposure to chemicals with LCA: examples of trichloroethylene and perchloroethylene in metal degreasing and dry cleaning. Environ Sci Technol 39(19):7741–7748

Hellweg S, Demou E, Bruzzi R, Meijer A, Rosenbaum RK, Huijbregts MAJ, McKone TE (2009) Integrating human indoor air pollutant exposure within life cycle impact assessment. Environ Sci Technol 43(6):1670–1679

Hertwich EG, McKone TE, Pease WS (1999) Parameter uncertainty and variability in evaluative fate and exposure models. Risk Anal 19(6):1193–1204

Hofstetter P (1998) Perspectives in life cycle impact assessment: a structured approach to combine models of the technosphere, ecosphere and valuesphere. Kluwer Academic Publishers, Boston

Huijbregts MAJ (1998) Application of uncertainty and variability in LCA Part I: a general framework for the analysis of uncertainty and variability in life-cycle assessment. Int J Life Cycle Assess 3(5):273–280

Huijbregts MAJ, Thissen U, Guinée JB, Jager T, Kalf D, van de Meent D, Ragas AMJ, Sleeswijk AW, Reijnders L (2000) Priority assessment of toxic substances in life cycle assessment. Part I: calculation of toxicity potentials for 181 substances with the nested multi-media fate, exposure and effects model USES-LCA. Chemosphere 41:541–573

Huijbregts MAJ, Rombouts LJA, Ragas AMJ, van de Meent D (2005a) Human-toxicological effect and damage factors of carcinogenic and noncarcinogenic chemicals for life cycle impact assessment. Integr Environ Assess Manag 1(3):181–244

Huijbregts MAJ, Struijs J, Goedkoop M, Heijungs R, Hendriks AJ, van de Meent D (2005b) Human population intake fractions and environmental fate factors of toxic pollutants in life cycle impact assessment. Chemosphere 61(10):1495–1504

Huijbregts M, Hauschild M, Jolliet O, Margni M, McKone T, Rosenbaum RK, van de Meent D (2010) USEtox TM User Manual Version 1.01, Substance database of the UNEP/SETAC model for the comparative assessment of chemicals released to air, water and soil and their toxic effects on the human population and ecosystems. http://www.usetox.org/sites/default/files/support-tutorials/user_manual_usetox.pdf

Humbert S, Manneh R, Shaked S, Wannaz C, Horvath A, Deschênes O, Jolliet O, Margni M (2009) Assessing regional intake fractions in North America. Sci Total Environ 407(17):4812–4820

IGHRC Interdepartmental Group on Health Risks from Chemicals (2003) Uncertainty factors: their use in human health risk assessment by UK Government. MRC Institute for Environment and Health, Leicester

Itsubo N, Inaba A (2012) LIME2 life-cycle impact assessment method based on endpoint modeling. Summary, Life Cycle Assessment Society of Japan, Tokyo

Jolliet O, McKone TE (2012) Rapid exposure-based prioritization of environmental chemicals using USEtox. Society of Toxicology (SOT) 51st annual meeting and ToxExpoTM, San Francisco, 11–15 Mar 2012

Jolliet O, Margni MD, Charles R, Humbert S, Payet J, Rebitzer G, Rosenbaum RK (2003) IMPACT 2002+: a new life cycle impact assessment methodology. Int J Life Cycle Assess 8(6):324–330

Jolliet O, Li DS, Ernstoff A (2012a) Direct consumer exposure during use of personal care products, plasticizers and flooring materials. In: Proceedings of the international conference on ecobalance, D2-05, Yokohama, Japan, November 2012

Jolliet O, Wannaz C, Fantke P, Shaked S (2012b) Multi-scale, multimedia modeling with Pangea – local to global human health impacts of emissions in multiple continents. In: Proceedings of the international conference on ecobalance, P-129, Yokohama, Japan, November 2012

Kijko G, Jolliet O, Margni M (2013) Impact of occupational exposures to organic chemicals in Life Cycle Assessment: a new framework based on measured concentrations, LCA XIII, Orlando, 2013/10/01. ACLCA, Orlando. http://lcacenter.org/lcaxiii/abstracts/797.htm

Kikuchi Y, Hirao M (2008) Practical method of assessing local and global impacts for risk-based decision making: a case study of metal degreasing processes. Environ Sci Technol 42 (12):4527–4533

Kounina A, Margni M, Shaked S, Bulle C, Jolliet O (2014) Spatial analysis of toxic emissions in LCA: a sub-continental nested USEtox model with freshwater archetypes. Environ Int 69:67–89

Li DS, Huijbregts MAJ, Jolliet O (2014) Life cycle health impacts of polycyclic aromatic hydrocarbon for source-specific mixtures. Int J Life Cycle Assess 29(1): 87–99

Lim SS et al (2012) A comparative risk assessment of burden of disease and injury attributable to 67 risk factors and risk factor clusters in 21 regions, 1990–2010: a systematic analysis for the Global Burden of Disease Study 2010. Lancet 380(9859):2224–2260

McKone TE (1993) CalTOX, a multimedia total exposure model for hazardous-waste sites, Part 1: executive summary. UCRL-CR--111456-Pt.1, Lawrence Livermore National Lab., CA, Livermore

McKone TE, Maddalena RL (2007) Plant uptake of organic pollutants from soil: bioconcentration estimates based on models and experiments. Environ Toxicol Chem 26(12):2494–2504

Mitragotri S, Anissimov YG, Bunge AL, Frasch HF, Guy RH, Hadgraft J, Kasting GB, Lane ME, Roberts MS (2011) Mathematical models of skin permeability: an overview. Int J Pharm 418:115–129

Pennington DW, Crettaz P, Tauxe A, Rhomberg L, Brand K, Jolliet O (2002) Assessing human health response in life cycle assessment using ED10s and DALYs Part 2: noncancer effects. Risk Anal 22(5):947–963

Pennington DW, Margni MD, Ammann C, Jolliet O (2005) Multimedia fate and human intake modeling: spatial versus nonspatial insights for chemical emissions in Western Europe. Environ Sci Technol 39(4):1119–1128

Potting J, Hauschild M (2005) Background for spatial differentiation in life cycle impact assessment. The EDIP2003 methodology. Danish Ministry of the Environment, Copenhagen

Rochat D, Margni M, Jolliet O (2006) Continent-specific intake fractions and characterization factors for toxic emissions: does it make a difference? Int J Life Cycle Assess 11(1):55–63

Rosenbaum RK, Margni MD, Jolliet O (2007) A flexible matrix algebra framework for the multimedia multipathway modeling of emission to impacts. Environ Int 33(5):624–634

Rosenbaum RK, Bachmann TM, Gold LS, Huijbregts MAJ, Jolliet O, Juraske R, Koehler A, Larsen HF, MacLeod M, Margni MD, McKone TE, Payet J, Schuhmacher M, van de Meent D, Hauschild MZ (2008) USEtox – The UNEP/SETAC toxicity model: recommended characterisation factors for human toxicity and freshwater ecotoxicity in life cycle impact assessment. Int J Life Cycle Assess 13(7):532–546

Rosenbaum RK, Huijbregts MAJ, Henderson AD, Margni M, McKone TE, van de Meent D, Hauschild MZ, Shaked S, Li DS, Gold LS, Jolliet O (2011) USEtox human exposure and toxicity factors for comparative assessment of toxic emissions in life cycle analysis: sensitivity to key chemical properties. Int J Life Cycle Assess 16(8):710–727

Sanborn M, Cole D, Kerr K, Vakil C, Sanin LH, Bassil K (2004) Systematic review of pesticide human health effects. The Ontario College of Family Physicians, Toronto

Scanlon KA, Gray GM, Francis RA, Lloyd SM, LaPuma P (2013) The work environment disability-adjusted life year for use with life cycle assessment: a methodological approach. Environ Health 12(21):21–36

Scanlon KA, Lloyd SM, Gray GM, Francis RA, LaPuma P (2014) An approach to integrating occupational safety and health into life cycle assessment: development and application of a work environment characterization factor. J Ind Ecol. doi:10.1111/jiec.12146

Udo de Haes HA, Jolliet O, Finnveden G, Hauschild MZ, Krewitt W, Müller-Wenk R (1999) Best available practice regarding impact categories and category indicators in life cycle impact assessment. Int J Life Cycle Assess 4(2):66–74

United States – Environmental Protection Agency (2012) Estimation Programs Interface Suite TM for Microsoft® Windows, v 4.11. Washington, DC

van Leeuwen CJ, Vermeire TG (2007) Risk assessment of chemicals: an introduction, 2nd edn. Springer, Dordrecht van Zelm R, Huijbregts MAJ, van de Meent D (2009) USES-LCA 2.0 – a global nested multimedia fate, exposure, and effects model. Int J Life Cycle Assess 14(3):282–284

van Zelm R, Stam G, Huijbregts MAJ, van de Meent D (2012) Making fate and exposure models for freshwater ecotoxicity in life cycle assessment suitable for organic acids and bases. Chemosphere 90(2):312–317

Wambaugh JF, Setzer RW, Reif DM, Gangwal S, Mitchell-Blackwood J, Arnot JA, Jolliet O, Frame A, Rabinowitz J, Knudsen TB, Judson RS, Egeghy P, Vallero D, Cohen Hubal EA (2013) High-throughput models for exposure-based chemical prioritization in the ExpoCast project. Environ Sci Technol 47(15):8479–8488

Wenger Y, Li D, Jolliet O (2012) Indoor intake fraction considering surface sorption of air organic compounds for life cycle assessment. Int J Life Cycle Assess 17(7):919–931

Wenzel H, Hauschild MZ, Alting L (1997) Environmental assessment of products, vol 1, Methodology, tools and case studies in product development. Chapman and Hall, London

Weschler CJ, Nazaroff WW (2012) SVOC exposure indoors: fresh look at dermal pathways. Indoor Air 22(5):356–377

Chapter 6
Particulate Matter Formation

Sebastien Humbert, Peter Fantke, and Olivier Jolliet

Abstract This chapter deals with the causes and consequences of exposure from emissions of primary particles and secondary particle precursors on human health and how to deal with them in life cycle impact assessment (LCIA).

Following a short introduction and literature review, the first part outlines the complete emission-to-damage pathway, from emissions of primary particles and secondary particle precursors to damage on human health, so called 'respiratory effects from particles'. It describes the assessment framework for quantifying respiratory effects from particles in the context of LCIA. The second part provides an overview of methods that have been available in LCA to address impact of particles on human health. We finally discuss variability and main sources of uncertainties, as well as future trends in modelling respiratory effects of particles in LCIA.

Keywords LCA • LCIA • Life cycle assessment • Life cycle impact assessment • Particulate matter (PM) • PM • Primary PM • Respiratory effects • Secondary PM

1 Impacts on Human Health from Exposure to Particles

Ambient particulate matter (PM) is considered to be one of the most important environmental stressors in terms of its contribution to the global human disease burden (Lim et al. 2012; Hänninen et al. 2014), and PM is a significant cause of

S. Humbert (✉)
Quantis, EPFL Innovation Park, Bât. D, CH-1015 Lausanne, Switzerland
e-mail: sebastien.humbert@quantis-intl.com

P. Fantke
Division for Quantitative Sustainability Assessment, Department of Management Engineering, Technical University of Denmark (DTU), Produktionstorvet 424, 2800 Kgs Lyngby, Denmark

O. Jolliet
School of Public Health, Department of Environmental Health Sciences, University of Michigan, 1415 Washington Heights, Ann Arbor, MI 48109-2029, USA

M.Z. Hauschild, M.A.J. Huijbregts (eds.), *Life Cycle Impact Assessment*,
LCA Compendium – The Complete World of Life Cycle Assessment,
DOI 10.1007/978-94-017-9744-3_6

adverse human health effects (Pope et al. 2009). The World Health Organisation (WHO 2008; Cohen et al. 2005) estimated that about 800,000 premature deaths worldwide occur each year due to exposure to outdoor particles emission. The burden of disease 2010 even estimates these effects at 3,200,000 deaths per year Lim et al. (2012).

Several epidemiological studies show that PM causes serious adverse health effects, including reduced life expectancy, lung cancer, chronic and acute respiratory and cardiovascular morbidity, chronic and acute mortality, diabetes, and adverse birth outcomes (Kuenzli et al. 2000; Chen et al. 2008; Lippmann and Chen 2009; Pelucchi et al. 2009; Pope et al. 2009; Brook et al. 2010; Hoek et al. 2013; Mehta et al. 2013; Straif et al. 2013). Toxicological studies also support the idea that exposure to PM can exert effects on key biological systems of the human body (Kelly and Fussell 2012). Ambient PM can be primary (i.e., directly emitted) or secondary (i.e., formed in the atmosphere from precursors). Precursors involved in secondary PM formation include sulphur oxides (SO_x), nitrogen oxides (NO_x), ammonia (NH_3), semivolatile and volatile organic compounds, of which the latter are most important for secondary organic aerosol formation.

Several life cycle impact assessment (LCIA) models and methods have been developed to evaluate the human health damage per mass of particles or precursors emitted (e.g., Hofstetter 1998; Bare et al. 2003; Jolliet et al. 2003; Van Zelm et al. 2008). Hofstetter (1998) created one of the first LCIA approaches evaluating damage factors for PM_{10} (respirable particles with aerodynamic diameter of less than 10 μm), based on a consistent integration of data from existing models and epidemiological studies. Since then, researchers have continued to develop fate and exposure models (e.g., Levy et al. 2002; Marshall et al. 2005; Greco et al. 2007; Rosenbaum et al. 2008) and revise epidemiological data (Pope et al. 2002; Kuenzli et al. 2000; Laden et al. 2006; Schwartz et al. 2008). Previous publications suggest that human health damage needs to be assessed in a regional context to increase the confidence in, accuracy of, and acceptance of LCIA results (Potting and Hauschild 2006; Sedlbauer et al. 2007; Reap et al. 2008).

Recently, within ongoing efforts of the UNEP-SETAC Life Cycle Initiative and the research project LC-IMPACT, significant review and consensual work has been done to evaluate the fate and exposure (Humbert et al. 2011) and effect and severity associated with particles (Gronlund et al. 2014) along with harmonising the assessment framework from emissions to health impacts (Fantke et al. 2014).

The present chapter lays down the state-of-the-art in assessing impacts on human health from particle emission in life cycle assessment. Information provided in this chapter is partly based on Humbert (2009), Humbert et al. (2011), and Fantke et al. (2014) where the reader may find additional information on the topic.

2 Emission-to-Damage Framework

In LCIA, impacts on human health from airborne pollutants can be expressed using different units from (i) 'non-health' based indicators such as the equivalent amount of a reference substance (e.g., kg of 1,4-DCB-eq or kg of $PM_{2.5}$-eq), to (ii) health-based indicators such as the number of cases of illness (e.g., expressed in USEtox as comparative toxic units with respect to human effects, CTU_h), the number of premature deaths, the reduction in life- expectancy expressed as quality-adjusted life years (QALY) or as disability-adjusted life years (DALYs – Murray and Lopez 1996). One of the advantages of the DALY is that it consistently accounts for several forms of burden such as mortality and morbidity, using internationally recognised disability weights. Another advantage of DALYs is to make damage results comparable across impact categories for all methods that use DALY for human health related impacts (e.g., Goedkoop et al. 2009; Jolliet et al. 2003; Bulle et al. 2013). DALY is especially appropriate for impacts of PM since available data used to determine dose-response is directly expressed in term of health outcomes. At midpoint, if desired, results can always be expressed in kg $PM_{2.5}$-equivalents by dividing the DALY by the amount of DALY per kg of $PM_{2.5}$, since most of the health effects are commonly attributed to the particles smaller than 2.5 μm.

Following the general framework for emitted atmospheric pollutants (Jolliet et al. 2003; Udo de Haes et al. 2002), the characterisation factor (CF, in DALY per kg emitted) for respiratory effects from particles or their precursors expresses the additional damage on human health per kg additional particulate matter or precursor emitted. As also done for the human and ecotoxicity impact categories, it can be determined as a function of an atmospheric fate factor, an exposure factor, a dose-response slope factor and a severity factor. The two former factors can be combined into an intake fraction (iF) and the two latter factors can be combined into an effect factor (EF):

$$CF = \underbrace{FF \cdot XF}_{iF} \cdot \underbrace{DRF \cdot SF}_{EF} \tag{6.1}$$

Figure 6.1 shows the emission-to-damage framework for impacts on human health from exposure to particles.

The fate factor (FF, [d]) represents the intermedia transfer and residence time in each media. It relates the quantity emitted ($kg_{emitted}$) to the mass in air as the exposure medium multiplied by its persistence ($kg_{in\ air} \cdot d$). The exposure factor (XF, d^{-1}) determines the fraction of the air that is inhaled every day by the population. It relates the mass in the environment to the dose taken in. The dose-response slope factor (DRF, disease incidence/$kg_{inhaled}$) indicates the change in morbidity or mortality per kg intake dose. Finally, the severity factor (SF, DALY/ disease incidence) is the human health damage per morbidity or mortality incidence.

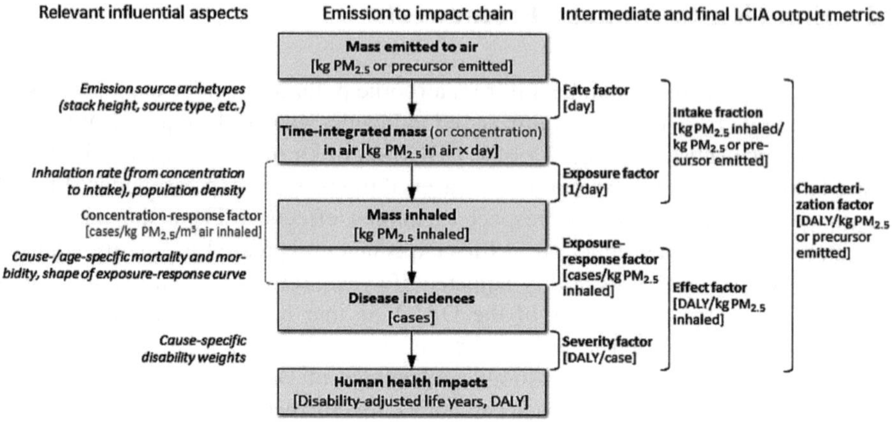

Fig. 6.1 Cause–effect pathway and impact assessment framework for impacts on human health from exposure to particles (Fantke et al. 2014)

2.1 Relevant Emissions

The pollutants considered to contribute to PM-related impacts in the different LCIA methodologies are typically total suspended particle, primary PM_{10}, primary $PM_{10\text{–}2.5}$, primary $PM_{2.5}$ and secondary PM formed from SO_2, NO_x, and NH_3. Direct exposure to SO_2, NO_x, NH_3 and CO is sometimes also considered in the so-called respiratory inorganic category. In accordance with the title of this chapter, the focus is exclusively on particulate-mediated effects, and we suggest to have non PM-related toxic effects of inorganic gases considered in the human toxicity category along with other organics and inorganics. Because of the lack of data, secondary PM from volatile organic compounds is often excluded but is recommended as an area of further research (Muller and Mendelsohn 2007; Ilacqua et al. 2007; Kanakidou et al. 2005).

2.2 Intake Fraction

The two terms XF and FF are often combined into the human intake fraction (iF, $kg_{inhaled}/kg_{emitted}$, which is expressed in this chapter in ppm, i.e., part per million or $mg_{inhaled}/kg_{emitted}$) (Bennett et al. 2002). The intake fraction for primary pollutants represents the fraction of the emission taken in (inhaled) by the (exposed) population. The intake fraction for secondary PM is the mass of PM attributable to a specific precursor inhaled per mass emission of the precursor. In cases secondary PM is composed of both nitrogen and sulphur such as in $(NH_4)_2SO_4$, an issue of double counting may appear when adding the damage from NH_3 emissions and SO_2

emissions. This is a limitation and would require further work to better address this issue of double counting, looking at the limiting pollutant in a specific area.

The inhalation intake fraction (iF) of a pollutant p is calculated according to Eq. 6.2 (Marshall et al. 2005; Bennett et al. 2002):

$$iF_p = \frac{\int_{time} N(t) \times BR(t) \times C_{p,air}(t) \times dt}{S_p} \tag{6.2}$$

where N (persons) is the number of persons exposed as a function of time t (s), BR (m^3/[person – d]) is the volumetric breathing rate, and $C_{p,air}$ (kg/m^3) is the incremental exposure concentration attributable to emission S_p (kg).

In Eq. 6.2 the intake fraction of secondary PM is calculated by dividing the mass of secondary PM inhaled by the mass of precursors emitted. It is assumed that SO_2 creates ammonium sulphate ($(NH_4)_2SO_4$), that NO_x creates ammonium nitrate (NH_4NO_3) and that NH_3 creates both ammonium sulphate ($(NH_4)_2SO_4$) and ammonium nitrate (NH_4NO_3). Further research is needed to better capture the composition and mass of secondary PM attributable to precursors which will also qualify the treatment of the potential double counting, e.g. in the case of ammonium sulphate $(NH_4)_2SO_4$, through better understanding of the influence of precursors already present in the air on the secondary PM formation of the precursor of interest.

The size distribution and the chemical composition may influence the dose-response relationship (Humbert 2009). They vary among emission sources and locations, and are the two main attributes of PM to consider when modelling their characterisation factors. Franklin et al. (2008) thus show that certain chemical species – that are, for example, adsorbed at the surface of primary PM at the point of emission – significantly modify the mortality of $PM_{2.5}$exposure, suggesting that mass alone may be an imperfect metric when evaluating the health effects of PM.

2.3 *Effect Factor*

The last two terms of Eq. 6.1, SF and DRF, can be combined into the effect factor (EF, DALY/$kg_{inhaled}$). LCIA studies often assume a linear, no-threshold dose response curve, an approach that for PM is supported by several studies (WHO 2006; Roman et al. 2008; Schwartz et al. 2008). However, in cases where PM concentrations are higher or lower than those observed in epidemiological studies of the U.S. or Europe (typically, around 10–35 μg/m^3 for $PM_{2.5}$), the linearity assumption may not apply, and non-linear relative risk functions have been proposed (Abrahamowicz et al. 2003; Pope et al. 2009; Burnett et al. 2013). Further studies are required to consider how these could be taken into account in the LCA framework and possibly locally linearised as a function of background $PM_{2.5}$ concentration.

At the present stage of knowledge, the human health impacts are primarily attributed to particulates smaller than 2.5 μm. Therefore the effect of PM_{10} is assumed by most methods to be equal to the effect of $PM_{2.5}$ multiplied by the fraction of $PM_{2.5}$ in PM_{10} inhaled.

The inhalation of PM can lead to many different health outcomes, and existing epidemiological studies show significant variations in the frequency and estimated damage of each outcome as a function of mass PM inhaled.

The total effect factor for PM exposure accounts for premature mortality and for some other endpoints such as asthma and restricted activity days. Premature mortality ('chronic mortality') is referring to the mortality associated with chronic diseases. Premature mortality also includes short-term increases in mortality ('acute mortality') from respiratory effects, as well as long-term mortality from carcinogenic effects. Acute data are based on time-series studies on daily mortality that measure the proportional increase in the daily death rate attributable to recent exposure to air pollution. Chronic data are based on cohort studies. Chronic data include those who died from chronic disease caused by long-term exposure, but also those whose death is advanced by recent exposure to air pollution (Kuenzli et al. 2000; WHO 2006; Van Zelm 2009).

By combining all considered outcomes, it is possible to find a final effect factor in DALY per kg PM inhaled. The effect factors thus represent endpoints where there is conclusive evidence of effects, but it should be acknowledged that present inconclusive epidemiological evidence for certain additional health endpoints potentially associated with PM exposure does not mean that these endpoints will not turn out to be relevant at a later stage.

2.4 Characterisation Factors as a Function of Source Type and Location

As explored by Humbert (2009), an important step towards reducing uncertainty in assessing the impacts on human health from exposure to particles is the capacity to account for variability in both source and location of emission. To illustrate the sensitivity to these parameters, Table 6.1 shows characterisation factors addressing impacts on human health from exposure to particles and precursors, expressed in a typical source-location matrix, as a function of the pollutant emitted, type of emission source (high-stack, low-stack, ground level and an emission-weighted average) and location of emission (indoor, in urban area with high population density of about 4,000 person per km^2, in rural area or in remote area with about 1 person per km^2). Note that the location (urban, rural and remote) influences the characterisation factor because of the differences in population density and therefore in exposure and not in terms of fate.

Table 6.1 Characterisation factors for impacts on human health from exposure to particles as a function of source type and location of emission

Pollutant emitted	Type of emission source	Characterisation factor for the respective location of emission						Population-weighted average (only for outdoor emissions)	Unit
		Indoor[a] (household)	Indoor[a] (office)	Indoor[a] (industrial)	Urban	Rural	Remote		
PM_{10}	High-stack	610,000	210,000	4,300	810	120	7.0	440	microDALY per kg PM_{10} emitted
	Low-stack				1,400	190	8.5	760	
	Ground-level				730	76	2.0	380	
	Emission-weighted average				780	88	3.2	410	
$PM_{10-2.5}$	High-stack	n/a[b]	n/a[b]	n/a[b]	n/a[b]	n/a[b]	n/a[b]	n/a[b]	microDALY per kg $PM_{10-2.5}$ emitted
	Low-stack				n/a[b]	n/a[b]	n/a[b]	n/a[b]	
	Ground-level				n/a[b]	n/a[b]	n/a[b]	n/a[b]	
	Emission-weighted average				n/a[b]	n/a[b]	n/a[b]	n/a[b]	
$PM_{2.5}$	High-stack	660,000	220,000	4,600	1,400	190	12	730	microDALY per kg $PM_{2.5}$ emitted
	Low-stack				1,900	260	12	1,000	
	Ground-level				6,000	630	17	3,100	
	Emission-weighted average				3,400	390	14	1,800	
SO_2	–	4,600	1,600	32	120	100	6.2	110	microDALY per kg SO_2 emitted
NO_x	–	2,200	740	15	14	13	0.79	13	microDALY per kg NO_x emitted
NH_3	–	4,000	1,300	28	120	120	7.6	120	microDALY per kg NH_3 emitted

Taken from Humbert (2009)

[a]The characterisation factor for indoor emissions includes only direct effects from indoor exposure. For the total characterisation factor for indoor emissions, one should add the characterisation factor for outdoor emissions (in general as a low-stack emission) of the archetype where the building is situated

[b]To be taken with care. Seems to be significantly lower than values for $PM_{2.5}$ and PM_{10}. Further research needed

3 Overview of Available Methods

Characterisation methods that address respiratory effects from particles all work with (part of) the above-mentioned framework. However, some LCIA methodologies consider impacts from particles within the 'human toxicity' impact category (e.g., Guinée et al. 2002), whereas most methods keep human toxicity and impacts from particles as two separate categories since the latter category is based on epidemiological data whereas toxicity impacts are mostly based on animal assays.

Table 6.2 provides an overview of the LCIA methodologies addressing respiratory effects from particles, with their main characteristics.

4 Variability and Uncertainty

There are many sources of variability and uncertainty along the emission-to-damage chain. Variability in the calculation of impacts contributes to uncertainty in the final impact results if it is not accounted for in the calculation of impacts. Therefore one way to reduce uncertainty is to account for spatial variability.

4.1 The Importance of Spatial Differentiation in Reducing Uncertainty

When looking at the characterisation factors at the damage level of methods presented in Table 6.2, one sees that the variation between characterisation methods is typically less than a factor 3. However, within characterisation methods, the characterisation factors can vary by more than 3 orders of magnitude for the same pollutant, depending on where it is emitted.

As evaluated in Humbert (2009), accounting for the emission source-specific population density reduces the variability (not the uncertainty) of the estimated intake fraction and the characterisation factor, which in turn reduces the uncertainty of the life cycle assessment results.

One of the main constraints in life cycle assessment regarding spatial differentiation is that most of the inventories of background processes do not give information (the format of the life cycle inventory databases often do not even provide the option to give information) on the country of emission. Certain specific processes may include the country of origin, but this information is often lost when the life cycle assessment software aggregates inventories before performing the impact assessment. In addition, for this impact category, variation within a country is often larger than between countries. Since inventory data can relatively easily be

Table 6.2 Life cycle impact assessment methodologies addressing respiratory effects from particles (including sometimes some inorganic gases) in the context of LCA

LCIA method	Substance coverage	Midpoint (named used in the method)	Unit of the characterisation factors	Endpoint (name used in the method)	Unit of the characterisation factors	Spatial resolution	References method [model]
IMPACT World+	$PM_{2.5}$, PM_{10}, NO, NO_2, SO_2, SO_3, SO_x, NH_3, CO	"Particulate matter/Respiratory inorganics midpoint"	kg $PM_{2.5}$ – eq/kg (note: the midpoint value is simply based on the endpoint divided by the characterisation factor, at endpoint, of $PM_{2.5}$)	"Respiratory effects on humans caused by inorganic substances"	DALY/kg	Global default; Version per continent	Bulle et al. (2013) [Combination of sources; taken from Humbert et al. (2011) for intake fractions and Gronlund et al. (2014) for effect factors]
ILCD LCIA	$PM_{2.5}$, PM_{10}, NO, NO_2, SO_2, SO_3, SO_x, NH_3, CO	"Particulate matter/Respiratory inorganics midpoint"	kg $PM_{2.5}$ – eq/kg (note: the midpoint value is simply based on the endpoint values calculated by Humber (2009) divided by the characterisation factor, at endpoint, of $PM_{2.5}$, also of Humbert (2009))	–	–	Global default	EC (2011), Hauschild et al. (2013) [Combination of sources; taken from Humbert (2009)]
ReCiPe	NO_x, SO_2, NH_3, PM_{10}	"Particulate matter formation"	kg PM_{10} – eq/kg	"Particulate matter formation"	DALY/kg	Europe	Goedkoop et al. (2009) [Van Zelm et al. 2008]
IMPACT 2002+/ LIME 2	$PM_{2.5}$, PM_{10}, TSP, NO, NO_2, NO_x, SO_2, SO_3, SO_x, NH_3, CO	"Respiratory effects"	kg $PM_{2.5}$ – eq/kg (note: the midpoint value is simply based on the endpoint divided by the characterisation factor, at endpoint, of $PM_{2.5}$)	"Respiratory effects"	DALY/kg	Europe	Jolliet et al. (2003), Itsubo and Inaba (2012) [Combination of sources, based on Hofstetter (1998) (note: the exact same values

(continued)

Table 6.2 (continued)

LCIA method	Substance coverage	Midpoint (named used in the method)	Unit of the characterisation factors	Endpoint (name used in the method)	Unit of the characterisation factors	Spatial resolution	References method [model]
							and therefore model as of Eco-indicator 99 are used – Goedkoop and Spriensma (1999)]
TRACI 2/TRACI 1	$PM_{2.5}$, PM_{10}, TSP, NO_x, SO_2	"Respiratory effects"	kg $PM_{2.5}$ – eq/kg	"Human health: criteria air pollutants"	DALY/kg	Global default, US at state level	Bare (2011), Bare et al. (2003) [Combination of sources, including Nishioka et al. (2002) for intake fraction and De Hollander et al. (1999) for severity]
EDIP 2003	NO_x, NO, NO_2, SO_2	"Human toxicity"	m^3/kg	–	–	Site-dependent	Hauschild and Potting (2003), Potting and Hauschild (2005) [Hauschild and Wenzel (1998)]
CML 2002	NO_2, SO_2, NH_3	"Human toxicity"	kg 1,4-DCB – eq/kg	–	–	–	Guinée et al. (2002) [Huijbregts et al. (2000)]

distinguished by archetype, even after aggregation, source-location matrices, such as suggested by Humbert (2009) and Humbert et al. (2011) and presented in Table 6.1 can be used to address spatial differentiation within current life cycle assessment constraints (including life cycle assessment software) and therefore, help reducing this source of uncertainty in the final results.

Variation in intake fraction is primarily caused by differences in population densities that can be up to two orders of magnitude between emissions in high population density areas (e.g., truck emissions in a city) and low density population areas (e.g., emissions from a truck crossing remote areas). Therefore, regionalising characterisation factors by considering variability in population density patterns is a high priority for the fate and exposure of primary and secondary PM. The intake fraction is also strongly influenced by the height of the emission source, with the highest intake fractions occurring for emissions occurring at ground-level – where people are breathing. Intake fraction also varies with local meteorological conditions, namely mixing height, wind speed and city width, approximated by the square root of the city area (Marshall et al. 2005). All other parameters considered equal, a city situated in a basin with low winds will experience a higher intra-urban intake fraction compared to a city situated near the coast with stronger winds towards the ocean. If the specific urban environment of an emission is known, Apte et al. (2012) provide continental, country and even city specific intra-urban intake fractions for ground level emissions, accounting for specific meteorological conditions and population densities of all cities with more than 100,000 inhabitants worldwide. These intra-urban intake fractions can be summed with the extra-urban rural *iF* used to replace the *iF* factor in Eq. 6.1 in the calculation of a characterisation factor that is representative of emissions into the considered city.

4.2 Other Types of Uncertainties

PM regulations and epidemiology studies typically focus on PM mass with particle size smaller than 2.5 μm. Present data do not enable to differentiate the impact as a function of particle size, and work in ultra-fine particles and nanoparticles suggests that effect on human health could be associated to the particle area rather than to particle mass. There is therefore a need to study area-based rather than mass-based dose-response relationships (Humbert 2009).

In addition, the use of epidemiological data means that PM-attributed impacts can in fact be caused by other pollutants whose concentrations could be correlated to PM (Reiss et al. 2007). Care must be taken to avoid double-counting the impacts of PM and the impact of other correlated variables in cases of common endpoints.

5 Future Trends

As outlined in Humbert (2009), the main challenges identified in LCIA modelling of respiratory effects from particles are better understanding of (i) the fate and exposure of PM as presented in the intake fractions, and of (ii) the effects.

(i) Intake fractions: First, further improvements are needed for the averaging method of wind speeds and mixing heights, which should be determined to best estimate intake fractions for rural and remote emissions. The influence of the season on the fate and exposure of PM should be evaluated, as the season has a large impact on mixing height, transport, and deposition (Ries et al. 2009). To account for meteorological specificities in the intra-urban intake fraction (such as the work of Apte et al. 2012), inventories should explore the possibilities to either capture the type of meteorological specificities associated with the city where the emission is occurring into archetypes or specify in which city the emission is occurring. The spatial differentiation of fate and exposure needs to be improved to also cover emissions in other types of environments (such as oceans or high altitudes), which would involve the evaluation of intake fractions and characterisation factors for different geographical regions in the world. Then, fate, exposure and effects from secondary PM from volatile organic compounds need to be quantified. Currently these precursors are not covered by LCIA. To further evaluate the influence of composition and size distribution, the intake fraction and effect factor should be differentiated depending on the PM source such as diesel, coal or road dust (Humbert 2009). Furthermore, the influence of buildings in modelling of the fate, exposure or effect factors of indoor exposure from outdoor emission should be considered. Buildings can change the particle size distribution and exposure to PM from outdoor origin (Riley et al. 2002; Liu and Nazaroff 2003) as well as the interpretation of epidemiological data. Another option that needs to be investigated is the possibility to determine industry specific intake fractions (e.g., for coal power plant in a country or for mining), especially in case most facilities in an industry sector are associated to an archetype (e.g. if all coal power plants would be situated in rural area and with high-stack emission). Finally, as pointed out by Fantke et al. (2014), additional areas that require further investigation include non-linearities in chemical formation of secondary PM.

(ii) Effect factors: Since chronic bronchitis in adults accounts for one-third of the PM effect factor (Humbert 2009), this aspect should be assessed with higher certainty. The influence of PM inhalation on low birth weight (Bell et al. 2008) and its expression in terms of DALYs also deserves further attention. Dose-responses from chronic exposure to precursors, CO, and secondary particulate matter from NH_3 require better understanding. Characterisation factors, both for PM smaller than 2.5 μm (e.g., PM_1 or $PM_{0.1}$) and for PM between 2.5 and 10 μm, should be studied further. On the dose-response side, there is a need to study the incorporation of the risk ratios proposed for PM in the 2010 burden of disease (Lim et al. 2012; Murray et al. 2013). Furthermore, because of the uncertain mode of action for PM, modelling of effect factors may consider surface area and number of particles instead of only mass as a proxy for adverse health effects (Humbert 2009). Similarly, fate and exposure modelling may consider the evolution of particle size distribution over time for an emission (Humbert 2009). Finally, as pointed out by Fantke et al. (2014), additional areas that require further investigation include non-linearity of the exposure-response for certain health endpoints, use of cause-specific vs. all-cause mortality data for calculation of effect factors, age- and cause-specific disability weights, and extension of the epidemiological assessment that are only mortality-based to also include morbidity.

While intake fractions, effect factors and characterisation factors for particles are still in need of further work, source-location matrices, as suggested by Humbert (2009) and Humbert et al. (2011) and presented in Table 6.1 provide a framework for life cycle assessment practitioners to improve their evaluations of adverse health effects caused by primary and secondary particulate matter. In many damage-oriented life cycle assessment studies, PM is responsible for a large or dominant fraction of the total human health damage. Harmonising the values used in life cycle assessment studies and making those values consistent with the characterisation of organics (Rosenbaum et al. 2008; Hauschild et al. 2008) will increase accuracy, consistency, and comparability among results for human health damage and strengthen the overall assessment of human health impacts in life cycle assessment.

To conclude, it is worth to note that in the ILCD recommendations (Hauschild et al. 2013), the '*particulate matter/respiratory inorganics*' human health impact category is the only impact category to get a classification 1 (i.e., classified as recommended and satisfactory) both at midpoint and endpoint level, so even though one can point to the need for further improvement, this expert assessment qualified it as satisfactory.

References

Abrahamowicz M, Schopflocher T, Leffondré K et al (2003) Flexible modeling of exposure-response relationship between long-term average levels of particulate air pollution and mortality in the American Cancer Society study. J Toxicol Environ Health 66:1625–1654

Apte JS, Bombrun E, Marshall JD, Nazaroff WW (2012) Global intra urban intake fractions for primary air pollutants from vehicles and other distributed sources. Environ Sci Technol 46 (6):3415–3423. doi:10.1021/es204021h

Bare JC (2011) TRACI 2. Available at http://www.epa.gov/ordntrnt/ORD/NRMRL/std/traci/traci.html

Bare JC, Norris GA, Pennington DW, McKone T (2003) TRACI: The tool for the reduction and assessment of chemical and other environmental impacts. J Ind Ecol 6(3–4):49–78

Bell ML, Ebisu K, Belanger K (2008) The relationship between air pollution and low birth weight: effects by mother's age, infant sex, co-pollutants, and pre-term births. Environ Res Lett 3(4):1–7. doi:10.1088/1748-9326/3/4/044003

Bennett DH, McKone TE, Evans JS, Nazaroff WW, Margni MD, Jolliet O, Smith KR (2002) Defining intake fraction. Environ Sci Technol 36(9):207A–211A

Brook RD, Rajagopalan S, Pope CA III, Brook JR, Bhatnagar A, Diez-Roux AV, Holguin F, Hong Y, Luepker RV, Mittleman MA, Peters A, Siscovick D, Smith SC Jr, Whitsel L, Kaufman JD (2010) Particulate matter air pollution and cardiovascular disease: an update to the scientific statement from the American Heart Association. Circulation 121(21):2331–2378

Bulle C, Jolliet O, Margni M, Rosenbaum R, Humbert S (2013) IMPACT World+. Available at http://www.impactworldplus.org

Burnett RT, Pope III CA, Ezzati M, Olives C, Lim SS, Mehta S, Shin H, Anderson HR, Smith KR, Cohen AJ (2013) An integrated exposure-response function for estimating the global burden of disease attributable to ambient PM2.5 exposure. In: 2013 Conference environment and health – bridging South, North, East and West, Joint meeting of the International Society for Environmental Epidemiology (ISEE), the International Society of Exposure Science (ISES), and the International Society of Indoor Air Quality and Climate (ISIAQ), Basel, 19–23 Aug 2013

Chen H, Goldberg MS, Villeneuve PJ (2008) A systematic review of the relation between long-term exposure to ambient air pollution and chronic diseases. Rev Environ Health 23(4):243–297

Cohen AJ, Anderson HR, Ostro B, Dev Pandey K, Krzyzanowski M, Kuenzli N, Gutschmidt K, Pope CA III, Romieu I, Samet JM, Smith K (2005) The global burden of disease due to outdoor air pollution. J Toxicol Environ Health A 68(13–14):1301–1307

De Hollander AEM, Melse JM, Lebret E, Kramers PGN (1999) An aggregate public health indicator to represent the impact of multiple environmental exposures. Epidemiology 10 (5):606–617

EC (2011) International Reference Life Cycle Data System (ILCD) handbook. Recommendations for Life Cycle Impact Assessment in the European context, 1st edn. European Commission-Joint Research Centre, Institute for Environment and Sustainability, Luxemburg

Fantke P, Jolliet O, Apte JS, Cohen AJ, Evans JS, Hänninen OO, Hurley F, Jantunen MJ, Jerrett M, Levy JI, Loh MM, Marshall JD, Miller BG, Preiss P, Spadaro JV, Tainio M, Tuomisto JT, Weschler CJ, McKone TE (2014) Health effects of fine particulate matter in life cycle impact assessment: findings from the Basel Guidance Workshop. Int J Life Cycle Assess. doi 10.1007/s11367-014-0822-2

Franklin M, Koutrakis P, Schwartz J (2008) The role of particle composition on the association between PM2.5 and mortality. Epidemiology 19(5):680–689

Goedkoop MJ, Spriensma R (1999) The eco-indicator'99: a damage-oriented method for life cycle impact assessment. Ministry of Housing, Spatial Planning, and Environment, The Hague

Goedkoop M, Huijbregts MAJ, Heijungs R, De Schryver A, Struijs J, Van Zelm R (2009) ReCiPe 2008: a life cycle impact assessment method which comprises harmonised category indicators at the midpoint and the endpoint level. Report I: characterisation, 1st edn. http://www.lcia-recipe.net

Greco SL, Wilson AM, Spengler JD, Levy JI (2007) Spatial patterns of low-stack source particulate matter emissions-to-exposure relationships across the United States. Atmos Environ 41 (5):1011–1025. doi:10.1016/j.atmosenv.2006.09.025

Gronlund C, Humbert S, Shaked S, O'Neil M, Jolliet O (2014) Characterizing the burden of disease of particulate matter for life cycle assessment. Air Qual Atmos Health. doi: 10.007/s11869-014-0283-6

Guinée JB, Gorée M, Heijungs R, Huppes G, Kleijn R, Koning AD, Oers LV, Wegener Sleeswijk A, Suh S, Udo de Haes HA, Bruijn HD, Duin RV, Huijbregts MAJ (2002) Handbook on life-cycle assessment: operational guide to the ISO standards. Kluwer Academic Publishers, Dordrecht

Hänninen O, Knol A, Jantunen M, Lim T-A, Conrad A, Rappolder M, Carrer P, Fanetti A-C, Kim R, Buekers J, Torfs R, Iavarone I, Classen T, Hornberg C, Mekel O (2014) Environmental burden of disease in Europe: estimates for nine stressors in six countries. Environ Health Perspect 122(5):439–446. doi: 10.1289/ehp.1206154. Epub 2014 Feb 26.

Hauschild M, Potting J (2003) Spatial differentiation in life-cycle impact assessment: the EDIP 2003 methodology. Guidelines from the Danish EPA. Institute for Product development, Technical University of Denmark, Copenhagen

Hauschild M, Wenzel H (1998) Environmental assessment of products, vol 2, Scientific background. Chapman & Hall, London

Hauschild MZ, Huijbregts M, Jolliet O, Macleod M, Margni M, Van de Meent D, Rosenbaum RK, McKone TE (2008) Building a model based on scientific consensus for life cycle impact assessment of chemicals: the search for harmony and parsimony. Environ Sci Technol 42 (19):7032–7037

Hauschild MZ, Goedkoop M, Guinée J, Heijungs R, Huijbregts M, Jolliet O, Margni M, De Schryver A, Humbert S, Laurent A, Sala S, Pant R (2013) Identifying best existing practice for characterization modeling in life cycle impact assessment. Int J Life Cycle Assess 18:683–697. doi:10.1007/s11367-012-0489-5

Hoek G, Krishnan RM, Beelen R, Peters A, Ostro B, Brunekreef B, Kaufman JD (2013) Long-term air pollution exposure and cardio-respiratory mortality: a review. Environ Health 12(1):43–57

Hofstetter P (1998) Perspectives in life cycle impact assessment: a structured approach to combine models of the technosphere, ecosphere and valuesphere. Kluwer Academic Publishers, Amsterdam

Huijbregts MAJ, Thissen U, Guinée JB, Jager T, Kalf D, van de Meent D, Ragas AMJ, Wegener Sleeswijk A, Reijnders L (2000) Priority assessment of toxic substances in life cycle assessment. Part I: calculation of toxicity potentials for 181 substances with the nested multi-media fate, exposure and effects model USES-LCA. Chemosphere 41:541–573

Humbert S (2009) Geographically differentiated life-cycle impact assessment of human health. Doctoral dissertation, University of California, Berkeley

Humbert S, Marshall JD, Shaked S, Spadaro JV, Nishioka Y, Preiss P, McKone TE, Horvath A, Jolliet O (2011) Intake fraction for particulate matter: recommendations for life cycle impact assessment. Environ Sci Technol 45(11):4808–4816

Ilacqua V, Hänninen O, Kuenzli N, Jantunen MF (2007) Intake fraction distributions for indoor VOC sources in five European cities. Indoor Air 17(5):372–383

Itsubo N, Inaba A (2012) LIME 2. Available at http://lca-forum.org/english/

Jolliet O, Margni M, Charles R, Humbert S, Payet J, Rebitzer G, Rosenbaum R (2003) IMPACT 2002+: a new life cycle impact assessment method. Int J Life Cycle Assess 8(6):324–330

Kanakidou M, Seinfeld JH, Pandis SN, Barnes I, Dentener FJ, Facchini MC, Van Dingenen R, Ervens B, Nenes A, Nielsen CJ, Swietlicki E, Putaud JP, Balkanski Y, Fuzzi S, Horth J, Moortgat GK, Winterhalter R, Myhre CEL, Tsigaridis K, Vignati E, Stephanou EG, Wilson J (2005) Organic aerosol and global climate modelling: a review. Atmos Chem Phys 5:1053–1123. doi:10.5194/acp-5-1053-2005

Kelly FJ, Fussell JC (2012) Size, source and chemical composition as determinants of toxicity attributable to ambient particulate matter. Atmos Environ 60:504–526

Kuenzli N, Kaiser R, Medina S, Studnicka M, Chanel O, Filliger P, Herry M, Horak F Jr, Puybonnieux-Texier V, Quenel P, Schneider J, Seethaler R, Vergnaud J-C, Sommer H (2000) Public-health impact of outdoor and traffic-related air pollution: a European assessment. Lancet 356(9232):795–801

Laden F, Schwartz J, Speizer FE, Dockery DW (2006) Reduction in fine particulate air pollution and mortality: extended follow-up of the Harvard Six Cities study. Am J Respir Crit Care Med 173(6):667–672. doi:10.1164/rccm.200503-443OC

Levy JI, Wolff SK, Evans JS (2002) A regression-based approach for estimating primary and secondary particulate matter iF. Risk Anal 22(5):895–904

Lim SS, Vos T, Flaxman AD et al (2012) A comparative risk assessment of burden of disease and injury attributable to 67 risk factors and risk factor clusters in 21 regions, 1990—2010: a systematic analysis for the Global Burden of Disease Study 2010. Lancet 380:2224–2260. doi:10.1016/S0140-6736(12)61766-8

Lippmann M, Chen L-C (2009) Health effects of concentrated ambient air particulate matter (CAPs) and its components. Crit Rev Toxicol 39(10):865–913

Liu D-L, Nazaroff WW (2003) Particle penetration through building cracks. Aerosol Sci Technol 37(7):565–573. doi:10.1080/02786820300927

Marshall JD, Teoh SK, Nazaroff WW (2005) Intake fraction of nonreactive vehicle emissions in US urban areas. Atmos Environ 39(7):1363–1371

Mehta S, Shin H, Burnett R, North T, Cohen AJ (2013) Ambient particulate air pollution and acute lower respiratory infections: a systematic review and implications for estimating the global burden of disease. Air Qual Atmos Health 6(1):69–83. doi:10.1007/s11869-011-0146-3

Muller NZ, Mendelsohn R (2007) Measuring the damages of air pollution in the United States. J Environ Econ Manage 54(1):1–14

Murray C, Lopez A (1996) The global burden of disease, a comprehensive assessment of mortality and disability from diseases, injuries, and risk factors in 1990 and projected to 2020. Global burden of disease and injury series, vols 1 & 2. Harvard School of Public Health, Boston, and World Health Organization, Geneva, and The World Bank, Washington, DC

Murray CJL, Abraham J, Ali MK et al (2013) The state of US health, 1990–2010: burden of diseases, injuries, and risk factors. J Am Med Assoc 310(6):591–608. doi:10.1001/jama.2013.13805

Nishioka Y, Levy JI, Norris GA, Wilson A, Hofstetter P, Spengler JD (2002) Integrating risk assessment and life cycle assessment: a case study of insulation. Risk Anal 22:1003–1017

Pelucchi C, Negri E, Gallus S, Boffetta P, Tramacere I, La Vecchia C (2009) Long-term particulate matter exposure and mortality: a review of European epidemiological studies. BMC Public Health 9:453–460. doi:10.1186/1471-2458-9-453

Pope CA III, Burnett RT, Thun MJ, Calle EE, Krewski D, Ito K, Thurston GD (2002) Lung cancer cardiopulmonary mortality and long-term exposure to fine particulate air pollution. J Am Med Assoc 287(9):1132–1141

Pope CA III, Burnett RT, Krewski D, Jerrett M, Shi Y, Calle EE, Thun MJ (2009) Cardiovascular mortality and exposure to airborne fine particulate matter and cigarette smoke: shape of the exposure-response relationship. Circulation 120(11):941–948

Potting J, Hauschild M (2005) Background for spatial differentiation in LCA impact assessment: the EDIP2003 methodology. Danish Ministry of the Environment, Environmental Project No. 996, Copenhagen

Potting J, Hauschild MZ (2006) Spatial differentiation in life cycle impact assessment: a decade of method development to increase the environmental realism of LCIA. Int J Life Cycle Assess 11 (Special issue 1):11–13

Reap J, Roman F, Duncan S, Bras B (2008) A survey of unresolved problems in life cycle assessment, part 2: impact assessment and interpretation. Int J Life Cycle Assess 13(5):374–388

Reiss R, Anderson EL, Cross CE, Hidy G, Hoel D, McClellan R, Moolgavkar S (2007) Evidence of health impacts of sulfate- and nitrate-containing particles in ambient air. Inhal Toxicol 19 (5):419–449

Ries FJ, Marshall JD, Brauer M (2009) Intake fraction of urban wood smoke. Environ Sci Technol 43(13):4701–4706

Riley WJ, McKone TE, Lai ACK, Nazaroff WW (2002) Indoor particulate matter of outdoor origin: importance of size-removal mechanisms. Environ Sci Technol 36:200–207

Roman HA, Walker KD, Walsh TL, Conner L, Richmond HM, Hubbell BJ, Kinney PL (2008) Expert judgment assessment of the mortality impact of changes in ambient fine particulate matter in the US. Environ Sci Technol 42(7):2268–2274

Rosenbaum RK, Bachmann TM, Swirsky Gold L, Huijbregts MA, Jolliet O, Juraske R, Köhler A, Larsen HF, MacLeod M, Margni M, McKone TE, Payet J, Schumacher M, Van de Meent D, Hauschild MZ (2008) USEtox—The UNEP-SETAC toxicity model: Recommended

characterisation factors for human toxicity and freshwater ecotoxicity in life cycle impact assessment. Int J Life Cycle Assess 13(7):532–546
Schwartz J, Coull B, Laden F, Ryan L (2008) The effect of dose and timing of dose on the association between airborne particles and survival. Environ Health Perspect 116(1):64–69
Straif K, Cohen A, Samet J (2013) Air pollution and cancer. IARC Scientific Publication No 161. International Agency for Research on Cancer, Lyon Cedex, p 229
Udo de Haes HA, Finnveden G, Goedkoop MJ, Hauschild M, Hertwich EG, Hofstetter P, Jolliet O, Klöpffer W, Krewitt W, Lindeijer E, Müller-Wenk R, Olsen SI, Pennington DW, Potting J, Steen B (2002) Life-cycle impact assessment: striving towards best practice. SETAC Press, Pensacola
Van Zelm R (2009) Personal communication. Radboud University, Nijmegen
Van Zelm R, Huijbregts MAJ, Den Hollander HA, Van Jaarsveld HA, Sautere FJ, Struijs J, Van Wijnen HJ, Van de Meent D (2008) European characterization factors for human health damage of PM10 and ozone in life cycle impact assessment. Atmos Environ 42(3):441–453
von Sedlbauer K, Braune A, Humbert S, Margni M, Schuller O, Fischer M (2007) Spatial differentiation in LCA: moving forward to more operational sustainability. Technikfolgenabschätzung Theorie Praxis 3(16):24–31
WHO (2006) Health risks of particulate matter from long-range transboundary air pollution. European Centre for Environment and Health. World Health Organization, Bonn
WHO (2008) The global burden of disease: 2004 update. World Health Organization, Geneva

Chapter 7
Photochemical Ozone Formation

Philipp Preiss

Abstract Anthropogenic ozone arises as the product of reactions in the atmosphere between OH-radicals, the anthropogenic air pollutants nitrogen oxides (NO_x) and different non-methane volatile organic compounds (NMVOC).

The photochemical oxidant of main interest within LCA is ground level ozone (O_3) caused by the emission of the air pollutants NO_x and NMVOC within life cycles of products and services. Several different LCIA methods have been developed in the last 20 years to characterise this impact category. Some provide midpoint and some endpoint characterisation factors, and there are site generic and spatially explicit methods. They all struggle with the highly non-linear dependence of ozone creation on background conditions regarding chemical substances and meteorology and also the fact that many response functions include thresholds and that resulting impacts depend on the ozone exposure, both for impacts on human health and on other living beings and even on materials. Hence, the modelled impacts due to ozone caused by anthropogenic emissions are subject to large variability and uncertainty.

Keywords Air pollutants • Ground level ozone (O_3) • Impacts on human health and vegetation • Life cycle impact assessment • Nitrogen oxides (NO_x) • NO_x • Non-methane volatile organic compounds (NMVOC) • NMVOC • O_3 • Ozone

1 Introduction

Photochemical oxidants arise as the product of reactions between OH-radicals, photochemical oxidants and the air pollutants nitrogen oxides (NO_x) and non-methane volatile organic compounds (NMVOC) in the atmosphere.

The photochemical oxidants are mainly ozone (O_3), peroxyacetyl nitrate and hydrogen peroxide. The main impact from photochemical oxidants on the natural environment is caused by an elevated O_3 concentration. Excessive concentrations of tropospheric O_3 have toxic effects on both plants (Davison and Barnes 1998;

P. Preiss (✉)
University of Stuttgart, Silberburgstrasse 150, D-70176 Stuttgart, Germany
e-mail: ppreiss44@gmx.de

M.Z. Hauschild, M.A.J. Huijbregts (eds.), *Life Cycle Impact Assessment*,
LCA Compendium – The Complete World of Life Cycle Assessment,
DOI 10.1007/978-94-017-9744-3_7

Ashmore 2002; Singh et al. 2009) and human health (WHO 2003; Bell et al. 2004; Anenberg et al. 2009, 2010).

Of main interest for the different life cycle impact assessment (LCIA) methods in this context is the impact assessment of the emissions of NO_x and NMVOC regarding their effects as a precursor of ground level ozone. NMVOC is a generic term for several substances or a group of substances consisting of alkanes, alkenes, aromatics, aldehydes and alcohols, etc. Sometimes, VOC is used synonymously with NMVOC, although NMVOC excludes methane. NO_x is the sum of NO and NO_2.

Ground level, or tropospheric ozone is especially known in connection with summer smog. Ozone also occurs in the ozone layer (stratosphere) where it is very beneficial for all living beings because it filters the harmful medium-frequency ultraviolet light from the sun.

The research on atmospheric chemistry on the one hand and impact assessment of ozone towards different receptors on the other hand has several motivations. Policy on air quality is concerned with absolute levels of ozone concentration and change of ozone concentration due to non-marginal change of emission of precursors. Tropospheric ozone also plays a role as a driver for climate change.

This chapter focuses on ground level ozone since it is the main contributor to photochemical oxidant impact. Therefore, the term ozone is used to refer to ground level ozone, unless explicitly stated otherwise.

1.1 Tropospheric Ozone

According to Amann and co-workers (2008), tropospheric ozone is a highly oxidative compound formed in the lower atmosphere from gases (originating to a large extent from anthropogenic sources) by photochemistry driven by solar radiation. From the perspective of LCIA, the most relevant substances which are contributing to ozone creation are NO_x and the different components of NMVOC.

1.2 Ozone Creation

The atmospheric chemistry of NMVOC and NO_x is described in detail e.g., in (Atkinson 2000). Ozone is formed photochemically from the photolysis of NO_2 as follows:

$$NO_2 + h\gamma \rightarrow NO + O \tag{7.1}$$

$$O + O_2 + M \rightarrow O_3 + M \tag{7.2}$$

M is a third 'body' – it can be actually any 'body' with mass, mostly nitrogen or oxygen molecules, but also particles, etc. It absorbs energy from the reaction as

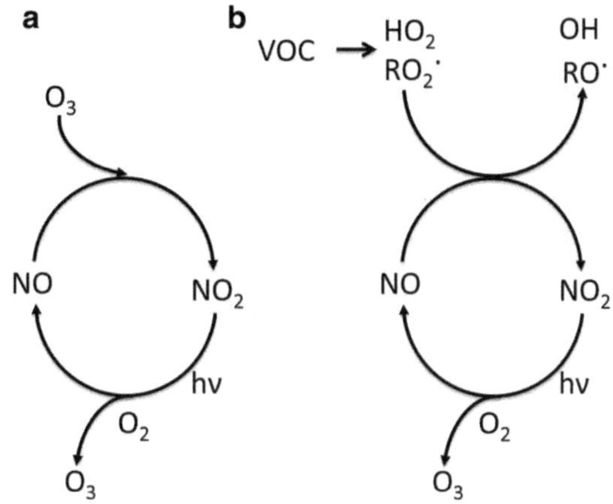

Fig. 7.1 Model of the reactions involved in NO to NO_2 conversion and O_3 formation. *Left* (**a**) NO-NO_2-O_3 systems in the absence of VOC. *Right* (**b**) NO-NO_2-O_3 systems in the presence of VOC (Taken from Atkinson 2000)

heat. Without this absorption, combining of O and O_2 into O_3 cannot be accomplished.

O_3 rapidly reacts with NO:

$$NO + O_3 \rightarrow NO_2 + O_2 \tag{7.3}$$

These reactions result in a photo equilibrium between NO, NO_2 and O_3 with no net formation or loss of O_3, as shown in Fig. 7.1a. However, in the presence of VOC (including methane) the degradation reactions of VOC lead to the formation of intermediate RO_2 and HO_2 radicals (hence, *R* stands for an 'organic remaining of a VOC after the corresponding reaction'). These HO_2 and RO_2 radicals react with NO, converting NO to NO_2

$$HO_2 + NO \rightarrow OH + NO_2 \tag{7.4}$$

$$RO_2 + NO \rightarrow RO + NO_2 \tag{7.5}$$

which then forms ozone through photolysis (Fig. 7.1b), hence, yielding a net formation of ozone.

Hydroxyl (OH) radicals are the key reactive species in the troposphere, because they are reacting with nearly all organic compounds. VOC are oxidised by a series of reactions, finally leading to formation of carbon monoxide (CO), carbon dioxide (CO_2) and water (H_2O).

The following reactions are involved:

$$\mathrm{RH + OH \rightarrow H_2O + R} \tag{7.6}$$

$$\mathrm{R + O_2 + M \rightarrow RO_2O + M} \tag{7.7}$$

$$\mathrm{RO_2 + NO \rightarrow NO_2 + RO} \tag{7.8}$$

The organic remaining 'R' from the VOC can also be generated by photolysis, i.e. caused by solar radiation. Usually, VOC with molecules containing the carbonyl (C = O) bond are involved. Most common is formaldehyde (HCHO).

$$OH + HCHO \rightarrow H_2O + HCO \tag{7.9}$$

Hydroperoxyl radical $\left(HO_{\downarrow}(2)\right)$ is then generated by the following reaction

$$HCO + O_2 \rightarrow HO_2 + CO \tag{7.10}$$

leading to the following reaction:

$$HO_2 + NO \rightarrow NO_2 + OH \tag{7.11}$$

Hence, the OH is available again and NO_2 is generated. Moreover, the carbon monoxide (CO) can finally create another hydroperoxyl radical.

Another effect of solar radiation can be described as follows

$$HCHO \overset{+hy_2}{\Rightarrow} H + HCO \tag{7.12}$$

The hydrogen atom and formyl radical H + HCO produced by this photolytic reaction yield two hydroperoxyl radicals via reaction with oxygen.

The reactions above comprise the simplest VOC oxidation combination. There are many other similar reactions taking place, but the shown schemes give an idea of the underlying mechanisms.

Hence, solar radiation is involved in the creation of ozone, mainly through regeneration of hydroxyl radicals, and hence, especially in summer on days with high ozone levels. The distribution of the emissions and background concentrations differ and change in time and space. However, NO_x is mainly emitted by combustion activities, such as thermal energy conversion (electricity and heat) and different transport-technologies such as aviation, shipping and vehicles, which use fossil fuels. NMVOC are mainly emitted by solvent applications but also by the domestic sectors. Moreover, there are many natural NMVOC emitted, e.g. by forests.

A difficulty in the assessment of NMVOC as one group of substances is that it can consist of very different substances in different composition, depending on the source of emission, with rather different potentials to contribute to the creation of ozone. For example, according to Theloke (2004) NMVOC emissions from solvent use in 2000 in Germany consisted of alcohols (31 %), alkanes and cycloalkanes (22 %), aromatics (15 %), esters (11 %), glycol derivatives (6 %) as well as ketones (5 %), terpenes (4 %), ethers and halogenated hydrocarbons (2 % each) and low amounts of organic acids, aldehydes, amines and amides. Approximately 2 % of the emissions could not be assigned to a substance class.

The photochemical ozone creation potential (POCP) indicates the potential capacity of an organic compound to create ozone in the troposphere. The value for ethene has been set as a reference. However, different lists and definitions of

Table 7.1 POCP-values for some NMVOC (sorted by declining POCP-values (Derwent et al. 1998))

Substance	POCP	Substance	POCP
1.3.5-Trimethylbenzene	138.1	1.3-Butadiene	85.1
1.2.4-Trimethylbenzene	127.8	2-Methyl-2-Butene	84.2
1.2.3-Trimethylbenzene	126.7	2-Methyl-1-Butene	77.1
cis-2-Butene	114.6	3-Methyl-1-Butene	67.1
trans-2-Butene	113.2	Toluene	63.7
Propene	112.3	Methyl propene	62.7
cis-2-Pentene	112.1	n-Butanol	61.2
trans-2-Pentene	111.7	n-Propanol	54.3
m-Xylene	110.8	2.3-Dimethylbutane	54.1
Isoprene	109.2	n-Heptane	49.4
1-Butene	107.9	n-Octane	45.3
trans-2-Hexene	107.3	n-Nonane	41.4
cis-2-Hexene	106.9	2-Methylhexane	41.1
o-Xylene	105.3	n-Decane	38.4
p-Xylene	101	3-Methylhexane	36.4
Ethene	**100**	Trichloroethylene	32.9
1-Pentene	97.7	Acetic acid ethyl ester	21.3
1-Hexene	87.4	Isopropanol	14
1.3-Butadiene	85.1	Ethane	12.3
2-Methyl-2-Butene	84.2	Acetone	9.4
1-Hexene	87.4	Perchloroethylene	2.9

POCP exist. The POCP is based on reaction rates with hydroxyl radical in the troposphere and hence, the spatial scope should be included, to reflect the local, regional and global effect on ozone creation (Table 7.1).

The POCP above is calculated for a certain time horizon, weather conditions, and for a certain area (Northwest Europe). The reactivity varies strongly between substances, hence, if the composition of the NMVOC is known, or when different specified solvents have to be compared, their actual ozone creation potential should be considered.

Sambat and co-workers (2005) explain that depending on the actual substance the photolysis can change one NMVOC (with high number of C atoms) to another one which again has certain reactivity. Hence, all NMVOC contribute to ozone generation depending on the spatial and time scales. Different approaches exist to the assessment of the reactivity of an NMVOC.

1.3 Photochemical Ozone Creation Potential

The value of the POCP as developed by Derwent and co-workers (1998) strongly depends on the considered time horizon. For a short integration time of some hours mainly very reactive NMVOC contribute to ozone formation – and therefore, on a local scale. If the integration time is increased to several days, even less reactive NMVOC contribute to a significant extent to ozone formation perhaps on a larger regional scale. In order to capture the full ozone formation burden which is of importance on the regional and even hemispheric scale, one has to consider a time horizon of weeks or even months.

1.4 Reactivity by OH

Sambat and co-workers point to the dependence of the ozone concentration on the atmospheric life time which is largely determined by its reaction with OH radicals. The lifetime thus depends both on the assumed atmospheric concentrations of OH radicals and the concentrations of the hydrocarbon (Sambat et al. 2005). Following this approach they make a distinction between NMVOC which contribute locally, regionally, supra-regionally or globally to the ozone formation.

1.5 Maximum Incremental Reactivity

With the incremental reactivity (IR) concept from Carter et al. (1995), the ozone formation potential is treated in terms of its incremental reactivity, i.e. the number of molecules of ozone formed per NMVOC carbon atom added to a certain atmospheric reaction mixture of NMVOC and NO_X. The peak IR value of an NMVOC is known as its maximum incremental reactivity (MIR).

In addition to the inherent reactivity of a specific compound, also specific conditions, emission and concentration of other pollutants and meteorological conditions are important for the resulting ozone formation. Hence, one of the main problems for the determination of ozone formation potential is the dependence on many parameters which vary with time, space and system boundaries.

1.6 Impact Pathway and Affected Area of Protection

As depicted in Fig. 7.1, the impact pathway leads from emission of NMVOC, CO and NO_x via creation of ozone to the effects on human health (acute and chronic effects), and effects on crops and natural vegetation, i.e. forestry and other plants, and therefore to damages on the ecosystem at large. In the diagram there are references made to certain LCIA-methods (circles) which address the effects at the corresponding level of the impact pathway. Therefore, these LCIA-methods provide characterisation factors for different midpoints or endpoints, and sometimes provide both (Fig. 7.2).

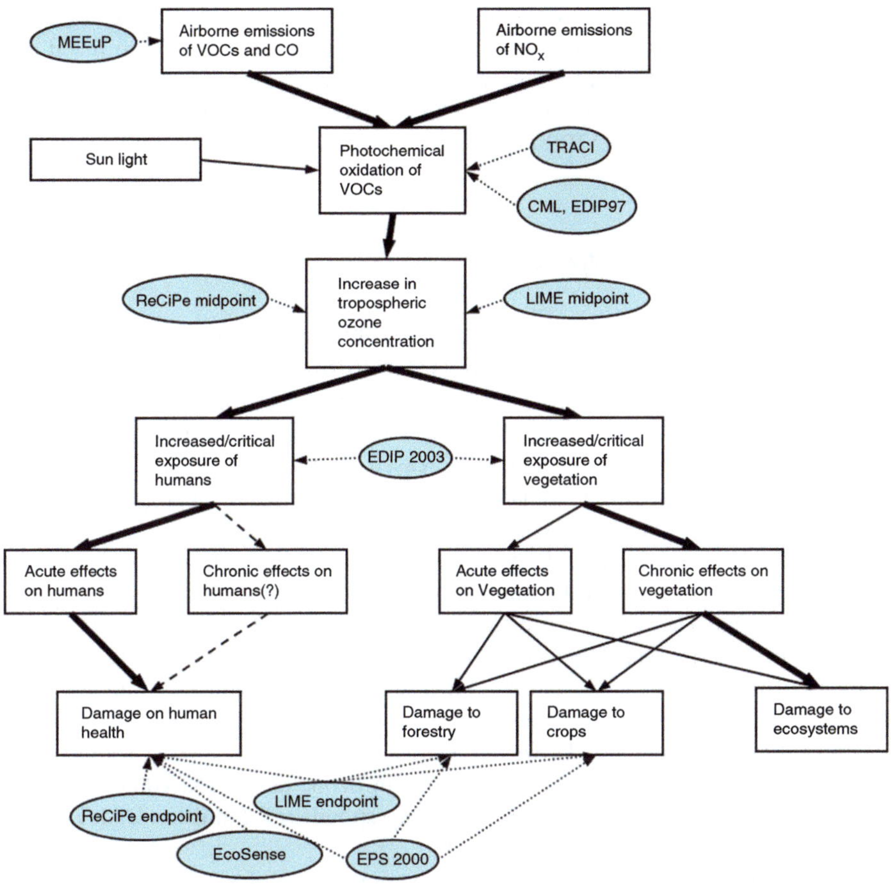

Fig. 7.2 Impact pathway for photochemical ozone formation and approach of different LCIA methods along the cause-effect chain (Taken from JRC-IES 2011)

There are different spatial scales of interest with regard to the environmental effects of the created ozone. Most popular is the so called summer smog (corresponding to very high concentration of ozone) which occurs only periodically and due to certain, relatively rare circumstances (e.g., high ambient temperature and intense solar radiation and low wind speed). This summer smog has a more local extent but relatively strong impacts on human health. The ozone formation may also lead to increasing tropospheric concentrations at the regional scale, and due to the larger area, the regional scale is more important for impacts on vegetation. Finally, it has to be mentioned that ozone acts as a direct greenhouse gas and that it affects the atmospheric lifetime of methane, another important greenhouse gas. This makes it relevant regarding radiative forcing, and, therefore, ozone is important on a global scale with regard to the impacts on global warming. However, this last issue is not further addressed in this chapter.

The pre-selected characterisation models which were considered for recommendation under the EU Commission's ILCD system (JRC-IES 2010) are listed and characterised in Table 7.2. The table shows the different levels of either including a midpoint and/or an endpoint model. A more detailed summary of the methods below can be found in JRC-IES (2011).

1.7 Impacts due to Ozone

Ozone is harmful to flora, fauna, human health and materials. However, the assessment of the corrosion of materials due to ozone has received only little attention within LCIA. Therefore, impact assessment has focused on impacts on crops, ecosystems and human health. Ozone can irritate the respiratory system and lead to different health effects and increased mortality. Due to its effect on vegetation, it may also affect human welfare due to its effects on crop productivity and biodiversity and hence, on ecosystem services.

The analysis by Amann and co-workers (2008) indicates that ozone pollution affects the health of most of the populations of Europe, leading to a wide range of health problems. The effects are estimated to include ca. 21,000 premature deaths annually in 25 European Union countries. Anenberg and co-workers (2010) estimate that anthropogenic O_3 is associated with an estimated 0.7 (range 0.4–1) million respiratory mortalities and correspondingly 6.3 (range 6.3–9.3) million years of life lost annually. However, Lim and co-workers (2012) estimate the global burden of disease in 2010 from ambient ozone as ca. 152,400 deaths with a range between 52,300 and 267,000. This corresponds to 2.5 million DALY with a range from 840,000 to 4.3 million DALY.

Van Dingenen and co-workers (2009) estimate the global impact of ozone on agricultural crop yields under current and future air quality legislation. Results indicated that present day global relative yield losses range between 7 and 12 % for wheat, between 6 and 16 % for soybean, between 3 and 4 % for rice, and between 3 and 5 % for maize.

1.8 Ozone Metrics and Levels

Ozone concentrations behave in a very dynamic way. The precursors NO_x and NMVOC are not emitted continuously, neither by natural nor by anthropogenic sources, but emissions show variation during the day (day, night, rush hours), during the week and between the different seasons. In addition, the ozone creation and resulting concentration are very much influenced by meteorological factors which are also changing during the day, across seasons, as well as between locations of emission sources. The meteorological factors are, for example, direct sunlight (and hence, there is an influence by clouds), temperature and atmospheric

Table 7.2 Characterisation models regarding ozone formation and their specifications

Method	Area of protection	Midpoint/ Endpoint	Unit	Spatial resolution	Chemicals	Model	Reference
CML 2002	–	M	kg ethylene-eq. per kg emitted	Country, Europe	1, 3	POCP	Guinée et al. (2002), Huijbregts et al. (2000)
Eco-Indicator 99	Human Health	E	DALY per kg emitted	Europe	1, 3	–	Goedkoop and Spriensma (1999)
EcoSense/ EcoSenseWeb	Human Health, crops	M, E	Accumulated exposure, YOLL and DALY per kg emitted	Europe/ Northern Hemisphere	1, 2	EMEP chemical transport Model	Preiss and Klotz (2008)
EDIP2003	Human health and vegetation	M	kg ethylene-eq. per kg emitted and m^2 ecosystem *ppm*hours per g emitted; pers*ppm*hours per g emitted	Country, Europe	1, 2, 3, 4, 5	EMEP Chemical Transport Model	Potting et al. (1998), Hauschild and Potting (2003)
EPS 2000	Human Health, crops	E	DALY per kg emitted	Generic	1, 2, 3	based on EcoSense	Steen (1999)
Impact 2002+	–	M	kg ethylene eq. into air per kg emitted	–	1, 2, 3	POCP	–
LIME	Human health, crops, wood and primary production	M, E	kg ethylene eq. into air per kg emitted	Japan	1, 2, 3	–	Hayashi et al. (2004)
LUCAS	–	O	kg ethylene eq. into air per kg emitted	Provinces, Canada	2	Maximum Incremental Reactivities	Carter (1994, 1998)
MEEuP	–	M	VOC in mg	Generic	2, no distinction between NMVOC	–	Kemna et al. (2005)

(continued)

Table 7.2 (continued)

Method	Area of protection	Midpoint/ Endpoint	Unit	Spatial resolution	Chemicals	Model	Reference
ReCiPe	yes	M, E	kg NMVOC-eq. per kg emitted and DALY estimates for acute mortality related to ozone	Europe	1, 2, 3	dynamic model LOTOS-EUROS	TNO (2005–2013), Van Zelm et al. (2007), Goedkoop et al. (2012)
Swiss Ecoscarcity 07	–	o	NMVOC and Critical flow for NMVOC corresponding to Swiss political aims. i.e. emission flow in year 1960	–	2, no distinction between NMVOC, i.e. weight-based summation as in MEEuP	–	–
TRACI	–	M	g – NO_x-eq. per kg emission	North-America and Mexico	2	–	Norris (2003)

Level:
o: Available in the methodology, but not further investigated
M: Midpoint model available and further analysed
E: Endpoint model available and further analysed
Chemicals – Substance detail:
1: NO_x, 2: NMVOC, 3: POCP-factors, 4: CO, 5: methane
Ecosystem: only terrestrial

mixing layer height. Furthermore, ozone itself is a very reactive gas with a relatively short lifetime in the atmosphere. Since it is very dependent on the conditions and reaction partners, it is difficult to specify the lifetime. At ground level the lifetime of ozone ranges from hours to days, whereas in the upper troposphere the lifetime can be weeks and months.

Hence, the spatial and temporal distribution of ozone concentrations shows large variations. Therefore, different concentration-response-relationships have been derived in epidemiological studies. They find relationships between a certain exposure indicator for humans or vegetation and observed effects in order to estimate impacts. Some indicators are based on a threshold of ozone exposure. Implicitly, it is assumed that below this threshold no impact should occur.

Ozone concentration indicators can be average concentrations over time with or without consideration of a threshold, such as daily maximum 8-h or 6-h or 1-h ozone averages. Anenberg and co-workers (2010) use seasonal average (6-month) 1-h daily maximum O_3 concentrations. Another indicator is named SOMO0 which is the daily maximum of 8-h running average (without threshold).

However, Sum of Ozone Means Over 35 ppb(SOMO35) (EMEP 2011) is a state-of-the-art indicator that is widely applied in Europe. It is the indicator for health impact assessment recommended by the World Health Organisation (WHO). It is defined as the yearly sum of the daily maximum of 8-h running averages over 35 ppb. For each day the maximum of the running 8-h average for O_3 is selected and the values over 35 ppb are summed over the whole year. If we let A_8^d denote the maximum 8-h average ozone on day (d), during a year with N_y days ($N_y = 365$ or 366), then SOMO35 can be defined as:

$$SOMO35 = \sum_{d=1}^{d=N_y} max\left(A_8^d - 35ppb, 0.0\right) \quad (7.13)$$

where the max function ensures that only A_8^d values exceeding 35 ppb are included. The corresponding unit is ppb · days (abbreviated also as ppb · d, where 1 ppb O_3 is ca. 2 $\mu g/m^3$ of O_3).

The indicators for impact assessment towards crops and natural vegetation are AOT40 and AOT40c, respectively. AOT40 stands for 'Accumulated Ozone concentration above a Threshold of 40 ppbV'. It is used to calculate forests using estimates of O_3 concentration at forest-top, assuming a default growing season of April-September for the northern hemisphere. AOT40c calculated for agricultural crops uses estimates of O_3 at the top of the crop. This AOT40c is using a default growing season of May-July, and a default crop-height of 1 m for the northern hemisphere.

2 Impact Assessment

For the impact category 'photochemical ozone formation', the relevant life cycle inventory (LCI) results consist of emissions of NO_x and different NMVOC to air.

Since the group of substances NMVOC is consisting of a large number of substances, some LCIA methods have developed substance specific characterisation factors (CF).

CFs can represent the impact on an indicator defined at a midpoint or an endpoint of the impact category. In the context of ozone, the midpoint indicator is, for example, the ozone creation or ozone exposure; the endpoint or damage indicators are human health impacts and impacts on vegetation (natural vegetation and crops).

To model the impact pathway in the calculation of a CF, certain steps have to be followed. They are described in the following.

2.1 Fate Modelling

According to Hofstetter (1998) there are five types of fate models:

- A model that relates average residence times to average mixing volumes
- A model that relates measured average concentrations to measured emissions
- A model that relates modelled average concentrations to measured emissions
- A multi-media model that models the fate of a unit emission and
- A model that relates modelled reductions in concentrations to assumed reductions in emissions concentrations to measured emissions.

These approaches differ significantly in their complexity, their computational demands and hence, in the alleged accuracy. For the latter ones, chemical transport modelling is necessary.

Chemical transport modelling (CTM): Atmospheric transport models are applied for calculating the concentrations of air pollutants in certain areas or at certain points within an underlying receptor area. The models do not only account for dispersion but also for chemical transformation. For practical reasons a large receptor area is divided into grid cells and the concentration is calculated as an average for each of these grid cells. In order to attribute a concentration increment to an emission from a certain source or source area, first a modelling of the background concentrations caused by the background emissions is done. This is then called the reference scenario. Then a second calculation has to be conducted based on a change of emissions from a certain source (e.g. a coal fired power plant) or source area (e.g. 15 % of the NO_x emissions occurring in Belgium in a certain reference year). This is the assessment scenario. Assuming linearity within the range of the marginal (or quasi marginal) emission change, the concentration change in each grid cell, i.e. the differences between the background concentrations in the reference scenario and the concentrations in the assessed scenarios are calculated. These changes can then be attributed to a unit change of emission, e.g. per mega gram emission from the assessed source or location. Based on this approach, so called source receptor matrices (or blame matrices) are derived for different source regions following a spatially differentiated approach.

2.2 *Impact Assessment for Impacts and Damage to Human Health*

As illustrated in Fig. 7.3 and in accordance with the approach taken for other human health-related impact categories, a CF for human health impacts from photochemical ozone formation can be expressed as the product of three factors:

- the intake fraction (iF)
- the effect factor (EF)
- the damage factor (DF)

The calculation of characterisation factors is summarised in the following expression:

$$CF_{s,a,p,i} = \sum_{i}\left(iF_{i,p,s,a}\sum_{e}\left(EF_{i,d} \times DF_{d}\right)\right) \tag{7.14}$$

where,

s is the source region (country of region)
a is a certain archetypical characteristic (e.g. high stack)

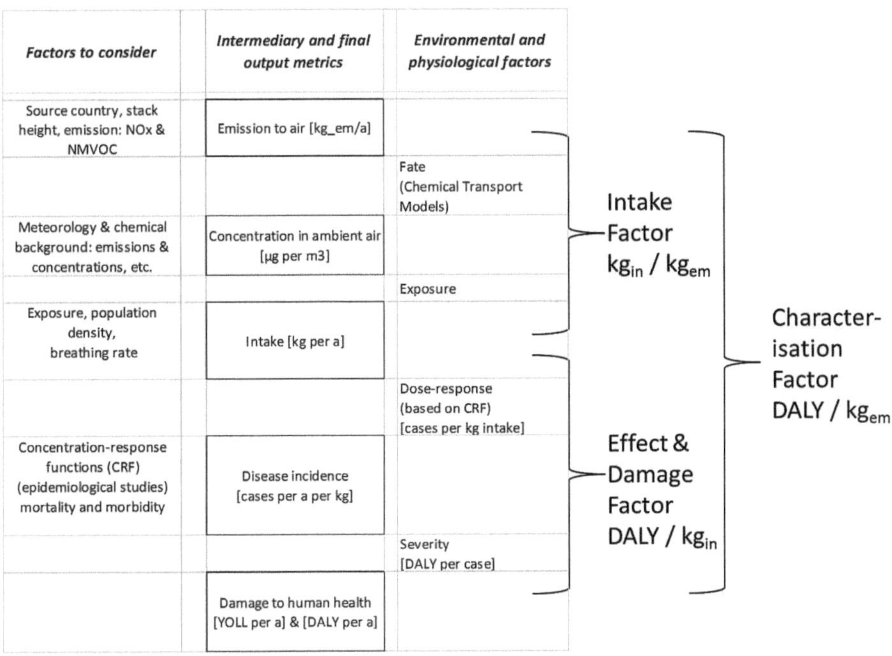

Fig. 7.3 Model to derive characterisation factors regarding human health impacts caused by ozone (Taken from Preiss et al. 2012)

p is the emitted pollutant (e.g. NO_x or NMVOC)
I is the pollutant taken in (in this case ozone)
iF is the intake fraction, i.e. the mass of ozone [g] taken in in relation to the mass [g] precursor pollutant (NO_x and NMVOC) emitted. The iF is dimensionless, because it represents actually a fraction taken in of a pollutant emitted.
EF is the effect factor due to ozone regarding different disease d, i.e. number of disease per unit of ozone taken in [g^{-1}] and finally,
DF is a so called damage factor which weights different disease in order to be able to sum them up, e.g. days lost per incident of illness [d].

The calculation of intake fraction iF, i.e. the amount of the photochemical oxidant i taken in by the exposed human population in relation to the emitted pollutant p from a source (or source region) *s* for certain archetypical characteristics is calculated as shown in

$$iF_{s,a,p,i} = \sum_{g=1}^{n} \left(\frac{dc_{i,g}}{dEm_{s,a,p}} \times BR \times N_g \right) \tag{7.15}$$

where

$dc_{i,g}$ is the concentration increment of photochemical oxidant i in grid cell g[g/m^3]
dEm is the delta emission (of pollutant p) at the defined conditions [g/day]
BR is the average breathing rate per capita of the exposed population (13 [m^3/day])
N_g is the number of people in grid cell g.

The receptor region consists of n grid cells. For example, the EMEP50 grid (EMEP 2008) for Europe consists of $132 \times 111 = 14{,}620$ grid cells of $0.5° \times 0.5°$, i.e. ca. 50×50 km^2 (at 60° North).

The effect factor $EF_{d,i,g}$ [g^{-1}] is based on concentration-response-functions, which relates the additional number of a certain disease (morbidity or mortality) to an ambient concentration increment.

In other words, the effect factor EF is calculated as the change of attributable burden *dAB* due to disease d caused by photochemical oxidant i and the total intake dI_i within 1 year, in grid cell g, according to Eq. 7.16.

$$EF_{d,i,g} = \frac{dAB_{d,i,g}}{dI_i} \tag{7.16}$$

where

*d*AB is attributable burden (e.g. days lost)
d is a disease or a cause of death
i is the photochemical oxidant causing the disease
dI_i is the intake within 1 year [g/year]
g is a certain grid cell.

The damage factor (DF) for a certain disease is calculated according to

$$DF_d = \frac{dDALY_d}{dAB_d} \tag{7.17}$$

CFs for human health damage caused by emitted pollutant p are defined as the change in the disability adjusted life years (DALY) of the total population in the concerned receptor area per unit of emission released in a certain source region.

The DALY is the sum of the YLD, i.e. year equivalents lost due to morbidity and YOLL, i.e. years of lifetime lost due to premature death.

DALYs are calculated according to Eq. 7.18.

$$DALY = YLD + YOLL \tag{7.18}$$

Year equivalents lost due to morbidity (YLD) are calculated according to

$$YLD = \sum_{i=1}^{n} ND_d \times DW_d \times L_d \tag{7.19}$$

Where

ND_d = is the number of cases of disease d occurring in the time frame of 1 year
DW_d = is the disability weight of disease d
L_d = is the duration of disease d (in years or fraction of a year).

The years of lifetime lost due to premature death (YOLL) are calculated according to

$$YOLL = \sum_{i=1}^{n} NM_d \times T_d \tag{7.20}$$

Where

NM_d = is the number of deaths due to cause of death, i.e. disease d
T_d = is the average expectancy of life minus the age at the time of death.

2.3 Assessment of Impacts and Damage to Plants

Ozone is an air pollutant affecting vegetation, including crops, trees and grassland species, i.e. the primary producers of terrestrial ecosystems and hence, the ecosystem at large.

Within EcoSenseWeb (Preiss et al. 2008) for the assessment of ozone impacts, a linear relation is assumed between yield loss and the AOT40 value for crops calculated for the growth period of crops (May to June). The relative yield change is calculated using linear equations together with sensitivity factors derived, e.g. by Fuhrer (1996). The damage (yield loss) to rice, tobacco, sugar beet, potato, sunflower and wheat can be evaluated. The underlying source receptor matrices are based on the EMEP chemical transport model (Simpson et al. 2012).

Hauschild and co-workers (2006) applied results of the RAINS model. The RAINS model is also based on EMEP source receptor matrices. EDIP 2003 uses a site-dependent regression equation derived for the RAINS model (for NMVOC and NO_x) corrected for substance properties using POCP factors from Derwent and Jenkins (Hauschild and Potting 2005).

Within the LC-IMPACT project spatially, explicit CFs for tropospheric ozone damage on natural vegetation have been derived for 65 European source regions. The CFs were defined as the area-integrated increase in the potentially affected fraction (PAF) of trees and grassland species due to a change in emission of NO_x and NMVOC. The area-integrated effect factors quantify the relationship between ozone exposure (AOT40) and the damage to natural vegetation. The relationships describing the ecological effects of a pollutant were based on a log normal relationship between the PAF and ground level ozone concentration.

3 LCIA Regarding Ozone: A Short Historical Overview

Comprehensive overviews of different LCIA methods are available, e.g. in Frischknecht et al. (2007) and Goedkoop et al. (2008), etc. This section describes some important stages in the development of impact assessment for the precursors NO_x and NMVOC regarding ozone formation in LCA.

The very first approaches used already existing approaches like POCP (CML92, EDIP97) or MIR as midpoint characterisation models. EDIP97 took it a step further and developed regression models to estimate substance specific CF for missing individual VOC based on classification into substance groups (Hauschild and Wenzel 1998).

Others made use of statistical methods which compared measured concentrations and emissions. By top-down allocation of emission and including factors like residence-time, volume of dilution, etc., Jolliet and Crettaz (1997) derived fate factors for air pollutants. Hofstetter (1998) covered ozone regarding fate analysis relating the primary emission to the concentration and provided 118 fate factors for precursors of ozone by making use of the ‘umbrella principle’ and applying the EMEP modelling for Europe which had a resolution of 150 km $\times$ 150 km per grid cell at that time.

3.1 *Krewitt and Co-workers*

Krewitt and co-workers (2001) derived CFs regarding health and crops yield loss for 15 European countries and a corresponding average. Estimates were also derived for Asia and South America. For Europe, emission scenarios for the years 1990 and 2010 were considered to analyse the influence of changing background conditions on the resulting impacts. The results showed variations in the impact per unit emission depending on the source region and emission scenarios. The results where expressed as exposure of people and finally, as impact in form of YOLL. The exposure is the so called 'accumulated exposure' per unit of emission, i.e. the sum over all grid cells' product of concentration and population resulting from the emission of pollutants (NO_x and NMVOC) from the respective source countries. 'Accumulated exposure' is expressed as person * $\mu g/m^3$.

The emission scenario years 1990 and 2010 showed an average range EU-15 from 191 to 198 (53 for Finland to 288 in Belgium in 1990) per million gram of NMVOC and −157 to −32 for NO_x (−429 for The Netherlands and +68 for Finland in the 1990). The negative CF values for NO_x for some countries reflect the complex chemistry behind the ozone formation, where additional emission of NO_x may decrease the ozone concentration near the source (due to reaction with NO as illustrated in?) but overall on the continental scale lead to increased ozone formation when ozone is formed elsewhere through the reduction of NO_2 to NO (see?).

Krewitt and co-workers also made a distinction between different source sectors (e.g. traffic or solvent use). The results regarding NMVOC have a range of 1–1.7 YOLL per kilo tonne of NMVOC due to O_3 formation whereas the values for NO_x have a range from −1.5 to 0.1 for the different sectors.

3.2 UNEP/SETAC LCI Task Force IV

In 2002 the UNEP/SETAC Life Cycle Initiative formed the LCIA Task Force IV on 'Transboundary impacts'. This task force aimed to establish recommended practice and guidance for use in transboundary categories, such as climate change or acidification, and coordinated with Task Force 3 on toxicity of photo oxidant formation and respiratory inorganic. The task force addressed midpoint categories and their relation to damage categories human health and biotic natural environment. The outcome was a summary list of the LCIA methods (UNEP/SETAC 2004).

In Jolliet et al. (2004) it is stated that two types of models have mostly been used to analyse midpoint indicators for smog – a Northern European model based on the calculated photochemical ozone creation potential and a model used in the United States based on the MIR, measured in units of ozone. The task force stressed that care should be taken to include the impact of NO_x appropriately.

3.3 Hauschild and Co-workers

Hauschild and co-workers (2006) further developed spatial differentiation regarding characterisation of photochemical ozone formation for the EDIP2003 methodology. According to JRC-IES (2011), EDIP 2003 is based on the RAINS model and meets the science based criteria. This model respects non-linearity of photochemical ozone formation and addresses both human health and vegetation impacts. It provides spatially differentiated CFs as well as overall site-generic factors for Europe for both human health and vegetation impacts. Adaptation to other continents is, however, not straightforward due to the reliance on the European RAINS model. Site-dependent CFs are provided for 41 European source regions, including Russia and sea regions, and for emission scenarios for the years 1990, 1995 and 2010. Moreover, an assessment of impacts on vegetation has been included.

3.4 ReCiPe

As described in van Zelm et al. (2008) and Goedkoop et al. (2012), the dynamic model LOTOS-EUROS (TNO 2005–2013) was applied to calculate human intake fractions for ozone due to emissions of NO_x and NMVOC. The intake fraction is $1.2 \cdot 10^{-7}$ for both substances. The impact considered regarding human health is acute mortality, and the endpoint CF (year $\cdot$ kg^{-1}) is 3.9×10^{-8}, also for both substances.

Regarding midpoint characterisation the ozone formation potential is expressed as NMVOC-eq./kg. In order to derive CFs for individual NMVOC, the different reactivity of the substances is taken into account. Following the approach developed in EDIP2003, the following equation is used to recalculate the CF for a specific hydrocarbon (x):

$$CF_x = \frac{POCP_x}{POCP_{NMVOC}} \times CF_{NMVOC} \tag{7.21}$$

As described in JRC-IES (2011), ReCiPe and EDIP2003 are the recommended default method at the midpoint level. Moreover, ReCiPe is also the recommended default method at endpoint level for human health impacts. For impacts on vegetation at endpoint level it is suggested to build on the EDIP2003 midpoint model (Hauschild et al. 2006), since it already models the time and area-integrated exposure above a critical level.

3.5 LC-IMPACT

Within the LC-IMPACT project, the assessment of NO_x and NMVOC regarding impacts due to ozone has been improved over several steps (Goethem et al. 2013; Preiss et al. 2012).

Firstly, results of the global chemical transport model TM5 (Krol et al. 2005) have been applied in order to assess the whole world. This has been achieved on a level of 56 source regions. Secondly, human health and natural vegetation have been assessed. The human health CFs are on a midpoint and endpoint level with a global coverage regarding the atmospheric dispersion and chemistry and regarding the receptor distribution. Analogous to Hauschild et al. (2006) and Goedkoop et al. (2012), CFs for different NMVOC have been derived according to their POCP.

The methodology for assessing impacts to natural vegetation is described in Goethem et al. (2013). It was first applied to source receptor matrixes for European areas. This enabled to derive country specific CFs expressing damage to natural vegetation by ozone whereas emissions and deposition of NO_x and NMVOC in 65 European regions were included. Subsequently the impact factors were also applied to global source receptor results from TM5. Hence, regionalised factors at a global level as well as global default value were developed.

4 Discussion on Variability, Uncertainty and Necessary Improvements and Future Research Needs

The main difficulties and research areas for potential improvement of the methodologies for impact assessment of ozone are:

1. To better quantify natural emission of NO_x, NMVOC (isoprene and various terpenes) and natural background concentration of ozone
2. To better quantify anthropogenic emission of NO_x, NMVOC and corresponding change of ozone
3. To better account for the chemistry influenced by meteorology and the amount of emission, i.e. is the background NO_x saturated or not
4. Further research regarding concentration response functions. The ozone concentration is fluctuating and changing during day/night and between the season because the ozone lifetime is short, hence:
   (a) Impacts on human health (probably) depend on level of concentration and corresponding duration
   (b) Impacts on plants (probably) also depend on level of concentration and corresponding duration and in addition on moisture and timing regarding growing season because of different stomata
5. Regionalisation – site dependent impact assessment is needed to derive aggregated impact per source region
6. Therefore, the height of release should also be differentiated leading to source sector differentiation
7. Ozone concentration in urban areas can be lower than outside of urban areas because of the so called NO_x titration – however, the spatial resolution of dispersion model regarding source regions, source sectors and receptor grid resolution is mostly too low to adequately reflect this phenomenon.

Considering the points above, regarding natural vegetation and crops, the exposure of the plant to the ozone based on AOT40 only takes into account the concentration and the exposure time (above a certain threshold). However, the effects on the plants depend on how much is taken in through the stomata. The stomata conductance is affected by several meteorological factors, especially humidity and soil water deficit and hence, these factors influence the amount of ozone absorbed. Therefore, this should be taken into account in specific future LCIA-methods.

Regarding human health, high ozone concentrations occur on relatively few days and the ozone creation is very much influenced by meteorological conditions. Moreover, the receptor sensitivity is probably different depending on meteorological conditions and depending on other stressors such as heat and air pollutant concentration (particulate matter, SO_2, CO, NO_x). Therefore, the annual and inter-annual variability leads to an uncertainty in the impact assessment.

Since the CTMs mainly have relatively coarse receptor resolution, it is difficult to cover the ozone concentration and human exposure adequately in urban areas (because of titration effects of NO_x which leads to lower concentration in urban areas than in the surrounding areas). Since most people live in urban areas, there is a need for more appropriate models with a higher spatial resolution in order to cover this issue.

In general, it has to be noted that the ozone creation in the context of NO_x and NMVOC emissions is a highly non-linear function, depending not only on location and meteorological conditions, but also on the absolute amount of the corresponding emissions of the pollutants. Therefore, it has to be kept in mind that the derived characterisation factors are always only an approximation for a marginal impact per unit of emission of the corresponding pollutants at a certain time and space.

5 List of Abbreviations and Symbols

Abbreviation	Explanation
AOT40	Accumulated exposure over a threshold of 40 ppb
AOT40f	Accumulated exposure over a threshold of 40 ppb for forests
AOT40f	Accumulated exposure over a threshold of 40 ppb for crops
c	Concentration
CF	Characterisation factor
CO	Carbon monoxide
CO_2	Carbon dioxide
CTM	Chemical transport model
DALY	Disability adjusted life years

(continued)

Abbreviation	Explanation
EF	Effect factor
EMEP	European monitoring and evaluation programme
FF	Fate factor
H_2O	Water
HCHO	Formaldehyde
HOx	Hydrogen oxide radicals
iF	intake fraction
IR	Incremental reactivity
LCA	Life cycle assessment
LCIA	Life cycle impact assessment
MIR	Maximum incremental reactivity
NMVOC	Non-methane volatile organic compounds
NO	Nitrogen oxide
NO_2	Nitrogen dioxides
NOx	Nitrogen oxides (refers to NO and NO_2)
O_3	Ozone
OH	Hydroxyl radical
PAF	Potentially affected fraction
POCP	Photochemical ozone creation potential
R	Organic rest 'R' from VOC
SOMO0	Sum of maximum 8-h ozone levels without a threshold
SOMO35	Sum of maximum 8-h ozone levels over 35 ppb (70 $\mu g/m^3$)
VOC	Volatile organic compound (NMVOC plus methane)
YLD	Years of lifetime lived disabled by a disease
YOLL	Years of lifetime lost due to premature death

References

Amann M, Derwent D, Forsberg B, Hänninen O, Hurley F, Krzyzanowski M et al (2008) Health risks of ozone from long-range transboundary air pollution, WHO. ISBN 978 92 890 42895

Anenberg SC, West JJ, Fiore AM, Jaffe DA, Prather MJ, Bergmann D et al (2009) Intercontinental impacts of ozone pollution on human mortality. Environ Sci Technol 43(17):6482–6487

Anenberg SC, Horowitz LW, Tong DQ, West JJ (2010) An estimate of the global burden of anthropogenic ozone and fine particulate matter on premature human mortality using atmospheric modeling. Environ Health Perspect 118:1189–1195. doi:10.1289/ehp.0901220

Ashmore M (2002) Effects of oxidants at the whole plant and community level. In: Bell JNB, Treshow M (eds) Air pollution and plant life, 2nd edn. Wiley, New York, pp 89–118

Atkinson R (2000) Atmospheric chemistry of VOCs and NO_x. Atmos Environ 34:2063–2101

Bell ML, McDermott A, Zeger SL, Samet JM, Dominici F (2004) Ozone and short-term mortality in 95 US urban communities, 1987–2000. J Am Med Assoc 292(19):2372–2378

Carter W (1994) Development of ozone reactivity scales for volatile organic compounds. J Air Waste Manage Assoc 44:881–899

Carter WPL (1998) Updated maximum incremental reactivity scale for regulatory applications. University of California, Riverside, p 73

Carter WPL, Pierce JA, Luo D, Malkina IL (1995) Environmental chamber study of maximum incremental reactivities of volatile organic compounds. Atmos Environ 29:2499–2511

Davison AW, Barnes JD (1998) Effects of ozone on wild plants. New Phytologist 139(1):135–151. doi:10.1046/j.1469-8137.1998.00177.x

Derwent RG, Jenkin ME, Saunders SM, Pilling MJ (1998) Photochemical ozone creation potentials for organic compounds in Northwest Europe calculated with a master chemical mechanism. Atmos Environ 32(14–15):2429–2441

EMEP (2008) Map of the 50 km grid used between 1997 and 2008. http://www.emep.int/grid/grid50.pdf

EMEP (2011) EMEP definitions & statistics used from http://www.emep.int/SR_data/definitions.pdf

Frischknecht R, Jungbluth N, Althaus H-J, Bauer C, Doka G, Dones R et al (2007) Implementation of life cycle impact assessment methods (No. 3). Swiss Centre for Life Cycle Inventories, Dübendorf

Fuhrer J (1996) The critical level for effects of ozone on crops and the transfer to mapping. testing and finalizing the concepts. UN-ECE Workshop, Department of Ecology and Environmental Science, University of Kuopio, Kuopio

Goedkoop MJ, Spriensma R (1999) The eco-indicator'99: a damage-oriented method for life-cycle impact assessment (no. 1999/36A). Ministry of Housing, Spatial Planning, and Environment, The Hague

Goedkoop M, Oele M, Schryver A, Vieira M (2008) SimaPro database manual: methods library. PRé Consultants, Amersfoort

Goedkoop M, Heijungs R, Huijbregts M, Schryver A, Struijs J, Zelm R (2012) ReCiPe 2008: a life cycle impact assessment method which comprises harmonised category indicators at the midpoint and the endpoint level. RIVM, Bilthoven

Goethem TMWJ, Preiss P, Azevedo LB, Roos J, Friedrich R, Huijbregts MAJ, Zelm R (2013) European characterization factors for damage to natural vegetation by ozone in life cycle impact assessment. Atmos Environ 77:318–324

Guinée JB, Gorée M, Heijungs R, Huppes G, Kleijn R, Koning AD, Oers LV, Wegener Sleeswijk A, Suh S, Udo de Haes HA, Bruijn HD, Duin RV, Huijbregts MAJ (2002) Handbook on life-cycle assessment: operational guide to the ISO standards. Kluwer Academic Publishers, Dordrecht

Hauschild M, Potting J (2003) Spatial differentiation in life-cycle impact assessment: the EDIP2003 methodology. guidelines from the Danish EPA. Institute for Product development, Technical University of Denmark, Copenhagen

Hauschild M, Potting J (2005) Photochemical ozone formation, chapter 7. In: Spatial differentiation in life cycle impact assessment – the EDIP2003 methodology. Environmental News no. 80, Danish Ministry of the Environment. Environmental Protection Agency, Copenhagen

Hauschild M, Wenzel H (1998) Environmental assessment of products, vol 2: scientific background. Chapman & Hall, London

Hauschild M, Potting J, Hertel O, Schöpp W, Bastrup-Birk A (2006) Spatial differentiation in the characterisation of photochemical ozone formation, the EDIP2003 methodology. Int J Life Cycle Assess 11(1):72–80

Hayashi K, Okazaki M, Itsubo N, Inaba A (2004) Development of damage function of acidification for terrestrial ecosystems based on the effect of aluminum toxicity on net primary production. Int J Life Cycle Assess 9:13–22

Hofstetter P (1998) Perspectives in life cycle impact assessment, a structured approach to combine models of the technosphere, ecosphere and value sphere. Springer Science + Business Media, New York

Huijbregts MAJ, Schöpp W, Verkuijlen E, Heijungs R, Reijnders L (2000) Spatially explicit characterization of acidifying and eutrophying air pollution in life-cycle assessment. J Ind Ecol 4:75–92

Jolliet O, Crettaz P (1997) Fate coefficients for the toxicity assessment of air pollutants. Int J Life Cycle Assess 2(2):104–110

Jolliet O, Müller-Wenk R, Bare J, Brent A, Goedkoop M, Heijungs R et al (2004) The LCIA midpoint-damage framework of the UNEP/SETAC life cycle initiative. Int J Life Cycle Assess 9:394–404

JRC-IES (2010) ILCD handbook: analysing of existing environmental impact assessment methodologies for use in life cycle assessment, 1st edn. Joint Research Centre, Ispra

JRC-IES (2011) ILCD handbook: recommendations for life cycle impact assessment (LCIA) in the European context based on existing environmental impact assessment models and factors, 1st edn. Joint Research Centre, Ispra

Kemna R, Van Elburg M, Li W, Van Holsteijn R (2005) MEEuP methodology report, final, 28.11.05. VHK for European Commission, Brussels

Krewitt W, Trukenmüller A, Bachmann TM, Heck T (2001) Country-specific damage factors for air pollutants: a step towards site dependent life cycle impact assessment. Int J Life Cycle Assess 6(4):199–210

Krol M, Houweling S, Bregman B, van den Broek M, Segers A, van Velthoven P et al (2005) The two-way nested global chemistry-transport zoom model TM5: algorithm and applications. Atmos Chem Phys 5(2):417–432. doi:10.5194/acp-5-417-2005

Lim SS, Vos T, Flaxman AD et al (2012) A comparative risk assessment of burden of disease and injury attributable to 67 risk factors and risk factor clusters in 21 regions, 1990—2010: a systematic analysis for the global burden of disease study 2010. Lancet 380:2224–2260. doi:10.1016/S0140-6736(12)61766-8

Norris GA (2003) Impact characterization in the tool for the reduction and assessment of chemical and other environmental impacts; methods for acidification, eutrophication and ozone formation. J Ind Ecol 6(3–4):79–101

Potting J, Schöpp W, Blok K, Hauschild M (1998) Site-dependent life-cycle impact assessment of acidification. J Ind Ecol 2(2):63–87

Preiss P, Klotz V (2008) Revised description of the updated and extended tool EcoSenseWeb for the detailed site-dependent assessment of external costs, NEEDS project, FP6, Rs1b_D7.1 – Project no: 502687. Stuttgart: Institut fürEnergiewirtschaft und Rationelle Energieanwendung (IER). Universitaet Stuttgart, Stuttgart

Preiss P, Friedrich R, Klotz V (2008) Report on the procedure and data to generate averaged/aggregated data, NEEDS project, FP6, Rs3a_D1.1 – Project no: 502687. InstitutfürEnergiewirtschaft und Rationelle Energieanwendung (IER). Universität Stuttgart, Stuttgart

Preiss P, Roos J, van Dingenen R, Dentener F, Friedrich R (2012) Photochemical ozone formation and effects on human health. In: LC-IMPACT, D3.5, Project no: 243827 FP7-ENV-2009-1

Sambat S, Theloke J, Friedrich R, Allemand N (2005) Bibliographic study concerning the speciation of NMVOC – within the project INTERREG III (Annex 2 – Point 2.1)

Simpson D, Benedictow A, Berge H et al. (2012) The EMEP MSC-W chemical transport model – technical description. Atmos Chem Phys 12:7825–7865. doi:10.5194/acp-12-7825-2012

Singh SN, Vandermeiren K, Harmens H, Mills G, Temmerman L (2009) Impacts of ground-level ozone on crop production in a changing climate. In: Singh SN (ed) Climate change and crops. Springer, Berlin/Heidelberg, pp 213–243

Steen B (1999) A systematic approach to environmental priority strategies in product development (EPS) Version 2000: models and data of the default method. Chalmers University of Technology, Göteborg

Theloke J (2004) NMVOC-Emissionen aus der Lösemittelanwendung und Möglichkeiten zu ihrer Minderung. Unpublished Promotion, Universität Stuttgart, Stuttgart

TNO (2005–2013) LOTOS-EUROS. http://www.lotos-euros.nl/documentation.php

UNEP/SETAC (2004) Life cycle impact assessment programme. LCIA methods and links, from http://lcinitiative.unep.fr/default.asp?site=lcinit&page_id=67F5A66D-9EB8-4E75-B663-297B7FD626B6

Van Dingenen R, Dentener FJ, Raes F, Krol MC, Emberson L, Cofala J (2009) The global impact of ozone on agricultural crop yields under current and future air quality legislation. Atmos Environ 43(3):604–618

Van Zelm R, Huijbregts MAJ, Van Jaarsveld JA, Reinds GJ, De Zwart D, Struijs J, Van de Meent D (2007) Time horizon dependent characterization factors for acidification in life-cycle assessment based on forest plant species occurrence in Europe. Environ Sci Technol 41:922–927

Van Zelm R, Huijbregts MAJ, Den Hollander HA, Van Jaarsveld HA, Sautere FJ, Struijs J, Van Wijnen HJ, Van de Meent D (2008) European characterization factors for human health damage of PM10 and ozone in life cycle impact assessment. Atmos Environ 42(3):441–453

WHO (2003) Health aspects of air pollution with particulate matter, ozone and nitrogen dioxide. Report on a WHO working group, 13–15 Jan 2003. World Health Organization, Bonn

Chapter 8
Ecotoxicity

Ralph K. Rosenbaum

Abstract Ecotoxicity impact assessment of chemicals in life cycle assessment (LCA) adheres to a number of underlying principles and boundary conditions: (1) a large number of emitted substances to cover (at least 100,000 potentially relevant elementary flows with current models covering around 2,500), (2) linearity of characterisation models, (3) conservation of mass and mass balance, (4) infinite time horizon, (5) additivity of toxicity, (6) assuming average conditions as best estimates to avoid bias in the comparison (including consideration of generic/average ecosystems and impacts). The cause-effect mechanism for ecotoxicity impacts of chemicals can be divided into four parts: (1) chemical fate (i.e. chemical behaviour/distribution in the environment), (2) exposure (i.e. bioavailability), (3) effects (i.e. affected species), and (4) severity (i.e. disappeared species). In terms of species represented, a freshwater ecosystem is described in this chapter by three trophic levels: (1) primary producers (e.g. algae), (2) primary consumers (i.e. invertebrates), and (3) secondary consumers (e.g. fish). Model uncertainty was estimated at about three orders of magnitude on top of important sources of parameter uncertainty such as degradation rates and effect factors. Current midpoint LCIA methodologies covering ecotoxicity include TRACI 2.0, and the ILCD recommended methodology, both employing the USEtox factors. Current LCIA methodologies covering midpoint and endpoint characterisation are ReCiPe, LIME, IMPACT 2002+, and IMPACT World+. Important research needs are (1) increasing substance coverage, (2) further developing marine and terrestrial ecotoxicity modelling for midpoint, (3) improving endpoint modelling for ecotoxicity towards biodiversity, (4) consideration of long-term emissions and impacts of metals, (5) importance of spatial and temporal variability, (6) mixture toxicity, and (7) decreasing model and parameter uncertainty.

R.K. Rosenbaum (✉)
IRSTEA, UMR ITAP, ELSA-PACT – Industrial Chair for Environmental and Social Sustainability Assessment, 361 rue Jean-François Breton, BP 5095, F-34196 Montpellier Cedex 5, France
e-mail: ralph.rosenbaum@irstea.fr

M.Z. Hauschild, M.A.J. Huijbregts (eds.), *Life Cycle Impact Assessment*,
LCA Compendium – The Complete World of Life Cycle Assessment,
DOI 10.1007/978-94-017-9744-3_8

Keywords Chemical fate • Comparative ecotoxicity • Concentration-response relationship • Ecosystem • LCA • LCIA • Life cycle assessment • Life cycle impact assessment • Potentially affected fraction of species (PAF)

1 Principles, Fundamentals, and Recommended Practice of Characterisation Modelling

In the early days of LCA, impact assessment was generally considered as unfeasible and only with the development of the first LCIA methodologies did this perception slowly start to change. Ecotoxicity is among those impact categories that, only within the last decade, started to be considered as becoming mature enough for application, although with considerable reservations by many practitioners until today. The origin for today's approaches in characterising this impact category is clearly in the field of environmental hazard and risk assessment (Pennington et al. 2006).

Environmental Risk Assessment (ERA) quantifies risks due to environmental changes (e.g. a chemical emission into an environmental compartment) affecting biological systems (e.g. animals, plants, or entire ecosystems). This risk depends on the toxicity of the substance and the level of exposure of living organisms. The toxicity is generally estimated based on toxicological tests relating adverse effects to concentrations of a substance, so called concentration-effect relationships. The level of exposure can be measured, or it can be estimated by modelling the fate of the substance from the emission to the relevant environmental media, resulting in environmental concentrations.

While LCA and ERA both aim at the evaluation of potentially toxic impacts on the environment, partly using common data and assumptions, there are also important differences between them (Pennington et al. 2006; Udo de Haes et al. 2002; Olsen et al. 2001; Barnthouse et al. 1997; Owens 1997). Some of those are:

1. ERA usually applies conservative estimates of toxicity and several other properties of a compound (e.g. biodegradability) and realistic conservative (worst-case) scenarios for the modelled environmental system while LCIA aims at best estimates for all parameters for a comparative assessment.
2. ERA is generally performed in a regulatory context ensuring that an emission at a given site poses no risk to the protection targets. LCIA aims to address all relevant environmental impacts anywhere in the world due to a product or service while currently not necessarily considering time and localisation of the emissions (Hauschild and Pennington 2003).
3. ERA only considers the potential impact, in terms of risk, of one compound or mixture on the environment. LCIA by definition assesses several, sometimes large sets of chemicals and has to ensure the compatibility of the toxicity impact indicator for each chemical and with indicators for other impact categories.

4. LCA considers impacts integrated over time and space at the ecosystem level, while regulatory ERA typically focuses on peak exposures to individual (most sensitive) species.

As opposed to ERA where actual risks are calculated, comparative assessments aim to estimate the impact of a chemical relative to other substances, typically represented by rankings of chemicals by a certain indicator, e.g. toxicity and/or persistence in the environment. These rankings are then used as the basis for decisions, e.g. regarding choices of chemicals as product compounds with the least environmental impact, or in the context of chemical policy identifying priority substances for regulation, etc. In ERA acceptable exposure levels in terms of regulatory thresholds are used to e.g. evaluate an emission or the acceptability of an industrial installation. It thus estimates the potential impact of a compound on ecosystem stability for a specific part of the life cycle and at a local scale, yielding results which are not necessarily comparable across different release sites or chemicals. Thus, risk assessment addresses different objectives, spatial scales and process chains.

Several important principles are common practice and required when developing a method for ecotoxicity impact assessment of chemicals in the framework of LCA:

- Large number of relevant substances emitted: Accounting for the often large number of potentially toxic elementary flows in a life cycle inventory requires coverage of a large number of substances in terms of available characterisation factors. For ecotoxicity this may range to hundreds of thousands potentially emitted substances.
- Linearity: As life cycle inventory (LCI) data are typically not spatially and/or temporally differentiated, integration of the impact over time and space is required. In ecotoxicity modelling in LCIA this leads to the use of characterisation models assuming steady-state conditions (i.e. no change in the relative distribution of a chemical between all environmental compartments over time), which implies a linear relationship between the increase in chemical emission into the environment and the consequent increase in chemical concentration in each environmental compartment. The effect model itself also assumes linearity (either for the full range of exposure or via different slopes for different ranges of exposure when considering marginal changes) between an increase in chemical exposure of an organism or ecosystem and observed toxic effects.
- Conservation of mass and mass balance: Mass cannot be created or disappear, it can only be transferred. Following this principle, the transport and transformation of a substance in the environment is modelled assuming that mass is conserved at all times.
- Time horizons: As another consequence of the steady-state assumption, most current ecotoxicity characterisation models essentially account for all potential ecotoxicity impacts, independently of their time and place of occurrence in the short or long-term future, which equals integration of the mass-balance differential equation system over infinity. A few exceptions exist, as some methods (e.g. ReCiPe and IMPACT World+), allow considering defined time horizons for metals.

- Additivity of toxicity: Current ecotoxicity models only characterise single substances, generally assuming that the toxicity of each substance can be added together. With three hypotheses possible, the toxicity of substances in mixtures may be (1) additive (i.e. response additive if independent toxicity mechanism, or concentration additive for chemicals with the same toxicity mechanism), (2) synergistic, or (3) antagonistic. Most likely, all three situations may occur to varying degrees, depending on which substances in which proportions are in the mixture and virtually endless combinations of substances are possible. Therefore, current research has no clear answer, and additive toxicity is commonly assumed in generic situations.
- Best estimates: A fundamental value choice in LCA is not to be conservative or protective, but to focus on avoiding any bias between compared scenarios by assuming average conditions, also referred to as best estimates. Products or services assessed in LCA are typically not representing one specific example (e.g. with a serial number or from a specific date), but an average, normally disregarding whether a specific life cycle process took place in summer or winter, during the day or night, etc. As discussed by Pennington et al. (2004b), LCA is a comparative assessment methodology. Direct adoption of regulatory methodology and data is often not appropriate. A conservative estimate of the ecotoxic effect of a substance is unwanted in a relative comparison. Best-estimates are desirable in LCA, with the need to account for uncertainties when making distinction amongst results. Furthermore, LCA is mostly used to compare competitive products (including services) for the same function. Avoiding a bias in the comparison, a best estimate of the potential risk of toxic releases associated with a product is needed, while risk assessment typically assumes conservative values in order to fulfil protective goals in line with prudent health or environmental policy thresholds.
- Generic/average ecosystems and impacts: Due to the limited information available on the sensitivity of species to toxic effects of substances and the local composition of an ecosystem (i.e. species present), the ecosystems assessed in LCIA are of a generic nature and currently do not consider variations in the composition of species present, or variability in their tolerance to toxic stressors. This may partly change with the introduction of terrestrial ecosystem assessment in LCA, where highly variable soil parameters can influence the toxicity of some substances. This level of detail (i.e. precision) may also not be required for each elementary flow in an LCA, making the generic/average character of models and data both a current methodological limitation (e.g. for very sensitive and uncertain elementary flows and impact indicators) as well as a principle (e.g. for the many elementary flows that are not sensitive to results/conclusions of an LCA).

Several working groups and initiatives have established criteria for good characterisation modelling for different impact categories. Of current relevance for ecotoxicity characterisation are notably the recommendations from the UNEP/SETAC Life Cycle Initiative and related activities:

1. The Lausanne review workshop in 2003 aimed to establish a framework for Life Cycle Toxicity Assessment and recommended a number of modelling elements

and choices based on input from ecotoxicity experts outside the LCA community (Jolliet et al. 2006).
2. The declaration of Apeldoorn from 2004 on LCIA of Non-Ferrous Metals (Ligthart et al. 2004) underlined the relevance of a number of aspects for a correct modelling of the fate and toxicity of essential elements.
3. The Clearwater consensus workshop in 2008 for the estimation of metal hazard in fresh water (Diamond et al. 2010) addressed 'inconsistencies in assumptions and approaches for organic substances and nonferrous metals'.
4. The recommendations from these workshops were implemented as far as possible in the UNEP/SETAC toxicity consensus model USEtox. Additionally, a large number of scientifically consensual model elements were identified by the USEtox team and hence became further recommended elements, also included in USEtox (Henderson et al. 2011; Rosenbaum et al. 2008).
5. Based on pre-defined criteria and requirements for good characterisation modelling practice (EC-JRC 2010) and supported by a team of experts (Hauschild et al. 2013), the European Commission established recommendations for LCIA as described in the ILCD Handbook on Recommendations for Life Cycle Impact Assessment in the European context (EC-JRC 2011).

2 Impact Pathway and Affected Areas of Protection

2.1 Overview

As shown in Fig. 8.1, the mechanism of toxic impacts of chemicals in LCA can be divided into four parts.

1. Fate modelling estimates the increase in concentration in a given medium due to an emission quantified in the life cycle inventory.
2. The exposure model quantifies the chemical's bioavailability in the different media by quantifying the bioavailable fraction of the total concentration.
3. The effect model relates the amount available to an effect on the ecosystem. This is typically considered a midpoint indicator in LCA, as no distinction between the severity of observed effects is made (e.g. a temporary/reversible decrease in mobility and death are given the same importance).
4. Finally, the severity (or damage) model translates the effects on the ecosystem into an ecosystem population (i.e. biodiversity) change integrated over time and space.

All four parts of this environmental mechanism are accounted for in the definition of the substance-specific and emission-compartment-specific characterisation factor *CF*:

$$CF = FF \times XF \times EF \times SF \tag{8.1}$$

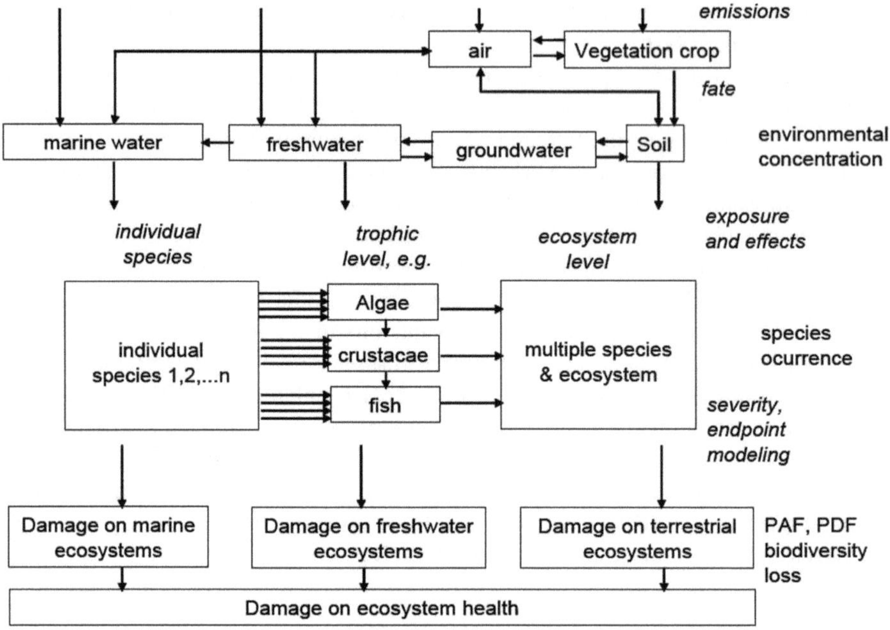

Fig. 8.1 General scheme of the environmental impact mechanism for ecotoxicity (EC-JRC 2011)

Where *FF* is the fate factor, *XF* the exposure factor, *EF* the effect factor, and *SF* the severity factor. Each of these four elements of the environmental mechanism of ecotoxicity, and thus its characterisation factor, is described in the following sections.

2.2 Fate

The fate model predicts the chemical behaviour/distribution in the environment accounting for multimedia (i.e. between environmental media and compartments) and spatial (i.e. between different zones but within the same compartment or medium) transport between environmental compartments (e.g. air, water, soil, etc.). This is accomplished via mass-balance-based modelling of (thermodynamic) exchange processes such as partitioning, diffusion, sorption, advection, convection – represented as arrows in Fig. 8.2 for the USEtox model – as well as biotic and abiotic degradation (e.g. biodegradation, hydrolysis, or photolysis), or burial in sediments. These processes are quantified in rate coefficients which are used to construct a differential equation system for all compartments. This system is solved assuming steady-state by employing matrix algebra (see (Rosenbaum et al. 2007). Further details on fate modelling principles in the USEtox model can be found in Henderson et al. (2011) and Rosenbaum et al. (2008).

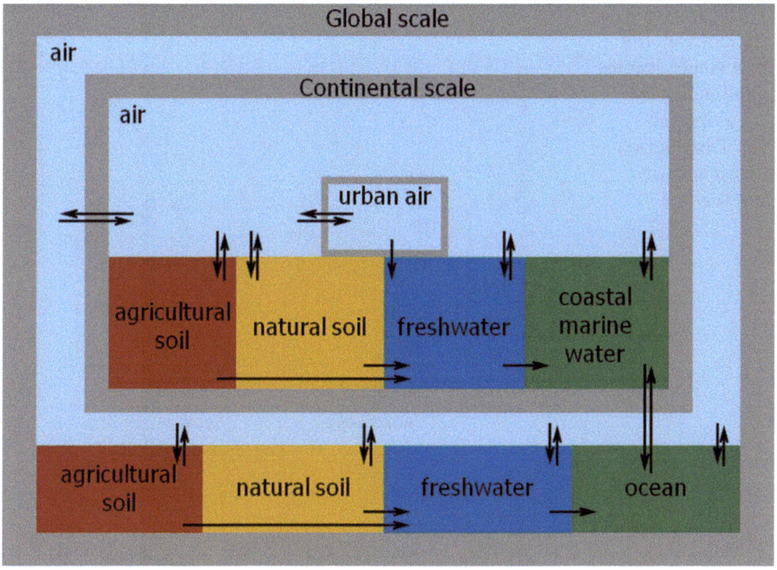

Fig. 8.2 The USEtox fate model (Rosenbaum et al. 2008)

2.3 Exposure

Exposure is the contact between a target and a pollutant via an exposure boundary for a specific duration and frequency (see detailed discussion by Duan et al. (1990)). The exposure model accounts for the fact that not necessarily the total ('bulk') chemical concentration present in the environment is available for exposure of organisms. Several factors and processes such as sorption, dissolution, dissociation and speciation may influence (i.e. reduce) the amount of chemical available for ecosystem exposure. According to Semple et al. (2004) such phenomena can be defined as bioavailability ('freely available to cross an organism's cellular membrane from the medium the organism inhabits at a given time'), and bioaccessibility ('what is actually bioavailable now plus what is potentially bioavailable'). Current LCIA methods consider exposure by calculating the dissolved concentration (Henderson et al. 2011; Rosenbaum et al. 2008), or the bioavailable fraction (Owsianiak et al. 2013; Gandhi et al. 2010, 2011a).

2.4 Effects

The effects model characterises the fraction of species within an ecosystem that will be affected by chemical exposure. Effects are described by lab-test derived concentration-response curves relating the concentration of a chemical to the fraction of a test group that is affected (e.g. 50 % of a group of individuals of the

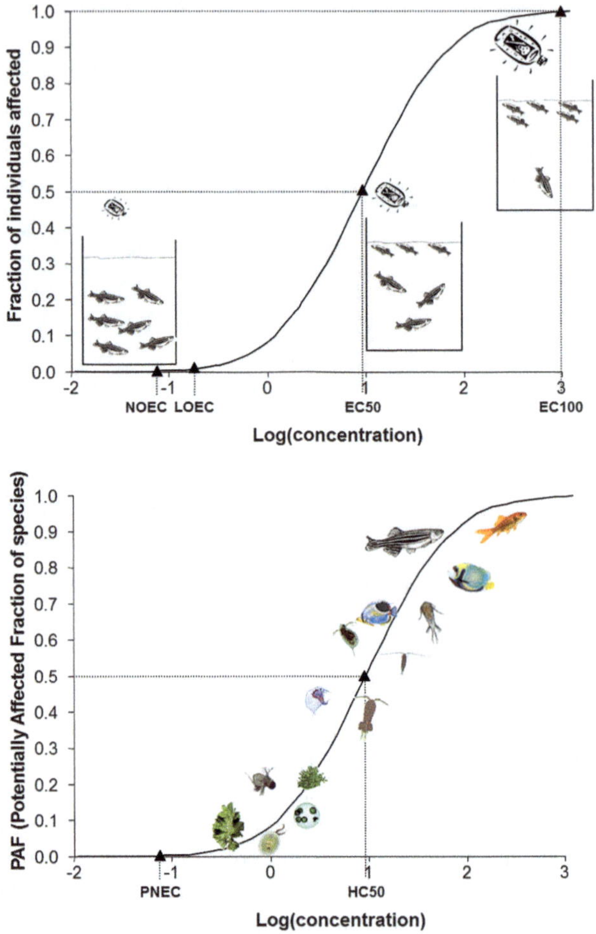

Fig. 8.3 Illustrative concentration-response curve for a single species (*above*) and multiple species in a Species Sensitivity Distribution (SSD) for an aquatic ecosystem (*below*)

same species compared to a control situation). Affected can mean various things, such as mortality, reduced mobility, reduced growth or reproduction rate, mutations, behavioural changes, changes in biomass or photosynthesis, etc. The toxicity tests are standardised and the results are specific for each substance and species. Toxic effects are further distinguished into acute, sub-chronic, and chronic toxicity (including further sub-groups like sub-acute, etc.). Acute toxicity describes an adverse effect after a short period of exposure, relative to the life-time of the animal (e.g. <7 days for vertebrates, invertebrates, or plants, and <3 days for algae). Chronic toxicity is based on exposure over a prolonged period of time covering at least one life cycle or one sensitive period (e.g. ≥ 32 days for vertebrates, ≥ 21 days for invertebrates, ≥ 7 days for plants, and ≥ 3 days for algae). A simplified (i.e. illustrative) dose-response curve for a single species is shown in Fig. 8.3 (above). Important toxicity measures typically determined and reported from the

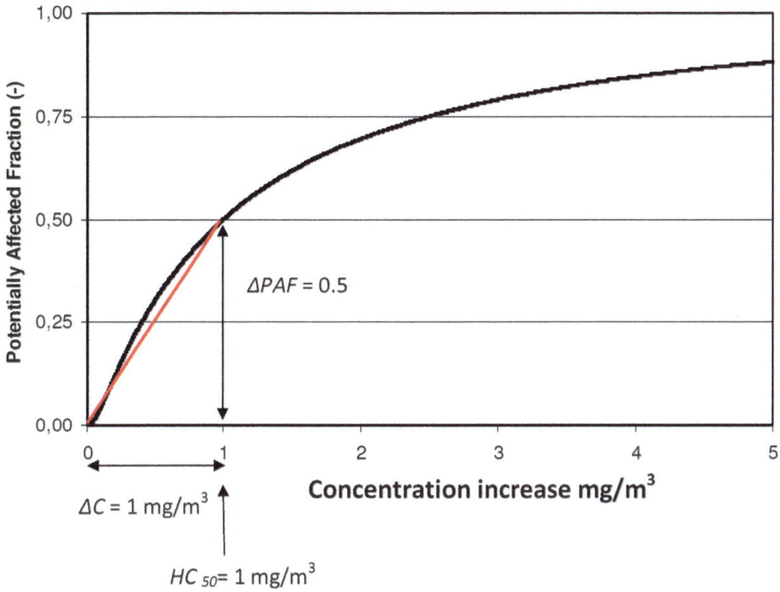

Fig. 8.4 Linearisation of the SSD, a simplification as applied e.g. in USEtox (Huijbregts et al. 2010)

tests are the NOEC – No Observed Effect Concentration (highest tested concentration without any observable effect), the LOEC – Lowest Observed Effect Concentration (lowest tested concentration at which an effect is observed), and the EC50 – Effect Concentration affecting 50 % of the individuals above background (if the observed effect is death, the reported parameter may be the LC50 – Lethal Concentration killing 50 % more of the individuals than in the background sample).

Between different species a large variation of sensitivity to a given substance can usually be observed. This is described by a species-sensitivity-distribution (SSD) curve, which hence represents the sensitivity of the entire ecosystem to a substance. The SSD is constructed using the respective geometric mean of all available and representative EC50 values for each species. This curve represents the range of sensitivity to exposure to a given substance among the different species from the most sensitive to the most robust species. As discussed above, LCA requires linear models, therefore the SSD curve is generally simplified to a linear regression between the origin (i.e. where x-axis and y-axis cross in their 0 values) along the concentration-response relationship up to the point where the PAF is 0.5 (Fig. 8.4). The ecotoxicity effect factor is then calculated using the HC50 – Hazardous Concentration at which 50 % of the species (in an aquatic ecosystem) are exposed to a concentration above their EC50. Based on Traas et al. (2002) and Klepper et al. (1998), the dimension of the effect factor is PAF – Potentially Affected Fraction of species, while the unit is typically m^3/kg. As visible from Fig. 8.3 (below), the HC50 employed in LCA is an average value (best estimate) and significantly different from the conservative choice of the PNEC – Predicted No

Effect Concentration used in Environmental Risk Assessment, which is based on the most sensitive species and thus on the lower end of the SSD curve. The use of the geometric mean to represent average toxicity for a population or an ecosystem is a standard approach in both ERA (Aldenberg et al. 2002; Forbes and Calow 2002; Versteeg et al. 1999; Newman and Dixon 1996) and LCA (Henderson et al. 2011; Larsen and Hauschild 2007b; Pennington et al. 2006; Pennington et al. 2004a; Payet 2004; Payet and Jolliet 2004). An important argument for its use in LCA is to avoid a bias in the comparative assessment of substances. As discussed by Henderson et al. (2011) as well as Larsen and Hauschild (2007b), the geometric mean is less sensitive to extremely high toxicity (low EC50) values and thus more representative for average toxicity to an ecosystem. In LCA this is very important, since a very well-studied substance, tested on many species, will sooner or later have been tested on a very sensitive species, whereas a substance that has only been tested on one or a few species will likely come out less toxic when using the PNEC approach (most sensitive species), simply because it has not yet been tested on a very sensitive species (the more sensitive species will differ among substances). Such a bias needs to be avoided in a comparative assessment.

According to the current scientific consensus, the ecotoxicological effect factor of a chemical is calculated as (Henderson et al. 2011):

$$EF = \frac{0.5}{HC50} \tag{8.2}$$

The log HC50 can be calculated as follows using the EC50 per species respectively:

$$\log\ HC50 = \frac{1}{n_s} \cdot \sum_s \log\ EC50_s \tag{8.3}$$

where n_s is the number of species.

2.5 Severity

A damage model, incorporating the severity of the effect, goes even further along the cause-effect chain and quantifies how many species are disappearing from a given ecosystem. Disappearance may be caused by mortality, reduced proliferation, or migration, for example. Currently, various approaches exist but none is sufficiently accepted by the scientific community to reach the status of recommendation. Several LCIA methodologies are available expressing the damage on ecosystems in PDF – Potentially Disappeared Fraction. IMPACT 2002+, for example, assumes a relation between PAF and PDF as: PDF = PAF/2, based on the assumption that 50 % of the affected species will disappear from the ecosystem (Jolliet et al. 2003). ReCiPe assumes that PAF(EC50) = PDF based on limited evidence from Posthuma and De Zwart(2006) that species loss due to mixture toxicity matches predicted risk with a maximum observed PDF equal to the EC50-based ecotoxicity predictor variable. Larsen and Hauschild (2007a) observed that 'the recovery time approach used as media recovery has been used in some attempts to include damage modelling in the Eco-indicator 99 method (Goedkoop et al. 1998, 2000), and most recently in IMPACT 2002+ (AMI) (Jolliet et al. 2003)'. Further details and a discussion on freshwater ecotoxicity damage modelling can be found in (Larsen and Hauschild 2007a).

2.6 *Affected Areas of Protection*

When relating to freshwater ecosystems, the question arises what exactly we mean by that. In LCIA, a freshwater ecosystem is typically described by at least the first three of the trophic levels (Henderson et al. 2011; Rosenbaum et al. 2008; Larsen and Hauschild 2007b; Pennington et al. 2004a, 2006; Payet 2004; Payet and Jolliet 2004):

1. Primary producers, converting sunlight into biomass via photosynthesis (i.e. phytoplankton, algae),
2. Primary consumers, living off primary producers (i.e. zooplankton, invertebrates, planktivorous fish),
3. Secondary consumers at the upper end of the aquatic food chain (i.e. piscivorous fish).

It should be noted that only impacts on cold-blooded species in freshwater ecosystems are currently considered. The latest state-of-the-art methods available in scientific literature, though only partially included in LCIA methodologies, also explore modelling the toxicity for warm-blooded freshwater predators (Golsteijn et al. 2012), as well as toxic impacts on terrestrial ecosystems (Owsianiak et al. 2013; Haye et al. 2007; Huijbregts 1999). However, there is no minimum requirement established, which trophic levels should be covered by a characterisation factor for terrestrial and marine ecosystems and available methods usually extrapolate from freshwater data or use the relatively few data available directly for these ecosystems.

Quantifying a potential reduction in species present in an ecosystem, ecotoxic impacts may contribute to damage to the Area of Protection sometimes called natural environment (e.g. ILCD Handbook (EC-JRC 2011)), and sometimes called ecosystem quality (e.g. IMPACT 2002+ (Jolliet et al. 2003), or ecosystems (e.g. ReCiPe (Goedkoop et al. 2012)).

3 Contributing Substances (Classification)

About 500 years ago Paracelsus stated that ‘All substances are poisons; there is none which is not a poison. The right dose differentiates a poison and a remedy’. Today’s toxicology science still agrees and adheres to this principle and in consequence any substance emitted may lead to toxic impacts depending on a number of driving factors:

1. Emitted quantity (as determined in the LCI)
2. Mobility (as determined by in the fate factor)
3. Persistence (as determined by the fate factor)
4. Exposure patterns and bioavailability (as determined by the exposure factor)
5. Toxicity (as determined by the effect factor)

This shows that toxicity is not the only parameter that determines the potential ecotoxic impact of a chemical in the environment as it first has to reach a potential target organism. For example, a substance may be very toxic, but never reach any organism due to its short lifetime in the environment (e.g. fast degradation) or because it is not mobile enough to be transported to a target organism and ends up bound to soil matrix or buried in sediment, in which case it contributes little to ecotoxic impacts. On the other hand, a substance may not be very toxic, but if it is emitted in large quantities and over prolonged periods of time or has a strong environmental persistence, it may still cause an ecotoxic impact.

Ecotoxity is very different from any other (non-toxicity) impact category when it comes to the number of potentially relevant elementary flows. Whereas no other (non-toxicity) impact category – with the exception of photochemical ozone formation – exceeds 100 contributing elementary flows (characterisation factors), both toxicity categories are facing the challenge of having to characterise several tens of thousands of chemicals. The CAS registry currently contains more than 70 million unique organic and inorganic substances (www.cas.org/about-cas/cas-fact-sheets) of which roughly 100,000 may play an important industrial role as reflected by the more than 90,000 substances registered in the European Classification and Labelling Inventory Database which contains REACH (Registration, Evaluation, Authorisation and Restriction of Chemical substances) registrations and CLP (Classification, Labelling and Packaging of substances and mixtures) notifications so far received by the European Chemicals Agency (ECHA: http://ihcp.jrc.ec.europa.eu/our_databases/esis). Current LCIA models cover around 2,500 substances for aquatic ecotoxicity (Rosenbaum et al. 2008).

4 Scale, Spatial Variability, Temporal Variability

Ecotoxicity is often considered as a local impact category (i.e. potential impacts will be taking place relatively close – within a few hundred kilometres – to the source of emissions). However, this is not fully true as this may vary greatly depending on the local conditions, the persistence, and the physico-chemical properties of the substance emitted. A very persistent, mobile and bioaccumulating substance may travel many thousands of kilometres and accumulate in food chains around the entire globe, thereby being diluted in the transport medium. Spatial variability is therefore a non-negligible source of uncertainty and requires further insights. With some exceptions (Owsianiak et al. 2013; Gandhi et al. 2010, 2011b), literature mostly focuses on spatial variability in chemical fate and human exposure. Temporal variability is a largely unexplored issue, which may have potentially important influence on the characterisation for some substances and/or ecosystems. Concerning the time scale, most LCIA methodologies employ an infinite time horizon, except ReCiPe and IMPACT World+. ReCiPe provides an extra scenario with characterisation factors for a 100-year time horizon for metals. IMPACT World + provides characterisation factors for 100 years and >100 years respectively for metals. Both methodologies thus allow considering the influence of the time scale on the impact score. For metals this is very important because metals do not degrade and their impacts occur over a very long period of time, which leads to very high (typically dominating) ecotoxicity impact scores when integrating impacts over infinity. In such a case the choice of time horizon represents a trade-off between representing all impacts (when integrating over infinity) on the one hand, and representing impacts that may be large for current generations (when integrating over 100 years), but which are 'diluted' when integrating over infinity on the other hand.

Model uncertainty observed in model variability between harmonised versions of IMPACT 2002, USES-LCA, EDIP, and USEtox was estimated by Rosenbaum et al. (2008) as about three orders of magnitude. Important sources of parameter uncertainty are degradation rates for organic substances (Rosenbaum et al. 2008), neglecting bioavailability due to speciation in metals (Chapman 2008; Chapman et al. 2003), the effect factors due to the use of chronic and acute data as well as a linear dose-response curve (Henderson et al. 2011; Rosenbaum et al. 2008), and the lack of toxicity data for species from various trophic levels (van Zelm et al. 2007). As discussed in (Rosenbaum et al. 2008), despite their uncertainty, ecotoxicity impact scores can still be usefully interpreted when seen in the context of 12 (and in fact up to 17) orders of magnitude difference between the lowest and the highest possible (known and characterised) chemical impacts per unit emission. This means that for the LCA practitioner, these CFs can help identify the 10 or 20 most important chemicals (i.e. dominating the impact by contributing together more than 99 % of the impact score) for a given application, and, perhaps more importantly, to disregard hundreds of other substances whose impact is not significant for the considered application. Tørsløv et al. (2005) discussed that excluding spatial variability may be less influential on overall uncertainty than parameter uncertainty. IMPACT World + (impactworldplus.org) is the first fully spatially resolved LCIA methodology that provides quantified uncertainty estimates for all CFs together with separate estimates of spatial variability contributing to overall uncertainty depending on the level of spatial resolution applied.

5 Midpoint Methodologies

In the 1990s several early models aiming at comparatively assessing toxicity have been published (Steen 1999; Krewitt et al. 1998; Hauschild and Wenzel 1998; Jolliet and Crettaz 1997; Walz et al. 1996; Guinée and Heijungs 1993; Braunschweig and Müller-Wenk 1993). More details and overview of many of the early models is given by Udo de Haes et al. (2002). Several early LCIA methodologies, such as Eco-indicator 99 (Goedkoop et al. 1998) or TRACI 1.0 (Bare et al. 2003) essentially adopted models and measures coming from the assessment of chemical risks to a local environment (e.g. EUSES 1.0 (EC 1996) and CalTOX 4.0 (Hertwich et al. 2001; McKone 2001) respectively), a methodology that was already well established and applied by then. Over time, however, a growing community of researchers started to adapt these models and redefine measures more suitable for comparative assessment of chemical impacts. Both communities are closely linked and collaborating but have evolved individually in the last decade, resulting in a number of specialised LCIA methodologies for characterisation of ecotoxicity impacts that employ well distinguishable approaches relative to regulatory risk assessment (Pennington et al. 2006).

Characterisation methods like EDIP (Hauschild and Potting 2003; Hauschild and Wenzel 1998) account for fate and exposure relying on key properties of the chemical applied to empirical models. For example, the octanol-water partitioning coefficient is used to determine the accumulation of the compound in the food chain. Mechanistic models and methodologies have been published accounting for fate, exposure, and effects providing cardinal impact measures. Among these methods are IMPACT 2002 (Pennington et al. 2005; Jolliet et al. 2003), USES-LCA (van Zelm et al. 2009; Huijbregts et al. 2000), Eco-Indicator 99 (Goedkoop et al. 1998) and ecotoxicity potentials provided by Hertwich et al. (2001) and McKone (2001) using the CalTOX model (McKone et al. 2001). All these methods adopt environmental multimedia, multipathway models employing mechanistic cause-effect chains to account for the environmental fate, exposure, and effects processes. However, they do not necessarily agree on how these processes are to be modelled, leading to variations in results of LCA studies related to the choice of LCIA methodology (Pant et al. 2004; Dreyer et al. 2003). The scientific consensus model USEtox (UNEP/SETAC toxicity consensus model) was developed with the intention to solve this situation by representing a scientifically agreed consensus approach to the characterisation of human and freshwater ecotoxicity (Henderson et al. 2011; Rosenbaum et al. 2008; Hauschild et al. 2008a). Tables 8.1 and 8.2 give

Table 8.1 Current midpoint characterisation methods (extended from ILCD handbook on LCIA (EC-JRC 2011))

LCIA methodology	Characterisation model	Fate/exposure modelling	Effect modelling	Ecosystems considered	Time horizon	Region modelled	No. of substances
CML 2002 (Guinée et al. 2002)	USES-LCA 1.0 (Huijbregts et al. 2000)	Mechanistic, nested, multimedia LCA model	Most sensitive species	Freshwater, marine, terrestrial	Infinite	Europe	~170
EDIP 2003 (Hauschild and Potting 2003; Hauschild and Wenzel 1998)	EDIP1997, combined with site dependent factors (Tørsløv et al. 2005)	Key property, partial fate	Most sensitive species	Freshwater, terrestrial	Infinite	Generic	~190
TRACI 1.0 (Bare et al. 2003)	CalTOX 4.0 (McKone et al. 2001)	Mechanistic, closed, multimedia ERA model	Most sensitive species	Freshwater, terrestrial	Infinite	USA	~160
IMPACT 2002+ (Jolliet et al. 2003)	IMPACT 2002 (Pennington et al. 2005)	Mechanistic, nested, multimedia LCA model	Average toxicity	Freshwater, marine, terrestrial	Infinite	Europe	~430
LUCAS (Toffoletto et al. 2007)				Freshwater, terrestrial		Canada	
MEEuP (Kemna et al. 2005)	None (policy-based target emission limits)	None	None	n/a	n/a	Europe	30
Swiss Ecoscarcity 2006 (Frischknecht et al. 2009)	None (policy-based target emission limits)	None	None	n/a	n/a	Switzerland	25
ReCiPe 2008 (updating CML2002) (Goedkoop et al. 2012)	USES-LCA 2.0 (van Zelm et al. 2009)	Mechanistic, nested, multimedia LCA model	Average toxicity	Freshwater, marine, terrestrial	Infinite, 100 years for metals	Europe	~2,650
TRACI 2.0 (Bare 2011)	USEtox (Rosenbaum et al. 2008)	Mechanistic, nested, multimedia LCA model	Average toxicity	Freshwater	Infinite	Global generic	~2,550
ILCD (EC-JRC 2011)							
IMPACT World + (impactworldplus.org) updating IMPACT 2002+, LUCAS, EDIP				Freshwater, marine, terrestrial	Infinite, ≤100 y, >100 y for metals	Global generic + 9 sub-continents	

Table 8.2 Current endpoint characterisation methods (extended from ILCD handbook on LCIA (EC-JRC 2011))

LCIA methodology	Characterisation model	Fate/exposure modelling	Effect modelling	Ecosystems considered	Time horizon	Region modelled	No. of substances
EPS 2000 (Steen 1999)	None (red list species pot. threatened by chemicals)	None	None	n/a	n/a	Generic	~45
Eco-Indicator99 (Goedkoop et al. 1998)	EUSES 1.0 (EC 1996)	Mechanistic, nested, multimedia ERA model	Most sensitive species	Freshwater, terrestrial	Infinite	Europe	~45
IMPACT 2002+ (Jolliet et al. 2003)	IMPACT 2002 (Pennington et al. 2005)	Mechanistic, nested, multimedia LCA model	Average toxicity	Freshwater, marine, terrestrial	Infinite	Europe	~430
LIME 1.0 (Itsubo and Inaba 2003)				Freshwater, terrestrial		Japan	
ReCiPe 2008 (updating EI99) (Goedkoop et al. 2012)	USES-LCA 2.0 (van Zelm et al. 2009)	Mechanistic, nested, multimedia LCA model	Average toxicity	Freshwater, marine, terrestrial	Infinite, 100 years for metals	Europe	~2,650
IMPACT World + (impactworldplus.org) updating IMPACT 2002+, LUCAS, and EDIP	USEtox (Rosenbaum et al. 2008)	Mechanistic, nested, multimedia LCA model	Average toxicity	Freshwater, marine, terrestrial	Infinite, ≤100 years, >100 years for metals	Generic + 9 parameterised sub-continents	~2,550

an overview of a number of LCIA methodologies and the ecotoxicity characterisation models they employ.

Among the existing LCIA methodologies on midpoint level, three main groups can be distinguished: (1) mechanistic, multimedia fate, exposure and effects models, (2) key property-based partial fate models, and (3) non-fate models (EC-JRC 2011). According to ISO 14044 (2006) "Characterisation models reflect the environmental mechanism by describing the relationship between the LCI results, category indicators and, in some cases, category endpoints. [...] The environmental mechanism is the total of environmental processes related to the characterisation of the impacts." Therefore, ecotoxicity characterisation models falling into categories (2) and (3), do not completely fulfil this criterion. Caution is advised regarding their use and most importantly the interpretation of their results, which should not be employed without prior in-depth study of their respective documentation. Having said that, depending on the goal and scope of the LCA, they may still be an adequate choice in some applications, and indeed agree quite well with the more sophisticated multimedia-based models as demonstrated for a harmonised version of the EDIP model (Rosenbaum et al. 2008). When deciding which LCIA methodology to use in an LCA, an overview of selected properties of these methodologies helps to identify suitable options. As a general recommendation for selecting an LCIA methodology (for ecotoxicity), the priority should be on the following three criteria: (1) substance coverage, (2) state-of-the-art mechanistic modelling, and (3) which ecosystems are considered. As science (and tools) is advancing visibly in this field, it is recommendable to choose a methodology not older than 5–10 years. Unless a methodology provides spatially variable characterisation factors, the regional focus is secondary and should not be a priority criterion for methodology selection. Table 8.1 provides an overview of current midpoint LCIA methodologies and several of their properties. A good overview and many further details can be found in the ILCD Handbook on LCIA recommendations (EC-JRC 2011).

6 Endpoint Methodologies

Ecotoxicity endpoint modelling is still in an early state and much research needs to be performed before maturity is reached. The authors of the ILCD LCIA handbook concluded that "For all the three evaluated endpoint methods (EPS2000, ReCiPe, IMPACT2002+), there is little or no compliance with the scientific and stakeholder acceptance criteria, as the overall concept of the endpoint effect factors is hardly validated and the endpoint part of the methods is not endorsed by an authoritative body. [...] No method is recommended for the endpoint assessment of ecotoxicity, as no method is mature enough" (EC-JRC 2011). Table 8.2 provides an overview of current endpoint LCIA methodologies and the ecotoxicity models employed, respectively.

7 New Developments and Research Needs

The principal objective and motivation behind any further research and development is reduction of uncertainty, notably parameter and model uncertainty, and ecotoxicity characterisation in LCIA is certainly no exception. As can be seen from the discussion above, substance coverage is an important area of further research needs (Hauschild et al. 2013; Finnveden et al. 2009; Rosenbaum et al. 2008). With at least 100,000 substances of potential industrial application and importance and current models covering some 2,500 substances, much remains to be done to improve the situation. The currently limiting factor is effect data availability, mainly driven by funding priorities focusing on substances "of highest political concern, [... while ...] coverage may not be for the most important chemical emissions in the life cycle of a specific product" (Finnveden et al. 2009). Even if a substance is already included in the list of available CFs, it may only be with significant uncertainties for some substances as reflected by the distinction between interim (i.e. with higher uncertainty) and recommended (i.e. with acceptable uncertainty) CFs in USEtox (Rosenbaum et al. 2008). Many substances are currently not characterised or only with insufficient accuracy. Important examples are ionic and amphiphilic substances, persistent bioaccumulating chemicals, persistent surface active compounds, pesticides and biocides, substituted musks/fragrances, biochemicals (i.e. antibiotics, nucleotides, proteins (including enzymes), peptides, polyamino acids, buffers, lipids, carbohydrates, antibodies), and metals. The same can be said for the robustness of the effect factor, which requires a minimum amount of ecotoxicological data for several species being available. These data should ideally represent chronic toxicity tests, but in reality most effect factors are based on acute effect data due to a lack of chronic data (which are significantly more expensive to obtain). Currently, availability and quality of ecotoxicological effect data are the most limiting factors for increasing substance coverage, as the physico-chemical properties required for the fate modelling are already available for several tens of thousands of substances.

Including marine and terrestrial ecosystems, in a scientifically more adequate way than currently, is another very important research need as they contain a significant biodiversity. Although some LCIA methodologies propose CFs for marine and terrestrial ecotoxicity, their application is not recommended neither at midpoint nor at endpoint level by EC-JRC (2011) and Hauschild et al. (2013). The same authors in accord with Rosenbaum et al. (2008) also conclude that for freshwater ecotoxicity, no viable approach is available to model impacts from midpoint to endpoint, corresponding to the step from accounting for the number of species affected by any kind of toxic impact to the number of species potentially disappearing from the ecosystem.

An important problem for some emissions is the time horizon of their potential impacts and how to meaningfully consider these in LCA and LCIA (Doka 2009; Zhao et al. 2009; Reid et al. 2009; Hauschild et al. 2008b; Doka and Hischier 2005; Hellweg and Frischknecht 2004; Hellweg et al. 2003; Finnveden and Nielsen 1999).

The most prominent examples are long-term emissions of metals from landfills or mine tailings. Small amounts of leachate or run-off containing very small concentrations of pollutants (especially heavy metals) are emitted from the landfill to the surrounding soil, aquifer, and eventually the surface water during thousands of years. In LCIA, impacts are modelled using steady-state conditions, applying integration over a defined time horizon. Integrating the impacts of long-term emissions over a relatively short time horizon like 500 years (as done for e.g. Global Warming Potentials), and thus neglecting impacts occurring later, leads to a strong underestimation of their impacts. On the other hand, their full consideration via integration over large or even infinite time horizons would lead to a strong overestimation, as the (perhaps small) impacts occurring over a long period of time would be fully attributed to the product as if they were occurring right now (as one large impact) and without considering the possibility of future technological solutions to the problem. While the latter approach does not account for the 'dilution of the impact over time', the first approach completely neglects these impacts. These two extremes represent a dilemma for which a meaningful solution is needed in LCIA.

Further insights into spatial variability – the influence of the place of an emission on the impact – will help reducing uncertainty due to neglecting locally specific ecotoxicity problems and thus help increase accuracy for this impact category. The influence of temporal variability, such as seasonal behaviour of species or weather patterns, remains to be examined. Mixture toxicity is a research area of high importance and complexity. One prominent example, potentially relevant for many LCA studies, concerns effluents from industrial and waste water treatment processes, which are mixtures of varying composition containing many different substances. These further research needs are thereby not a matter of increasing the complexity of characterisation models and their application to an impractical level, but rather about establishing a parsimonious balance between necessary complexity and maximised simplicity. In order to simplify, scientists have to first explore complexity, which will lead to finding a meaningful balance including the consideration of practical needs on the level of application and with full conscience about the uncertainties introduced.

References

Aldenberg T, Jaworska J, Traas TP (2002) Normal species sensitivity distributions and probabilistic ecological risk assessment. In: Posthuma L, SuterII GW, Traas TP (eds) Species sensitivity distribution in ecotoxicology. Lewis, Boca Raton, pp 49–102

Bare J (2011) TRACI 2.0: the tool for the reduction and assessment of chemical and other environmental impacts 2.0. Clean Technol Environ Policy 13(5):687–696. doi:10.1007/s10098-010-0338-9

Bare JC, Norris GA, Pennington DW, McKone T (2003) TRACI: the tool for the reduction and assessment of chemical and other environmental impacts. J Ind Ecol 6(3–4):49

Barnthouse LW, Fava JA, Humphreys K, Hunt R, Laibson L, Noesen S, Norris GA, Owens JW, Todd J, Vigon B, Weitz K, Young JS (1997) Life-cycle impact assessment: the state of the art, 2nd edn. SETAC Press, Pensacola

Braunschweig A, Müller-Wenk R (1993) Oekobilanzen für Unternehmungen; eine Wegleitung für die Praxis. Eine Wegleitung für die PraxisVerlag Paul Haupt. Verlag Paul Haupt/BUWAL, Bern

Chapman PM (2008) Environmental risks of inorganic metals and metalloids: a continuing, evolving scientific odyssey. Hum Ecol Risk Assess 14:5–40

Chapman PM, Wang F, Janssen CR, Goulet RR, Kamunde CN (2003) Conducting ecological risk assessments of inorganic metals and metalloids: current status. Hum Ecol Risk Assess 9:641–697

Diamond ML, Gandhi N, Adams WJ, Atherton J, Bhavsar SP, Bulle C, Campbell PGC, Dubreuil A, Fairbrother A, Farley K, Green A, Guinee J, Hauschild MZ, Huijbregts MAJ, Humbert S, Jensen KS, Jolliet O, Margni M, McGeer JC, Peijnenburg WJGM, Rosenbaum RK, van de Meent D, Vijver MG (2010) The clearwater consensus: the estimation of metal hazard in fresh water. Int J Life Cycle Assess 15(2):143–147

Doka G (2009) Life cycle inventories of waste treatment services. Ecoinvent report no 13 part II: landfills – underground deposits – landfarming. Swiss Centre for Life Cycle Inventories, Dübendorf

Doka G, Hischier R (2005) Waste treatment and assessment of long-term emissions. Int J Life Cycle Assess 10(1):77–84

Dreyer LC, Niemann AL, Hauschild MZ (2003) Comparison of three different LCIA methods: EDIP97, CML2001 and eco-indicator 99: does it matter which one you choose? Int J Life Cycle Assess 8(4):191–200

Duan N, Dobbs A, Ott W (1990) Comprehensive definitions of exposure and dose to environmental pollution. Department of Applied Earth Sciences, Stanford University, Stanford, California, Stanford

EC (1996) EUSES, the European Union System for the Evaluation of Substances. National Institute of Public Health and the Environment (RIVM), Bilthoven

EC-JRC (2010) Framework and requirements for LCIA models and indicators. ILCD handbook – International Reference Life Cycle Data System, vol EUR24571EN. European Union, Ispra

EC-JRC (2011) International Reference Life Cycle Data System (ILCD). Handbook-recommendations for life cycle impact assessment in the European context, 1st edn. Luxemburg

Finnveden G, Nielsen PH (1999) Long-term emissions from landfills should not be disregarded. Int J Life Cycle Assess 4(3):125–126

Finnveden G, Hauschild MZ, Ekvall T, Guineé J, Heijungs R, Hellweg S, Koehler A, Pennington DW, Suh S (2009) Recent developments in life cycle assessment. J Environ Manage 91:1–21

Forbes VE, Calow P (2002) Species sensitivity distribution revisited: a critical appraisal. Hum Ecol Risk Assess 8(3):473–492

Frischknecht R, Steiner R, Jungbluth N (2009) The ecological scarcity method – eco-factors 2006: a method for impact assessment in LCA. Federal Office for the Environment (FOEN), Bern

Gandhi N, Diamond ML, Van de Meent D, Huijbregts MAJ, Peijnenburg WJGM, Guinée J (2010) New method for calculating comparative toxicity potential of cationic metals in freshwater: application to copper, nickel, and zinc. Environ Sci Technol 44(13):5195–5201

Gandhi N, Diamond M, Huijbregts MJ, Guinée J, Peijnenburg WGM, Meent D (2011a) Implications of considering metal bioavailability in estimates of freshwater ecotoxicity: examination of two case studies. Int J Life Cycle Assess 16(8):774–787. doi:10.1007/s11367-011-0317-3

Gandhi N, Huijbregts MAJ, van de Meent D, Peijnenburg WJGM, Guinée J, Diamond ML (2011b) Implications of geographic variability on comparative toxicity potentials of Cu, Ni and Zn in freshwaters of Canadian ecoregions. Chemosphere 82:268–277

Goedkoop M, Müller-Wenk R, Hofstetter P, Spriensma R (1998) The eco-indicator 99 explained. Int J Life Cycle Assess 3(6):352–360

Goedkoop M, Effting S, Collignon M (2000) The Eco-indicator 99, a damage oriented method for life cycle impact assessment. Methodology Annex 2nd edn. Amersfoort, Pré Consultants, B.V.

Goedkoop M, Heijungs R, Huijbregts MAJ, De Schryver A, Struijs J, van Zelm R (2012) ReCiPe 2008 – a life cycle impact assessment method which comprises harmonised category indicators at the midpoint and the endpoint level. Volume report I: characterisation, 1st (revised) edn. Ministry of Housing, Spatial Planning and Environment (VROM), Den Haag

Golsteijn L, van Zelm R, Veltman K, Musters G, Hendriks AJ, Huijbregts MAJ (2012) Including ecotoxic impacts on warm-blooded predators in life cycle impact assessment. Integr Environ Assess Manag 8(2):372–378. doi:10.1002/ieam.269

Guinée J, Heijungs R (1993) A proposal for the classification of toxic substances within the framework of life cycle assessment of products. Chemosphere 26(10):1925–1944

Guinée JB, Gorrée M, Heijungs R, Huppes G, Kleijn R, van Oers L, Wegener Sleeswijk A, Suh S, Udo de Haes HA, de Bruijn H, van Duin R, Huijbregts MAJ (2002) Handbook on life cycle assessment: operational guide to the ISO standards. Eco-efficiency in industry and science. Kluwer Academic Publishers, Dordrecht

Hauschild M, Pennington DW (2003) Chapter 6: Indicators for ecotoxicity in life-cycle impact assessment. In: Udo de Haes H (ed) Life-cycle impact assessment: striving towards best practice. SETAC Press, Pensacola, pp 149–176

Hauschild MZ, Potting J (2003) Spatial differentiation in life cycle impact assessment: the EDIP 2003 methodology. Institute for Product Development, Technical University of Denmark, Lyngby

Hauschild M, Wenzel H (1998) Environmental assessment of products, vol 2, Scientific background. Thomson Science, London

Hauschild MZ, Huijbregts MAJ, Jolliet O, MacLeod M, Margni M, Van de Meent D, Rosenbaum RK, McKone TE (2008a) Building a model based on scientific consensus for life cycle impact assessment of chemicals: the search for harmony and parsimony. Environ Sci Technol 42 (19):7032–7037

Hauschild MZ, Olsen SI, Hansen E, Schmidt A (2008b) Gone...but not away—addressing the problem of long-term impacts from landfills in LCA. Int J Life Cycle Assess 13:547–554

Hauschild M, Goedkoop M, Guinée J, Heijungs R, Huijbregts M, Jolliet O, Margni M, Schryver A, Humbert S, Laurent A, Sala S, Pant R (2013) Identifying best existing practice for characterization modeling in life cycle impact assessment. Int J Life Cycle Assess 18(3):683–697. doi:10.1007/s11367-012-0489-5

Haye S, Slaveykova VI, Payet J (2007) Terrestrial ecotoxicity and effect factors of metals in life cycle assessment (LCA). Chemosphere 68(8):1489–1496

Hellweg S, Frischknecht R (2004) Evaluation of long-term impacts in LCA. Int J Life Cycle Assess 9(5):339–341

Hellweg S, Hofstetter TB, Hungerbühler K (2003) Discounting and the environment. Should current impacts be weighted differently than impacts harming future generations? Int J Life Cycle Assess 8(1):8–18

Henderson A, Hauschild M, Van de Meent D, Huijbregts MAJ, Larsen HF, Margni M, McKone TE, Payet J, Rosenbaum RK, Jolliet O (2011) USEtox fate and ecotoxicity factors for comparative assessment of toxic emissions in life cycle analysis: sensitivity to key chemical properties. Int J Life Cycle Assess 16:701–709. doi:10.1007/s11367-011-0294-6

Hertwich E, Matales SF, Pease WS, McKone TE (2001) Human toxicity potentials for life-cycle assessment and toxics release inventory risk screening. Environ Toxicol Chem 20(4):928–939

Huijbregts M (1999) Ecotoxicological effect factors for the terrestrial environment in the frame of LCA. University of Amsterdam, Amsterdam

Huijbregts MAJ, Thissen U, Guinée JB, Jager T, Kalf D, van de Meent D, Ragas AMJ, Wegener Sleeswijk A, Reijnders L (2000) Priority assessment of toxic substances in life cycle assessment. Part I: calculation of toxicity potentials for 181 substances with the nested multi-media fate, exposure and effects model USES-LCA. Chemosphere 41(4):541–573

Huijbregts M, Hauschild MZ, Jolliet O, Margni M, McKone TE, Rosenbaum RK, van de Meent D (2010) USEtox user manual, http://www.usetox.org/sites/default/files/support-tutorials/user_manual_usetox.pdf

ISO 14044 (2006) International standard. Environmental management – life cycle assessment – requirements and guidelines. International Organisation for Standardisation, Geneva

Itsubo N, Inaba A (2003) A new LCA method: LIME has been completed. Int J Life Cycle Assess 8(5):305

Jolliet O, Crettaz P (1997) Critical surface time 95: a life cycle assessment methodology including fate and exposure. Swiss Federal Institute of Technology, Institute of Soil and Water Management, Lausanne

Jolliet O, Margni M, Charles R, Humbert S, Payet J, Rebitzer G, Rosenbaum RK (2003) IMPACT 2002+: a new life cycle impact assessment methodology. Int J Life Cycle Assess 8(6):324–330

Jolliet O, Rosenbaum RK, Chapmann P, McKone T, Margni M, Scheringer M, van Straalen N, Wania F (2006) Establishing a framework for life cycle toxicity assessment: findings of the Lausanne review workshop. Int J Life Cycle Assess 11(3):209–212

Kemna R, Van Elburg M, Li W, Van Holsteijn R (2005) MEEUP – Methodology report. Final version, 28-11-2005. EC, Brussels

Klepper O, Bakker J, Traas TP, Van de Meent D (1998) Mapping the potentially affected fraction (PAF) of species as a basis for comparison of ecotoxicological risks between substances and regions. J Hazard Mater 61:337–344

Krewitt W, Mayerhofer P, Trukenmüller A, Friedrich R (1998) Application of the impact pathway analysis in the context of LCA. Int J Life Cycle Assess 3(2):86–94

Larsen HF, Hauschild M (2007a) Evaluation of ecotoxicity effect indicators for use in LCIA. Int J Life Cycle Assess 12(1):24–33

Larsen HF, Hauschild MZ (2007b) GM-troph: a low data demand ecotoxicity effect indicator for use in LCIA. Int J Life Cycle Assess 12(2):79–91

Ligthart T, Aboussouan L, Van de Meent D, Schönnenbeck M, Hauschild M, Delbeke K, Struijs J, Russel A, Udo de Haes H, Atherton J, van Tilborg W, Karman C, Korenromp R, Sap G, Baukloh A, Dubreuil A, Adams W, Heijungs R, Jolliet O, De Koning A, Chapmann P, Verdonck F, van der Loos R, Eikelboom R, Kuyper J (2004) Declaration of Apeldoorn on LCIA of non-ferrous metals. http://lcinitiative.unep.fr/includes/file.asp?site=lcinit&file=38D1F49D-6D64-45AE-9F64-578BA414E499

McKone TE (2001) Ecological toxicity potentials (ETPs) for substances released to air and surface waters. Environmental Health Sciences Division, School of Public Health, University of California, Berkeley, 94720

McKone T, Bennett D, Maddalena R (2001) CalTOX 4.0 Technical Support Document, vol 1. Lawrence Berkeley National Laboratory, Berkeley

Newman MC, Dixon PM (1996) Ecologically meaningful estimates of lethal effect in individuals. In: Newman MC, Jagoe CH (eds) Ecotoxicology – a hierarchical treatment. Lewis, Boca Raton, pp 225–253

Olsen SI, Christensen FM, Hauschild M, Pedersen F, Larsen HF, Tørsløv J (2001) Life cycle impact assessment and risk assessment of chemicals – a methodological comparison. Environ Impact Assess Rev 21(4):385

Owens JW (1997) Life-cycle assessment in relation to risk assessment: an evolving perspective. Risk Anal 17(3):359

Owsianiak M, Rosenbaum RK, Huijbregts MAJ, Hauschild MZ (2013) Addressing geographic variability in the comparative toxicity potential of copper and nickel in soils. Environ Sci Technol 47(7):3241–3250. doi:10.1021/es3037324

Pant R, Van Hoof G, Schowanek D, Feijtel TCJ, De Koning A, Hauschild M, Olsen SI, Pennington DW, Rosenbaum RK (2004) Comparison between three different LCIA methods for aquatic ecotoxicity and a product environmental risk assessment: insights from a detergent case study within OMNIITOX. Int J Life Cycle Assess 9(5):295

Payet J (2004) Assessing toxic impacts on aquatic ecosystems in life cycle assessment (LCA). Ph. D. Diss., Ecole Polytechnique Fédérale de Lausanne (EPFL), Lausanne
Payet J, Jolliet O (2004) Comparative assessment of the toxic impact of metals on aquatic ecosystems: the AMI Method. In: Dubreuil A (ed) Life cycle assessment of metals: issues and research directions. SETAC, Pensacola, FL, pp 172–175
Pennington DW, Payet J, Hauschild M (2004a) Aquatic ecotoxicological indicators in life-cycle assessment. Environ Toxicol Chem 23(7):1796–1807
Pennington DW, Rydberg T, Potting J, Finnveden G, Lindeijer E, Jolliet O, Rebitzer G (2004b) Life cycle assessment part 2: current impact assessment practice. Environ Int 30(5):721–739
Pennington DW, Margni M, Ammann C, Jolliet O (2005) Multimedia fate and human intake modeling: spatial versus nonspatial insights for chemical emissions in Western Europe. Environ Sci Technol 39(4):1119–1128
Pennington DW, Margni M, Payet J, Jolliet O (2006) Risk and regulatory hazard based toxicological effect indicators in life cycle assessment (LCA). Hum Ecol Risk Assess 12(3):450–475
Posthuma L, De Zwart D (2006) Predicted effects of toxicant mixtures are confirmed by changes in fish species assemblages in Ohio, USA, rivers. Environ Toxicol Chem 25(4):1094–1105. doi:10.1897/05-305r.1
Reid C, Bécaert V, Aubertin M, Rosenbaum RK, Deschênes L (2009) Life cycle assessment of mine tailings management in Canada. J Clean Prod 17:471–479
Rosenbaum RK, Margni M, Jolliet O (2007) A flexible matrix algebra framework for the multimedia multipathway modeling of emission to impacts. Environ Int 33(5):624–634
Rosenbaum RK, Bachmann TK, Gold LS, Huijbregts MAJ, Jolliet O, Juraske R, Koehler A, Larsen HF, MacLeod M, Margni M, McKone TE, Payet J, Schuhmacher M, Van de Meent D, Hauschild MZ (2008) USEtox – The UNEP/SETAC-consensus model: recommended characterisation factors for human toxicity and freshwater ecotoxicity in life cycle impact assessment. Int J Life Cycle Assess 13 (7):532–546. doi:10.1007/s11367-008-0038-4
Semple KT, Doick KJ, Jones KC, Burauel P, Craven A, Harms H (2004) Defining bioavailability and bioaccessibility of contaminated soil and sediment is complicated. Environ Sci Technol 38 (12):228A–231A
Steen B (1999) A systematic approach to environmental priority strategies in product development (EPS). Version 2000 – Models and data of the default method. Centre for Environmental assessment of products and material systems. Chalmers University of Technology, Technical Environmental Planning, Gothenburg
Toffoletto L, Bulle C, Godin J, Reid C, Deschênes L (2007) LUCAS – a new LCIA method used for a Canadian-specific context. Int J Life Cycle Assess 12(2):93–102
Tørsløv J, Hauschild MZ, Rasmussen D (2005) Ecotoxicity. From Hauschild M, Potting J: spatial differentiation in life cycle impact assessment – The EDIP2003 methodology. Environmental News no 80. The Danish Ministry of the Environment, Environmental Protection Agency, Copenhagen
Traas TP, Van de Meent D, Posthuma L, Hamers THM, Kater BJ, De Zwart D, Aldenberg T (2002) Potentially affected fraction as measure of toxic pressure on ecosystems. In: Posthuma L, Suter GWI, Traas TP (eds) Species-sensitivity distributions in ecotoxicology. Lewis, Boca Raton, pp 315–344
Udo de Haes H, Jolliet O, Finnveden G, Goedkoop M, Hauschild M, Hertwich E, Hofstetter P, Klöpffer W, Krewitt W, Lindeijer E, Mueller-Wenk R, Olson S, Pennington D, Potting J, Steen B (2002) Life-cycle impact assessment: striving towards best practice. SETAC Press, Pensacola
van Zelm R, Huijbregts MAJ, Harbers JV, Wintersen A, Struijs J, Posthuma L, Van de Meent D (2007) Uncertainty in msPAF-based ecotoxicological effect factors for freshwater ecosystems in life cycle impact assessment. Integr Environ Assess Manag 3(2):203–210
van Zelm R, Huijbregts MAJ, Van de Meent D (2009) USES-LCA 2.0-a global nested multi-media fate, exposure, and effects model. Int J Life Cycle Assess 14(3):282–284

Versteeg DJ, Belanger SE, Carr GJ (1999) Understanding single species and model ecosystem sensitivity. Data-based comparison. Environ Toxicol Chem 18:1329–1346

Walz R, Herrchen M, Keller D, Stahl B (1996) Impact category ecotoxicity and valuation procedure, ecotoxicological impact assessment and the valuation step within LCA: pragmatic approaches. Int J Life Cycle Assess 1(4):193–198

Zhao W, van der Voet E, Huppes G, Zhang Y (2009) Comparative life cycle assessments of incineration and non-incineration treatments for medical waste. Int J Life Cycle Assess 14:114–121

Chapter 9
Acidification

Rosalie van Zelm, Pierre-Olivier Roy, Michael Z. Hauschild, and Mark A.J. Huijbregts

Abstract This chapter outlines the cause-effect pathway of terrestrial and aquatic acidification from air emissions to ecosystem damage. Carbon dioxide is the main cause of (coastal) marine acidification, while nitrogen and sulfur inputs are underlying the damage due to freshwater and terrestrial acidification. Various life cycle impact assessment (LCIA) methods address parts of the impact pathway. Terrestrial acidification, caused by base cation leaching, has been addressed by a number of midpoint methods and several methods determining impacts to biodiversity. To decrease uncertainty in the ecological effect predictions, more insight needs to be gained in the stressor-response curves for many regions of the world. Moreover, research is needed regarding other indicators related to biodiversity than relative species richness as such. For freshwater acidification, only one midpoint and one endpoint method are available, with substantial options for improvement. To address ocean acidification in LCIA in the future, a carbon cycle model needs to be used to make the link to ocean acidification and stressor-response curves that assess impacts on marine biodiversity.

Keywords Aquatic acidification • Cause-effect pathway • Endpoint • LCA • LCIA • Life cycle assessment • Life cycle impact assessment • Midpoint • Terrestrial acidification

R. van Zelm (✉) • M.A.J. Huijbregts
Department of Environmental Science, Institute for Water and Wetland Research, Radboud University, Nijmegen, The Netherlands
e-mail: r.vanzelm@science.ru.nl; M.Huijbregts@science.ru.nl

P.-O. Roy
CIRAIG, Department of Chemical Engineering, École Polytechnique de Montréal, P.O. Box 6079, Montréal, QC H3C 3A7, Canada
e-mail: pierre-olivier-3.roy@polymtl.ca

M.Z. Hauschild
Division for Quantitative Sustainability Assessment, Department of Management Engineering, Technical University of Denmark (DTU), Produktionstorvet, Building 424, Lyngby 2800, Denmark
e-mail: mzha@dtu.dk

M.Z. Hauschild, M.A.J. Huijbregts (eds.), *Life Cycle Impact Assessment*, LCA Compendium – The Complete World of Life Cycle Assessment,
DOI 10.1007/978-94-017-9744-3_9

1 Introduction

This chapter deals with the causes and consequences of acidification in terrestrial and aquatic ecosystems and how to deal with them in life cycle impact assessment. Section 2 will outline the complete cause-effect pathway, from emissions of acidifying substances to damage to the natural environment. Subsequently, Sect. 3 will outline the framework for acidification for life cycle impact assessment (LCIA). Section 4 will then provide an overview of all methods that have been available in LCA to address (parts of) the cause-effect pathway. Current methods still include uncertainties in the final characterisation factor outcomes, which will be discussed. Finally, future trends will be outlined in acidification modelling for LCIA.

2 Cause-Effect Pathway

Atmospheric deposition of inorganic substances on the Earth's surface, such as oxides of sulfur and nitrogen, are the basis of acidification in terrestrial and freshwater ecosystems, and to a lesser extent in (coastal) marine ecosystems. Their dissociation products alter alkalinity, pH, and inorganic carbon storage in oceans. While these acidifying compounds have natural sources in volcanic eruptions and emissions from the oceans (e.g. volatile sulfur gases), most derive from anthropogenic activities, such as the combustion of fossil fuels at power stations and industrial plants, vehicle exhausts, and agriculture.

Terrestrial acidification is mainly caused by nitrogen and sulfur depositions (Psenner 1994; Bouwman et al. 2002) resulting from nitrogen oxides (NO_x), ammonia (NH_3), and sulfur dioxide (SO_2), and to a lesser extent pyrite (FeS_2) and hydrogen sulfide (H_2S) or strong inorganic acids like hydrogen fluoride (HF), hydrogen chloride (HCl), or monocalcium phosphate ($Ca(H_2PO_4)_2$).

SO_2 for example, which is the acidic anhydride of sulfurous acid, H_2SO_3, can absorb water from the atmosphere to form sulfurous acid, which can release two hydrogen ions:

$$SO_2 + H_2O \rightarrow H_2SO_3 \leftrightarrow 2H^+ + SO_3{}^{2-} \quad (9.1)$$

Ammonia, on the other hand, is a substance that releases a hydrogen ion on mineralisation. In itself, ammonia is a base (absorbing hydrogen ions via the reaction $NH_3 + H^+ \rightarrow NH_4^+$), but it is oxidised on bacterial mineralisation through nitrite, NO_2^-, to nitrate, NO_3^-:

$$NH_3 + 2O_2 \rightarrow H^+ + NO_3{}^- + H_2O \quad (9.2)$$

$Ca(H_2PO_4)_2$ is often used as a component in fertilizers and will react with water to form the acidifying phosphoric acid (H_3PO_4):

$$Ca(H_2PO_4)_2 + H_2O \rightarrow CaHPO_4 + H_3PO_4 \tag{9.3}$$

Areas that are most susceptible to terrestrial acidification have an unreactive geology such as granite and a base-poor soil. After (wet) deposition, biogeochemical processes can delay chemical response to acid deposition in soil and subsequent runoff to freshwater. The amount and place of deposition depends on atmospheric climate conditions (i.e. wind, temperature, precipitation, etc.), chemical interactions with the atmosphere and topography. There are landscapes, such as mine spoil and mangrove reclamation areas, in which sulfur content is naturally high. If pyrite is present, acidification can form a serious problem, as 2 hydrogen ions are produced for every sulfide ion oxidised:

$$2FeS_2 + 6H_2O + 7O_2 \rightarrow 4SO_4{}^{2-} + 8H^+ + 2Fe(OH)_2 \tag{9.4}$$

Atmospheric input of acidifying substances can be neutralised by a number of buffer reactions in an ecosystem (Blaser et al. 1999). The buffer systems are dependent on the chemical status of an aquatic or terrestrial system and vary around the globe (Clair et al. 2007; Dangles et al. 2004). The concept of critical load was developed to be able to set standards and targets for emission reduction policy. A definition was set by Nilsson and Grennfelt (1988) as: "A critical load for acid deposition is the highest deposition of acidifying compounds that will not cause chemical changes leading to long term harmful effects on ecosystem structure and function." Critical loads are defined for specific combinations of pollutants, effects, and receptors. There are a number of indicators for ecosystem acidity, e.g. base saturation, base cations (BC, i.e., Ca^{2+}, K^+, Mg^{2+}) to aluminium (Al) ratio (i.e., BC/Al), the aluminium to calcium (Ca) ratio (i.e., Al/Ca), soil solution pH, dissolved Al concentration (Posch et al. 2001).

Delays in responses occur due to biogeochemical processes as well as biological processes in plants and organisms (Van Zelm et al. 2007). As a result of changes in nutrient regulation in the soil, terrestrial species may suffer from a decrease in phosphorus and magnesium content in tissue, an increase in tissue yellowing, a reduction in biomass, coverage, and root growth, unsuccessful germination and regeneration, and competitive exclusion by acid-tolerant species (Falkengren-Grerup 1986; Zvereva et al. 2008; Roem and Berendse 2000). Metabolism of plants is dependent on optimal pH values as well, and resistance to acidification differs from one plant species to another (Scholz and Reck 1977). Acidic environments also enhance aluminium toxicity. Trivalent cationic Al^{3+} present as $Al(H_2O)_6^{3+}$ in acid environments is the most relevant toxic form to plants, and research on plant resistance to aluminium toxicity has been ongoing (Poschenrieder et al. 2008).

In the end, biodiversity will be altered. The acidifying effects that chemicals cause on the environment can be modelled along the cause-effect pathway up to biodiversity damage, which is the endpoint for this impact category. All previous points on the cause-effect pathway may serve as midpoints for this impact category.

Freshwater acidification is mainly caused by protons resulting from the mineralisation of nitrogen and sulfur deposition as well. The decrease in the pH of the Earth's oceans, on the other hand, is mainly caused by the accelerated dissolution of carbon dioxide (CO_2) from the atmosphere following the increasing concentrations of CO_2 in the atmosphere due to anthropogenic activities (Caldeira and Wickett 2003). A large share of the carbon dioxide released by humans into the atmosphere dissolves into the oceans (and some in rivers and lakes), which increases the hydrogen ion concentration in the ocean, and thus decreases ocean pH, as follows:

$$CO_{2\,(aq)} + H_2O \leftrightarrow H_2CO_3 \leftrightarrow HCO_3^{-} + H^{+} \leftrightarrow CO_3^{2-} + 2H^{+} \tag{9.5}$$

On a global scale, the alterations in surface water chemistry from anthropogenic nitrogen and sulfur deposition are only a few percent of the ocean acidification. However, their impacts on seawater chemistry can be much more substantial in coastal waters, i.e. 10–50 % or more of the anthropogenic changes caused by CO_2 uptake near the major source regions and in marginal seas (Doney et al. 2007).

Aluminium toxicity also plays a role in freshwater acidification. Furthermore, aquatic acidification particularly affects sensitive processes like calcification. Ocean acidification enhances growth of autotrophs and reduces fertility and metabolic rates (Hendriks et al. 2010).

Figure 9.1 shows the cause-effect pathway for acidifying impacts. Initial emissions to air will affect a variety of species in the end.

3 Framework

The cause-effect pathway is translated into a series of numerical indicators resulting in a characterisation factor (CF). Following the general framework for emitted atmospheric pollutants (Udo de Haes et al. 2002), a CF for acidification can be expressed as a function of an atmospheric fate factor (FF), a receiving environment exposure factor (XF) and an effect factor (EF).

$$CF_{i,x} = \sum_{j} \left(FF_{i,j,x} \cdot XF_{j,x} \cdot EF_{j} \right) \tag{9.6}$$

The atmospheric fate factor (FF) describes the source-receptor relationship i.e. the atmospheric impact pathway from the emission location i of pollutant x to the corresponding deposition location in the receiving environment j. The receiving environment exposure factor (XF) evaluates the ability of the receiving environment to withstand acidic deposition due to buffer reactions. The effect factor

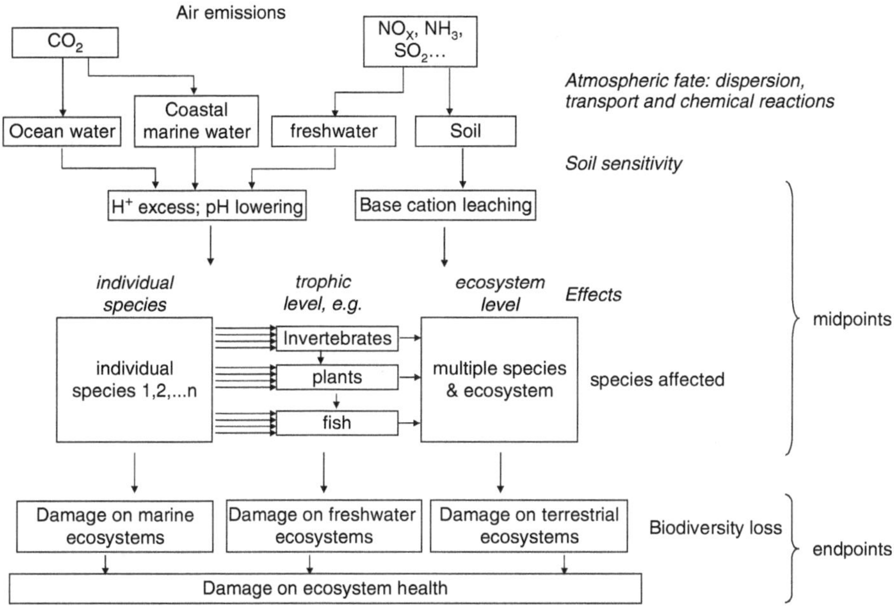

Fig. 9.1 Cause–effect pathway for acidification (Adapted from EU-JRC (2011))

assesses the effects or damage caused by the acid deposits to species, such as changes in biodiversity.

As local conditions, e.g. climate, topography, and biogeochemistry, influence the CFs, spatial differentiation is of importance for acidification. Potting et al. (1998) were the first to indicate spatial differentiation related to acidification impacts. Emission location (i) and receiving environment (j) can be a continent, country, province or grid. They even do not necessarily need to be of the same spatial differentiation level, i.e. an emission change in Europe can be considered and deposition modelled on a grid level (Van Zelm et al. 2007).

Midpoint indicators can be set anywhere along the cause-effect chain, linking, for example, emissions to the deposition of acidifying substances or change in receiving environment sensitivity. For terrestrial acidification, a midpoint indicator is, for example, the accumulated exceedance, which integrates both the area exceeded and amount of exceedance of the critical load (Seppälä et al. 2006). Midpoint indicators express the acidification impacts of atmospheric pollutants on a common scale so they can be compared quantitatively. They are referred to as acidification potentials (AP), and SO_2 is generally used as a reference substance and therefore has an AP of 1. The midpoint characterisation factors can be used in comparison studies to understand which industrial process(es) contribute most to acidification to optimise a system.

Endpoint indicators link the emissions to the consequences of acidification on the ecosystem (e.g. biodiversity or plant productivity loss). Endpoint indicators have been addressed in terms of relative species richness, i.e. the change in species composition in a community. For terrestrial acidification vascular plants are generally included (Goedkoop and Spriensma 1999; Hayashi et al. 2004; Van Zelm et al. 2007). As such, the endpoint level provides information on the ecosystem quality which directly matters to society.

In LCIA, interest generally lies in CFs that can be applied to small emission changes, a so-called marginal approach (Udo de Haes et al. 1999). This approach is based on the assumption that addition of a certain stressor introduces marginal changes to a background situation (Huijbregts et al. 2011). The background situation is defined by 'current' emission and deposition levels. The marginal change is generally set between 1 and 10 % of current emissions (Huijbregts et al. 2000; Krewitt et al. 2001; Potting et al. 1998; Seppälä et al. 2006; Van Zelm et al. 2007; Hettelingh et al. 2005).

4 Methods

LCIA methodologies and methods that address acidification, namely CML 1992 and 2002 (Heijungs et al. 1992; Guinée et al. 2002), Eco-indicator '99 (Goedkoop and Spriensma 1999), MEEuP (Kemna et al. 2005), EDIP97 (Hauschild and Wenzel 1998; Wenzel et al. 1997), EDIP2003 (Potting et al. 1998), TRACI (Norris 2003), EPS (Steen 1999), Accumulated Exceedance (Seppälä et al. 2006), LIME (Hayashi et al. 2004), ReCiPe (Van Zelm et al. 2007), LUCAS (Fréchette-Marleau et al. 2008) and Roy and Azevedo (Azevedo et al. 2013; Roy et al. 2012a, b, 2014a, b) all work with (part of) the above mentioned framework. Table 9.1 provides an overview of the methods, with their main characteristics.

LCIA methods for terrestrial acidification are relatively well developed, while methods for freshwater and marine acidification are lagging.

4.1 Terrestrial Acidification

The first available method, CML 1992 (Heijungs et al. 1992), evaluates the AP in term of H^+ releases. The AP equals the number of H^+ ions that can be formed per molecular weight of a chemical relative to SO_2. EDIP97 also determines the theoretical maximum quantity of hydrogen ions which can be released into the environment compared to SO_2 (Hauschild and Wenzel 1998). Apart from the environmental oxidation, the chemical fate in air and in soil is neglected in these approaches. Kemna et al. (2005) adopted the factors in the MEEuP methodology. Steen (1999), in the EPS methodology, applied the same method, but assuming that a part of the H^+ is not deposited on land.

Potting et al. (1998) were the first to include emission dispersion, subsequent deposition and sensitivity of the receiving environment with help of the integrated

Table 9.1 Acidification methods and their specifications

Method	Ecosystem	Midpoint	Endpoint	Unit	Spatial resolution	Chemicals	Model	Reference
CML 1992	Terrestrial, freshwater	H^+ ions formed	–	kg-SO_2 eq·kg^{-1}	Generic	NO_x, SO_2, NH_3	–	Heijungs et al. (1992)
EDIP97	Terrestrial, freshwater	H^+ ions formed	–	kg-SO_2 eq·kg^{-1}	Generic	NO_x, SO_2, SO_3, NH_3, HCl, HNO_3, H_2SO_4, HF, H_3PO_4, H_2S	–	Wenzel et al. (1997), Hauschild and Wenzel (1998)
Eco-indicator '99	Terrestrial	–	Plant species occurrence due to deposition	m^2·year·kg^{-1}	Europe	NO_x, SO_2, NH_3	SMART2 MOVE	Goedkoop and Spriensma (1999)
MEEuP	Terrestrial	H^+ ions formed	–	kg-SO_2 eq·kg^{-1}	Generic	NO_x, SO_2, NH_3	–	Kemna et al. (2005)
TRACI	Terrestrial	H^+ ions deposited on land	–	kg-SO_2 eq·kg^{-1}	State, North-America and Mexico	NO_x, SO_2, HF, HCl	ASTRAP	Norris (2003)
EDIP 2003	Terrestrial	Affected ecosystem area	–	m^2 -UES·kg^{-1}	Country, Europe	NO_x, SO_2, NH_3	RAINS	Potting et al. (1998)
CML 2002	Terrestrial	Relative risk ratio	–	kg-SO_2 eq·kg^{-1}	Country, Europe	NO_x, SO_2, NH_3	RAINS	Guinée et al. (2002), Huijbregts et al. (2000)
Seppälä	Terrestrial	Accumulative critical load exceedance	–	Eq·$year^{-1}$	Country, Europe	NO_2, SO_2, NH_3	EMEP	Seppälä et al. (2006)
EPS	Terrestrial	H^+ ions deposited on land	–	mole-SO_2 eq·kg^{-1}	Generic	NO_x, SO_2, NH_3, HF, HCl	–	Steen (1999)

(continued)

Table 9.1 (continued)

Method	Ecosystem	Midpoint	Endpoint	Unit	Spatial resolution	Chemicals	Model	Reference
LIME	Terrestrial	H^+ ions deposited on land	Net primary productivity	kg-SO_2 eq·kg^{-1} (mid); kg·$year^{-1}$ (end)	Japan	NO_x, SO_2, NH_3, HCl	Eulerian, effect	Hayashi et al. (2004)
ReCiPe	Terrestrial	Soil acidity change	Plant species occurrence	kg-SO_2 eq·kg^{-1} (mid); m^2·year·kg^{-1}(end)	Europe	NO_x, SO_2, NH_3	EUTREND, SMART2, effect	Van Zelm et al. (2007), Goedkoop et al. (2009)
LUCAS	Freshwater	Critical load exceedance	–	Eq·kg^{-1}	Provinces, Canada	NO_x, SO_2	ASTRAP	Fréchette-Marleau et al. (2008)
Roy/ Azevedo	Terrestrial	Soil acidity change	Plant species occurrence	m^2·year·kg^{-1}	$2° \times 2.5°$, worldwide	NO_x, SO_2, NH_3	GEOS-Chem, Pro-file, effect	Azevedo et al. (2013), Roy et al. (2012a, b, 2014a)
Roy	Freshwater	Lake acidity change	Fish species occurrence	m^2·year·kg^{-1}	$2° \times 2.5°$, worldwide	NO_x, SO_2, NH_3	GEOS-Chem, Exposure, effect	Roy et al. (2014b)

assessment model RAINS that describes the pathways of emissions of SO_2, NO_x, and NH_3. Their method includes 44 emission regions in Europe, and 150 × 150 km receiving grids. Potting et al. (1998) were also the first to use critical loads as a concept in LCIA. Their CF equals the change in area of unprotected ecosystems (UA) following a 10 % reduction of the emission of an acidifying chemical. The UA refers to an area which receives a deposition load that is above its critical load. Their method was implemented in the EDIP 2003 methodology. Krewitt et al. (2001) applied the method of Potting and colleagues, applying a similar atmospheric fate model and assuming a national emission increase of 10 % in each European country. Huijbregts et al. (2000) slightly changed the method by Potting et al., applying the same model, but taking into account the acidifying load above the threshold on top of the area exposed. Their method was implemented in the CML 2002 methodology. The country-specific CFs for Europe from Seppälä et al. (2006) evaluate the absolute acidifying load above the threshold, referring to it as the accumulated exceedance. They applied the Lagrangian model of long-range transport of air pollution EMEP, which is also used in RAINS to assess the atmospheric fate of NO_x, NH_3 and SO_2 emissions.

TRACI (Norris 2003) determined spatial specific midpoint characterisation factors per North-American state, equaling the deposited H^+ ions per kg of emitted substance. They did not go further in the cause-effect chain modelling because of the lack of the North American receiving environment sensitivity databases at that time. Hayashi et al. (2004) did the same for Japanese regions. Source-receptor relationships (SRR) were determined with an empirically calibrated atmospheric chemistry and transport model for SO_2 and NO_x, while SRRs for HCl and NH_3 were created from existing data on emission and deposition from different sources. Hayashi et al. (2004) subsequently determined soil acidity change, assuming the H^+ load completely contributes to pH decrease and the exposure factor equals the change in aluminium (Al^{3+}) concentration on a grid-specific deposition level. They determined effect factors based on the net primary productivity of one Japanese pine tree to finally come up with a Japanese CF.

For the Eco-indicator '99, Goedkoop and Spriensma (1999) applied a dynamic soil model to determine soil acidity (pH) based on deposition data for Europe. This was the first time the biogeochemical processes occurring in the soil were directly taken into account, assuming all acidification changes can lead to effects. Subsequently they based effect factors on target species from a Dutch database on the occurrence of plant species versus pH. Final CFs, however, combine acidification and eutrophication effects. Van Zelm et al. (2007) expanded the method by Goedkoop and Spriensma for the ReCiPe methodology by including a fate modelling step assuming a European generic marginal emission increase and grid-specific deposition increase. An atmospheric fate model, dynamic soil exposure model and effects on plant species occurrence were coupled to obtain a European CF for acidification only. The dynamic model allowed determining CFs for 20, 50, 100 and 500 year time horizons.

Latest developments in terrestrial acidification modelling for LCIA build on the obtained knowledge throughout the years and includes spatially explicit (2° × 2.5°)

atmospheric fate modelling, soil fate modelling and plant species occurrence on a worldwide scale. Atmospheric fate, soil exposure, and plant species effect modelling were evaluated separately (Roy et al. 2012a, b; Azevedo et al. 2013). The CFs evaluate the change in relative species richness present following a change in H^+ concentration in worldwide receiving environments due to a marginal emission change (Roy et al. 2014a). The pH was chosen as soil acidity indicator as most data on species occurrence was available together with soil pH data. From this work, CFs can be derived on a grid-scale or any scale larger (e.g. ecoregions, biomes, countries, or continents).

4.2 *Freshwater Acidification*

Freshwater acidification impacts are typically evaluated with generic characterisation factors (CFs) also applicable to terrestrial acidification (Heijungs et al. 1992; Hauschild and Wenzel 1998). Without spatially explicit midpoint or endpoint CFs, impacts of freshwater acidification are typically disregarded in LCA (EU-JRC 2011).

Fréchette-Marleau et al. (2008) proposed a method for aquatic acidification in Canadian provinces for the LUCAS methodology. CFs were based on the critical load exceedance approach. Atmospheric fate modelling was done with a long-range air emission model, and the aquatic ecosystem was assumed to be exposed to all the chemicals reaching the destination environment.

Recently, Roy et al. (2014b) proposed a set of spatially explicit worldwide endpoint CFs for freshwater lake acidification. These CFs assess the potential impacts of acidifying emissions to lakes. The spatially explicit ($2° \times 2.5°$) modelling includes atmospheric fate, lake exposure (deposition to H^+ concentration), and effects on relative fish species richness on a worldwide scale.

4.3 *Marine Acidification*

No publications were found in the peer reviewed literature reporting on attempts to include marine acidification in LCIA, caused by CO_2 uptake in the seas and oceans.

5 Uncertainty

Three different types of uncertainty are particularly relevant in LCIA modelling: (1) uncertainty due to lack of knowledge of the 'true' value of a model input parameter (parameter uncertainty), (2) uncertainty caused by arbitrary choices into a model (decision rule uncertainty), and (3) uncertainty caused by the loss of information resulting from the simplification of reality by the use of models (model structure uncertainty).

Parameter uncertainties were only quantitatively estimated very recently. Roy et al. (2014a, b) showed the influence, through Monte Carlo simulation, of the combination of fate, exposure and effects to the overall uncertainty of endpoint CFs. In a 95 % confidence interval, they showed that for terrestrial as well as freshwater acidification the uncertainty in the CF is dominated by uncertainty in the effect factor. This is due to the low availability of data on species occurrence and acidity on a worldwide scale. Whereas fate and exposure can quite adequately be modelled nowadays, effect modelling is still in an early phase of development.

Decision rule uncertainties related to acidification that we identified are the time horizon and the species protection level, and the adaptation of ecosystems/species to acidification and the dispersal or migration of species (De Schryver et al. 2009; Goedkoop et al. 2009). Time horizon is important as the acidification in soil can be delayed due to biogeochemical processes, which was illustrated by Van Zelm et al. (2007). Although studies looked at either including effects to all species (Hayashi et al. 2004; Van Zelm et al. 2007; Azevedo et al. 2013), or target species only (Goedkoop and Spriensma 1999), no acidification study so far looked at the difference between the two types of effect estimations. This aspect, as well as possible adaptation and migration of species still need more investigation.

Regarding model structure uncertainty, it can be seen that throughout the years efforts have been made to decrease this uncertainty type when new knowledge on the acidification cause-effect pathway was gained. LCIA acidification methods have become more sophisticated, related to spatial explicitness and the inclusion of fate, exposure as well as effect modelling.

6 Future Trends

The main challenges identified in LCIA modelling of acidification are (i) fully including marine and freshwater acidification, and (ii) reducing uncertainty in effect factor estimations.

For aquatic acidification, only two methods are available, one on midpoint addressing freshwater systems and one on endpoint addressing freshwater lakes only, neglecting the acidification of other freshwater environments. Improvements needs of the midpoint method relate to the low model resolution and the lack of temporal differentiation (Fréchette-Marleau et al. 2008). A number of shortcomings were identified with respect to the derived endpoint CFs for lake freshwaters (Roy et al. 2014b). The main issues that need to be addressed in the future are (i) the current assumption of an even mixing within the lake, while lakes pH are known to be highly heterogeneous, and (ii) effect factor derivation. Currently, species richness was estimated from fish species only, with region specific regressions that showed poor correlation (R^2 between 0.01 and 0.42) with available data.

To address ocean acidification, a carbon cycle model can be used to calculate fate factors, such as the model described by Montenegro et al. (2007). Subsequently, effect factors can be determined in the same way as for terrestrial acidification, i.e. based on occurrence of species following an acidification indicator gradient (e.g. pH).

To decrease uncertainty in effect factor estimation, more insight needs to be gained in the stressor-response curves for many regions of the world. There is a large number of parameters that influences species richness such as temperature, environmental area, compartment depth (soil/lake), altitude, and pH. The influence of all these factors related to world regions needs further exploration. Moreover, relative species richness as such is not the only indicator related to biodiversity. Further research should focus on, e.g., exploring a stressor-response curve for a specific region for which many data are available on a number of different species, from genes level to ecosystem level (Curran et al. 2011), and on various effect types, e.g. growth, reproduction, lethality. This way, a better balanced and focused upscaling can take place geographically, and at a species level that is relevant for use in LCIA.

References

Azevedo LB, Van Zelm R, Hendriks AJ, Bobbink R, Huijbregts MAJ (2013) Quantitative effects of soil pH on plant species richness: a global assessment. Environ Pollut 174:10–15

Blaser P, Zysset M, Zimmermann S, Luster J (1999) Soil acidification in Southern Switzerland between 1987 and 1997: a case study based on the critical load concept. Environ Sci Technol 33:2383–2389

Bouwman AF, Van Vuuren DP, Derwent RG, Posch M (2002) A global analysis of acidification and eutrophication of terrestrial ecosystems. Water Air Soil Pollut 141(1–4):349–382

Caldeira K, Wickett ME (2003) Anthropogenic carbon and ocean pH. Nature 425(6956):365–365

Clair TA, Dennis IF, Scruton DA, Gilliss M (2007) Freshwater acidification research in Atlantic Canada: a review of results and predictions for the future. Environ Rev 15:153–167

Curran M, de Baan L, De Schryver AM, van Zelm R, Hellweg S, Koellner T, Sonnemann G, Huijbregts MAJ (2011) Toward meaningful end points of biodiversity in life cycle assessment. Environ Sci Technol 45(1):70–79

Dangles O, Malmqvist B, Laudon H (2004) Naturally acid freshwater ecosystems are diverse and functional: evidence from boreal streams. Oikos 104(1):149–155

De Schryver AM, Brakkee KW, Goedkoop MJ, Huijbregts MAJ (2009) Characterization factors for global warming in life cycle assessment based on damages to humans and ecosystems. Environ Sci Technol 43(6):1689–1695

Doney SC, Mahowald N, Lima I, Feely RA, Mackenzie FT, Lamarque JF, Rasch PJ (2007) Impact of anthropogenic atmospheric nitrogen and sulfur deposition on ocean acidification and the inorganic carbon system. Proc Natl Acad Sci U S A 104(37):14580–14585

EU-JRC (2011) International Reference Life Cycle Data System (ILCD) handbook- recommendations for life cycle impact assessment in the European context. Volume 1st edn. November 2011. European Commission-Joint Research Centre, Institute for Environment and Sustainability, Luxemburg

Falkengren-Grerup U (1986) Soil acidification and vegetation changes in deciduous forest in southern Sweden. Oecologia 70(3):339–347

Fréchette-Marleau S, Bécaert V, Margni M, Samson R, Deschênes L (2008) Evaluating the variability of aquatic acidification and photochemical ozone formation characterization factors for Canadian emissions. Int J Life Cycle Assess 13(7):593–604

Goedkoop MJ, Spriensma R (1999) The Eco-indicator '99: a damage-oriented method for life-cycle impact assessment. Ministry of Housing, Spatial Planning, and Environment, The Hague

Goedkoop M, Huijbregts MAJ, Heijungs R, De Schryver A, Struijs J, Van Zelm R (2009) ReCiPe 2008: a life cycle impact assessment method which comprises harmonised category indicators at the midpoint and the endpoint level. Ministry of Housing, Spatial Planning and the Environment (VROM), Amersfoort, The Netherlands

Guinée JBE, Gorrée M, Heijungs R, Huppes G, Kleijn R, De Koning A, Van Oers L, Wegener Sleeswijk A, Suh S, Udo de Haes HA, De Bruijn JA, Van Duin R, Huijbregts MAJ (2002) Handbook on life cycle assessment: operational guide to the ISO standards. SeriesL eco-efficiency in industry and science. Kluwer Academic Publishers, Dordrecht

Hauschild M, Wenzel H (1998) Environmental assessment of products vol 2: scientific background. Chapman & Hall/Kluwer Academic Publishers, London/Hingham, 1997. ISBN 0-412-80810-2

Hayashi K, Okazaki M, Itsubo N, Inaba A (2004) Development of damage function of acidification for terrestrial ecosystems based on the effect of aluminium toxicity on net primary production. Int J Life Cycle Assess 9:13–22

Heijungs R, Guinee JB, Huppes G, Lankreijer RM, Udo de Haes HA, Wegener Sleeswijk A, Ansems AMM, Eggels PG, Van Duin R, De Goede HP (1992) Environmental life cycle assessment of products: guide and backgrounds. Centre of Environmental Science, University, Leiden, Leiden

Hendriks IE, Duarte CM, Alvarez M (2010) Vulnerability of marine biodiversity to ocean acidification: a meta-analysis. Estuar Coast Shelf Sci 86(2):157–164

Hettelingh JP, Posch M, Potting J (2005) Country-dependent characterization factors for acidification in Europe; a critical evaluation. Int J Life Cycle Assess 10:177–183

Huijbregts MAJ, Schöpp W, Verkuijlen E, Heijungs R, Reijnders L (2000) Spatially explicit characterization of acidifying and eutrophying air pollution in life-cycle assessment. J Ind Ecol 4:75–92

Huijbregts MAJ, Hellweg S, Hertwich EG (2011) Do we need a paradigm shift in life cycle impact assessment? Environ Sci Technol 45:3833–3834

Kemna R, Van Elburg M, Li W, Van Holsteijn R (2005) MEEuP methodology report, final. 28 Nov 2005. VHK for European Commission, Brussels

Krewitt W, Trukenmüller A, Bachmann TM, Heck T (2001) Country-specific damage factors for air pollutants. Int J Life Cycle Assess 6:199–210

Montenegro A, Brovkin V, Eby M, Archer D, Weaver AJ (2007) Long term fate of anthropogenic carbon. Geophys Res Lett 34(19):L19707

Nilsson J, Grennfelt P (1988) Report from a workshop held at Skokloster, Sweden, 19–24 March 1988. NORD Report, Copenhagen, 418 pp

Norris GA (2003) Impact characterization in the tool for the reduction and assessment of chemical and other environmental impacts; methods for acidification, eutrophication and ozone formation. J Ind Ecol 6(3–4):79–101

Posch M, De Smet PAM, Hettelingh JP, Downing RJ (2001) Modelling and mapping of critical thresholds in Europe: status report 2001. National Institute for Public Health and the Environment, Bilthoven

Poschenrieder C, Gunse B, Corrales I, Barcelo J (2008) A glance into aluminium toxicity and resistance in plants. Sci Total Environ 400(1–3):356–368

Potting J, Schöpp W, Blok K, Hauschild M (1998) Site-dependent life-cycle impact assessment of acidification. J Ind Ecol 2(2):63–87

Psenner R (1994) Environmental impacts on fresh-waters: acidification as a global problem. Sci Total Environ 143(1):53–61

Roem WJ, Berendse F (2000) Soil acidity and nutrient supply ratio as possible factors determining changes in plant species diversity in grassland and heathland communities. Biol Conserv 92 (2):151–161

Roy PO, Deschenes L, Margni M (2012a) Life cycle impact assessment of terrestrial acidification: modeling spatially explicit soil sensitivity at the global scale. Environ Sci Technol 46 (15):8270–8278

Roy PO, Huijbregts M, Deschenes L, Margni M (2012b) Spatially-differentiated atmospheric source-receptor relationships for nitrogen oxides, sulfur oxides and ammonia emissions at the global scale for life cycle impact assessment. Atmos Environ 62:74–81

Roy PO, Azevedo LB, Margni M, Van Zelm R, Deschênes L, Huijbregts MAJ (2014a) Characterization factors for terrestrial acidification at the global scale: a systematic analysis of spatial variability and uncertainty. Sci Tot Environ 500:270–276

Roy PO, Deschênes L, Margni M (2014b) Uncertainty and spatial variability in characterization factors for aquatic acidification at the global scale. Int J Life Cycle Assess 19:882–890

Scholz F, Reck S (1977) Effects of acids on forest trees as measured by titration in vitro, inheritance of buffering capacity in Picea abies. Water Air Soil Pollut 8(1):41–45

Seppälä J, Posch M, Johansson M, Hettelingh JP (2006) Country-dependent characterisation factors for acidification and terrestrial eutrophication based on accumulated exceedance as an impact category indicator. Int J Life Cycle Assess 11(6):403–416

Steen B (1999) A systematic approach to environmental priority strategies in product development (EPS). Version 2000 – models and data of the default method. Chalmers University of Technology, Göteborg

Udo de Haes HA, Jolliet O, Finnveden G, Hauschild M, Krewitt W, Müller-Wenk R (1999) Best available practice regarding impact categories and category indicators in life cycle impact assessment. Int J Life Cycle Assess 4:66–74

Udo de Haes HA, Finnveden G, Goedkoop MJ, Hauschild M, Hertwich EG, Hofstetter P, Jolliet O, Klöpfer W, Krewitt W, Lindeijer E, Müller-Wenk R, Olsen SI, Pennington DW, Potting J, Steen B (2002) Life-cycle impact assessment: striving towards best practice. SETAC, Pensacola

Van Zelm R, Huijbregts MAJ, Van Jaarsveld JA, Reinds GJ, De Zwart D, Struijs J, Van de Meent D (2007) Time horizon dependent characterization factors for acidification in life-cycle assessment based on forest plant species occurrence in Europe. Environ Sci Technol 41:922–927

Wenzel H, Hauschild M, Alting L (1997) Environmental assessment of products vol 1: methodology, tools and case studies in product development. Chapman & Hall/Kluwer Academic Publishers, London/Hingham, 1997. ISBN 0 412 80800

Zvereva EL, Toivonen E, Kozlov MV (2008) Changes in species richness of vascular plants under the impact of air pollution: a global perspective. Global Ecol Biogeogr 17(3):305–319

Chapter 10
Eutrophication

Andrew D. Henderson

Abstract Anthropogenic increases in nitrogen and phosphorus inputs to terrestrial and aquatic ecosystems have driven increases in eutrophication, the occurrence of ecosystem changes due to over-supply of nutrients. Eutrophic water bodies exhibit changes in species composition that often include algal blooms and oxygen depletion, with occasionally arresting images of fish kills or dead zones. Though dramatic and subtle consequences of eutrophication itself have been described for over 100 years, understanding of nutrients as the main drivers for this phenomenon is more recent. Modelling nutrient fate has reached a basic level of operability, with a general rule that freshwaters are limited in phosphorus (and hence respond to its addition), and terrestrial and marine systems are nitrogen limited. However, understanding of ecosystems responses such as species shifts or changes in primary productivity is still growing. Future work should incorporate more comprehensive metrics to quantify impacts of eutrophication on ecosystems – and the human systems that depend on them.

Keywords Freshwater eutrophication • LCA • LCIA • Life cycle assessment • Life cycle impact assessment marine eutrophication • Nutrient enrichment • Terrestrial eutrophication

1 Introduction

1.1 Historical Perspective

Eutrophication is the result of supplying nutrients to ecosystems in excess of natural rates, which may drive a cascade of changes, including alterations in species composition, biomass, or productivity. Nutrient cycling varies across ecosystems, and eutrophication does naturally occur to some degree, but anthropogenic

A.D. Henderson (✉)
Division of Epidemiology, Human Genetics and Environmental Sciences, School of Public Health, The University of Texas Health Science Center at Houston, 1200 Herman Pressler, RAS W634, Houston, TX 77030, USA
e-mail: andrew.d.henderson@uth.tmc.edu

M.Z. Hauschild, M.A.J. Huijbregts (eds.), *Life Cycle Impact Assessment*,
LCA Compendium – The Complete World of Life Cycle Assessment,
DOI 10.1007/978-94-017-9744-3_10

emissions of nutrients and organic matter have increased nutrient cycling, disturbing the natural dynamic (Bouwman et al. 2009). This increased cycling, and often over-supply, of nutrients can directly affect humans: fish kills or red tides in aquatic systems may have direct health effects, and changing crop yields in terrestrial systems affects food production. Algal blooms continue to be of concern in many areas (e.g., Liu et al. 2013). Comprehensive overviews of nutrient-related ecosystem alterations are provided by Schindler (2006) and Smith et al. (1999).

The response of terrestrial systems to nutrient supply was documented in the mid-nineteenth century (von Liebig 1855). Excess nutrients have been a concern for over a century; a qualitative description of lakes as eutrophied, based on hypolimnetic oxygen depletion, the changed occurrence of benthic macroinvertebrates, and visual appearance of water, was first published in the early 1900s (Weber 1907). However, attention intensified in the middle of the last century (NRC 1992; OECD 1982), largely due to increased nitrogenous atmospheric emissions and synthetic fertilizer use, and the corresponding increase in the visibility of eutrophication and its impacts (MEA 2005; Tilman 1999; Vitousek et al. 1997). There was uncertainty about the causes of aquatic eutrophication into the 1960s, with a variety of substances, including phosphorus (P) and nitrogen (N), carbon, vitamins, amino acids, and trace elements identified as possible causes (NAS 1969). Even after phosphorus was identified as a critical substance in eutrophication (Vollenweider 1968), it took time – and demonstration via whole-lake manipulation – to overcome counter-claims and resistance from the scientific community and the soap and detergent industries (Schindler 2006).

Initial efforts to incorporate eutrophication into LCIA were largely based on biomass production (Lindfors et al. 1995); for aquatic eutrophication, this was connected to nutrient input via the Redfield ratio, the average stoichiometric ratios of carbon, nitrogen, and phosphorus(C_{106}:N_{16}:P_1) found in plankton (Redfield 1934). This ratio indicates that algae and other aquatic organisms require 16 mol of N for every one mole of P. A lack of one of these nutrients can limit biomass production; such systems are said to be N- or P-limited. A SETAC working group on LCIA noted that nearly all LCIA aquatic eutrophication methods based their characterisation factors on the Redfield ratio, with the corresponding impact indicator being algal growth (Udo de Haes et al. 2002). Alternately, some methods considered oxygen depletion in water, which has the advantage of providing a direct way to include organic matter, linking these to oxygen depletion via the biological or chemical oxygen demand (BOD or COD) (e.g., Guinée et al. 2002). The oxygen-consuming degradation of organic matter by bacteria forms the basis for the BOD; COD is a measure of all substances that can be oxidised. Newer LCIA methods have begun to move beyond the Redfield ratio and algae to consider changes in species composition, which itself is an interim step in the progression towards more complete assessments.

2 Principles of Characterisation Modelling

Modelling eutrophication requires capturing interrelationships between hydrology, ecosystem biology, microbiology, and chemistry. Modelling choices reflect differences both in scientific opinion about the cause-effect chain and in societal values about what merits protection (Hertwich et al. 2000; Udo de Haes et al. 2002; UNEP/SETAC 2005). Capturing complexity while creating a usable approach is a thread that runs through eutrophication modelling in LCIA.

One modelling choice regards the use of average or marginal impacts: if an LCIA is being conducted to assess a change or comparison, then marginal modelling is recommended; average impacts are suitable for information-gathering (Udo de Haes et al. 2002). To date, LCIA models for eutrophication have relied on marginal changes; e.g., Struijs et al. (2010a) used a 1 % increase above 1995 emission levels. However, this conceptual framework does discount impacts when the marginal increase occurs in an already-stressed receiving area (Huijbregts et al. 2011). Another modelling challenge, for both the average and marginal approaches, is the possibility of non-linear responses to nutrient loading. Loading may reduce ecosystem resilience; beyond critical levels, ecosystems may undergo radical shifts to alternate, metastable states (Scheffer et al. 2001).

2.1 *Criteria for Good Characterisation Models*

The variety of modelling approaches for eutrophication and other categories in LCIA has led to ongoing efforts to create a framework to objectively compare models (e.g., Udo de Haes et al. 2002; Margni et al. 2008). The most comprehensive effort to date has been the International Reference Life Cycle Data System (ILCD) (EC-JRC 2010a). In the ILCD framework, scientifically sound LCIA eutrophication models include a complete of scope, environmental relevance, and scientific robustness. These criteria require accurately capturing the cause-effect chain, to the extent made possible by current knowledge. For aquatic transport of eutrophying substances, major transport phenomena include precipitation and sedimentation for N and P, as well as oxidation, specifically denitrification, for N. For atmospheric emissions, these include oxidation and deposition. At the damage level, sound models should include discrimination between receiving areas based on sensitivity, possibly including a critical level, and a dose-response relationship (EC-JRC 2011). The application of these criteria to eutrophication models is discussed in the following sections.

3 Impact Pathway and Affected Areas of Protection

Damage categories represent changes to those components of the environment that are valued by human society (UNEP/SETAC 2005); however, valuing and quantifying ecosystem qualities is challenging, as society may value components ranging from the subjective (e.g., aesthetic quality) to the concrete (e.g., fish production for food supply). The Natural Environment Area of Protection, as defined by the UNEP/SETAC working group, encompasses both ends of this spectrum. The intrinsic values of the existence and stability of the environment is captured via biodiversity, which can be measured as a species loss. The functional values of natural resources are captured via net primary productivity, which can be measured financially (Margni et al. 2008). Some methods have considered human health, since direct human impacts are possible: algal blooms may be toxic to humans and have resulted in beach closures (Anderson 1989; Paerl et al. 2001). However, human health has rarely been considered directly for eutrophication in LCIA. Developing a set of metrics to fully capture the myriad aspects of the biotic environment is ongoing (see Sect. 8).

The addition of nutrients or organic matter to terrestrial or aquatic ecosystems can affect cell synthesis or energy supply. Organisms that can take advantage of changing inputs of either nutrients or organic matter will be able to outcompete other species, resulting in dynamic changes to the steady-state ecosystem composition, changing biodiversity and productivity. Figure 10.1 provides an overview of the terrestrial and aquatic eutrophication processes, which are discussed below.

3.1 Terrestrial Eutrophication

To date, LCIA has focused on changes to terrestrial vegetation, as interactions of other components of the ecosystem with changing nutrient cycling and changing plant communities are not yet well understood. Plants in terrestrial systems are usually nitrogen limited; i.e., there is sufficient P for growth, but not adequate N for the typical cellular nutrient ratio of N and P (see Sect. 1.1) (Grouzet et al. 2000; Hornung et al. 1994; Nilsson and Grennfelt 1988). The terrestrial N:P nutrient ratio, as well as the freshwater ratio, are similar to the Redfield ratio, although there are differences in the ratio of carbon to these nutrients (Elser et al. 2000; McGroddy et al. 2004). Excess nitrogen can change the structure and function of terrestrial, N-limited ecosystems by favoring a (typically) limited number of N-adapted species. This may change tolerance to disease or other stressors (e.g., drought, frost), resulting in overall biodiversity and productivity changes.

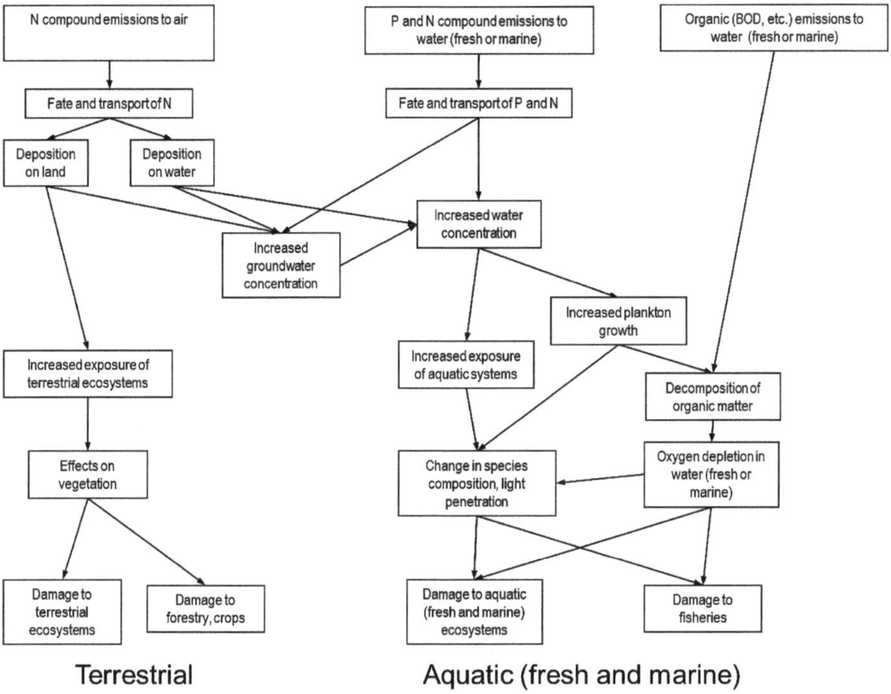

Fig. 10.1 Flow diagram of eutrophication impact pathway (Adapted from ILCD EC-JRC (2010b))

3.2 Aquatic Eutrophication

Increases of nitrogen and phosphorus, or inputs of respirable organic matter, in aquatic ecosystems can likewise change the structure and function of communities of plants and animals through a cause-effect chain that involves organisms in the benthos and water column. Excess nutrients first stimulate the growth of phytoplankton, increasing turbidity. This, in turn, affects plants in the light-dependent photic zone, as well as limiting predatory success of fish. The fish community shifts away from piscivorious toward zooplanktivorious species, removing zooplankton and favoring population increases in their prey, phytoplankton. Microbiological respiration of dead phytoplankton or other species (increased both due to direct nutrient input as well as the indirect ecosystem community changes) may lead to hypoxic conditions (i.e., low levels of dissolved oxygen in the water column), driving further changes (Kristensen and Hansen 1994). Such changes can also be caused by direct addition of excess organic matter (BOD or COD). To a certain degree, dissolved oxygen may be replenished via exchange with atmospheric oxygen at the lake surface.

However, some lakes undergo a seasonal stratification, limiting oxygen replenishment. The spring and summer warming of the surface layer, the epilimnion, creates a sharp temperature gradient, limiting mixing with the lower layer, the hypolimnion. In this circumstance, decomposition of organic matter that has settled to the lake bottom consumes oxygen which is not replenished, potentially creating hypoxic or anaerobic conditions in the hypolimnion (Boehrer and Schultze 2008).

Low-nutrient water inputs through lakes, estuaries, or coastal zones, can serve to flush nutrients. Point-source emissions to well-mixed lakes can be modeled, but for diffuse nutrient sources in a watershed, the relationship between flow, nutrients, and eutrophication is less clear and can be difficult to model (Schindler 2006). The assimilation and movement of phosphorus through water bodies varies spatially and temporally (Withers and Jarvie 2008). Nitrogen, as it exists in multiple redox states, some of which are volatile, is more complex to model. Among other complicating factors, previously deposited P and N can be re-introduced to the water column during periods of anoxia (Levine et al. 1986; Mortimer 1942).

3.3 Freshwater and Marine Systems

It is important to distinguish fresh and marine waters; marine water bodies' salinity is not affected by freshwater inputs. Current understanding is that freshwaters are often—but not always—limited by phosphorus (Carpenter et al. 1998; Schindler 1977). Since the mid-twentieth century investigations that began to identify nutrients as drivers of eutrophication, it has been known that nitrogen may also limit freshwater productivity (Schindler 2006). Indeed, nitrogen, and even iron, may be co-limiting over time scales relevant to biological cycles, though it is still likely that phosphorus is the controlling, i.e. limiting, nutrient over scales of multiple years (Schindler 2006; Sterner 2008). If nitrogen is also limited in a water body, species capable of fixing N from the atmosphere (e.g., cyanobacteria) may be favored, reducing the extent to which N limits productivity (Smith 1983). In addition, N-limited lakes receiving anthropogenic nitrogen inputs may be eventually transformed to P-limited (Goldman 1988).

In contrast, productivity and eutrophication of marine waters has a more complex relationship with the two major nutrients than do freshwater systems, as hydrodynamics and trace elements play a more significant role (Grouzet et al. 2000). As a general rule, though, marine systems are typically N-limited (Jørgensen and Richardson 1996), and this provides a useful starting point for LCIA. While coastal zones and the upper layers of the ocean typically have greater biodiversity and productivity than other zones, zone-specific modelling of oceans has proven difficult. However, zones are taken into account in the cross-category biodiversity comparisons of the ReCiPe method (Goedkoop et al. 2009).

The limiting nutrient concept should be interpreted cautiously; as noted above, nutrient relationships in a water body can exhibit long-term changes, as well as change on sub-annual time scales. However, P limitation of freshwater and N limitation of marine systems provides a convenient foundation upon which to build LCIA models.

4 Contributing Substances (Classification)

As shown in Fig. 10.1, the major categories of emissions driving eutrophication are nutrients (nitrogen and phosphorus) and organic matter emitted to water. For aquatic effects, LCIA methods can differ in their approaches to capturing these ecosystem changes, with some endpoints focusing on nutrient enrichment, and others on oxygen depletion. Table 10.1 presents a summary of LCIA characterisation methods for eutrophication.

Consistent with the development of ecological understanding of eutrophication as driven by the supply of limiting nutrients, many LCIA methods primarily consider emissions of phosphorus and nitrogen compounds. Life cycle inventory (LCI) data may include a variety of forms of P and N; typical inventory substances are elemental P and N, phosphate (PO_4^{3-}), NH_3 or NH_4^+ (ammonia and ammonium), aqueous NO_3^- (nitrate), and gaseous nitrogen oxides (NO_x, representing the sum of NO and NO_2).

Since the addition of limiting nutrients may lead to oxygen depletion (due to microbial respiration of excess biomass), some LCIA approaches include substances that lead directly to oxygen demand, either biological (BOD) or chemical (COD) (see Fig. 10.1). The respiration of organic matter by bacteria consumes oxygen and forms the basis for the BOD metric; compounds not easily respired by microbial communities are captured in the COD.

LCIA methods that assess oxygen depletion, e.g., LIME (Itsubo and Inaba 2003), include organic matter (BOD and COD) as contributing substances. Although this biological material will contain some P and N, the mechanisms of action of these organic substances are distinct from those of the limiting nutrients, so double-counting is not a concern. The methods that do measure oxygen depletion therefore require an approach to translate P and N emissions to changes in oxygen levels.

5 Scale and Variability (Spatial and Temporal)

5.1 Scale

Atmospheric emissions of NO_x and NH_3 can be transported over a continental scale; e.g., the established RAINS and EMEP models for Europe (Alcamo et al. 1990; EEA 2009), or ASTRAP for North America (Shannon and Clark 1985; Toffoleto et al. 2007), affecting broad terrestrial regions, as well as being deposited on aquatic systems, though the latter is a minor source of aquatic nutrient inputs, given the small fraction of surface area occupied by freshwater relative to land.

In the case of freshwater emissions, the region of impact is defined by the downstream path of a receiving water body and the biogeochemical processes affecting nutrient transport. Some fractions of P emissions may travel ~1,000 km

Table 10.1 Summary table of LCIA eutrophication methods

Method	Impacted ecosystem	Midpoint	Endpoint	Unit (per kg emission)	Coverage (spatial, temporal)	Substances	Underlying model	Notes	Reference
CML 1992	Terrestrial, aquatic (combined)	Nutrification potential	–	PO_4^{3-}-eq.	Constant	N, NO, NO_2, NO_x, NO_3–, NH_4^+, P, PO_4^{3-}, COD	–	Potential based on redfield ratio	Heijungs et al. (1992)
CML 2002	Terrestrial, aquatic (combined)	Eutrophication potential	–	PO_4^{3-}eq.	Generic	NH_3, NH_4^+, NO_3^-, HNO_3, N, NO_2, NO, NO_x, PO_4^{3-}, $H_3PO_4^{3-}$, P, P_2O_5, COD	–	European country factors available from Huijbregts and Seppälä (2000)	Guinée et al. (2002)
Eco Scarcity 2006		Ecopoints (EP) (assigned based on exceedance of critical load)		EP	Switzerland (+ regional waters)	Air: NO_x, NH_3; Water: N and P substances		Critical flow defined by political target	Frischknecht et al. (2009)
Eco-Indicator 99	Terrestrial		Affected target plants – probability of disappearance	PDF·m^2·year	Europe	NO_x, NH_3 (to air)	SMART2 (Kros et al. 1995)	Based on change in Dutch species (Latour et al. 1997), extrapolated to Europe. Acidification combined with eutrophication	Goedkoop and Spriensma (2000)
EDIP 97	Aquatic, terrestrial (aggregated)	Eutrophication potential		NO_3^-, N, or P-eq.;	generic	NO_3^-, NO_2, NO_2^-, NO_x, N_2O, NO, NH_3, CN-, total N, PO_4^{3-}, $P_2O_7^{2-}$, total P		factors can consider P or N-limited systems; uses Redfield ratio	Hauschild and Wenzel (1998), Wenzel et al. (1997)

EDIP 2003	Aquatic, terrestrial	Eutrophication potential	Ecosystem area that becomes unprotected (UES)	Aquatic: NO_3^-, N, or P-eq.; Terrestrial: m^2 UES (terrestrial)	Aquatic: European country; Terrestrial: Europe (44 regions or generic; 1990 or 2010 factors)	See EDIP 97	Aquatic: CARMEN; Terrestrial: RAINS	Aquatic adds fate factor to EDIP 97 site-generic factors	Potting and Hauschild (2005)
EPS	Aquatic		Fish and meat production (air emissions); Normalised EXtinction of species (NEX) (water emissions)	kg fish production; kg NEX	Global	To air: NO_x, N_2O, NH_3, N-tot; to water: BOD, COD, N-total, P-total		Convert BOD to O_2 demand using Lindfors et al. (1995)	Steen (1999a)
LIME	Marine	Dissolved oxygen concentration	Biodiversity, primary production, and economic value		Japan (specific bays modeled)	NO_x, NH_3		Limited documentation	Itsubo and Inaba (2003)
LUCAS	Aquatic (+ groundwater), terrestrial	Terrestrial: amount of emitted N that reaches ecosystem		kg N-eq.	Canada (15 ecozones)	N and P	ASTRAP (air), CARMEN (water)	Adapts EDIP 2003 to Canadian context	Toffoleto et al. (2007)
MEEuP	Aquatic	Eutrophication potential		PO_4-eq.	generic	N, NO_3-, NH_4^+, P, PO_4^{3-}, P_2O_5, COD, BOD, SS, DOC, TOC NO_x (to water)	–	Substance weighting based on European directives (Vermeire et al. 1997, 2007)	Kemna et al. (2005)
ReCiPe	Freshwater and marine	Eutrophication potential	Species disappearance	Freshwater: P-eq. (mid), PDF·year·m^3 (end): Marine: N-eq. (mid)	Europe	N and P substances	CARMEN (water), EUTREND (air) (van Jaarsveld 1995)		Goedkoop et al. (2009)
TRACI	Aquatic	Eutrophication potential		kg N-eq. or kg PO_4^{3-}eq.	US (state resolution)	N and P substances, BOD, COD	ASTRAP (air), exorheic fraction (water)	Eutrophication potential based on Redfield ratio	Bare et al. (2003)

Impact 2002+ (Jolliet et al. 2003) uses terrestrial eutrophication from Eco-Indicator 99 and aquatic eutrophication from CML 2002

(Helmes et al. 2012). For marine emissions, the scale of impact is related to the extent to which receiving bodies create partially-enclosed systems with limited mixing with the larger ocean. Scales of impact can be quite large, e.g., zones of hypoxia up to 80,000 km^2 in the Baltic Sea (Diaz 2001; Hansson et al. 2009).

The most important spatial distinction is between P and N-limited water bodies. Stoichiometric (e.g., NH_3 vs. NO_3^-) and bioavailability differences between eutrophying compounds are minor relative to correctly capturing the limiting nutrient or possible spatial transport differences. The exclusion of the latter may diminish the relevance of an impact assessment (Potting and Blok 1994), and inclusion of spatial differentiation was identified as a major challenge for LCIA early on (Potting 2000). Finnveden and Potting (1999) had begun working on spatial eutrophication in the early 1990s (UNEP 2003). Some of the first published spatially-differentiated transport factors for eutrophication were country-specific estimates of marine deposition of air emissions of NH_3 and NO_x, and these were included in a multi-scale (country, Europe, world) assessment that included runoff and leaching from agricultural lands (Huijbregts et al. 2000; Huijbregts and Seppälä 2001).

Several studies have pointed to the possible variation, by orders of magnitude, in transport between sources and receptors. Transport differences between European countries for airborne nitrogen compounds can be up to three orders of magnitude (Posch et al. 2008; Potting et al. 1998b). At the impact level, also comparing European countries, the eutrophication potentials of nitrogen air emissions were found to vary by up to 1.5 orders of magnitude (up to 3.5 orders of magnitude for acidification) (Huijbregts et al. 2000). Differences in aquatic transport between European countries and US states, according to previous work, were less than one order of magnitude (Norris 2003; Potting and Hauschild 2005). However, for P transport, recent modelling has suggested possible variations of 3 orders of magnitude between US states (Helmes et al. 2012).

At small spatial scales, variation in the transport and eventual impacts of P and N can therefore be significant. On the aquatic side, emissions to rivers and lakes have different fates and impacts; this is a level of detail not captured in impact models nor inventory data. UNEP/SETAC working groups formalised the call for archetypical situations (Margni et al. 2008), which represent deviations from the default, generic situation and could be used when spatial differences cause a variation above some threshold factor (e.g., 2–10, depending on the study). Margni et al. (2008) recommended continental-level resolution as preliminary step. More nuanced divisions are also possible, provided that inventory data are sufficiently detailed: for freshwater transport, the presence or absence of large lakes downstream from emission sources is a possible archetypical division point (Helmes et al. 2012).

Comparing freshwater bodies or terrestrial areas, there can be differences driven by variation in climate, species composition, underlying geology, or previous environmental stresses, among others. Many of these factors control existing levels of nutrients; this may affect the limiting nutrient. However, within one nutrient limitation regime, the response of a water body to nutrient input may vary depending on existing levels of that nutrient. As a threshold concept, this presented challenges to developers of LCIA (e.g., UNEP (2003), for the existence of a cutoff value can imply the existence of an infinite sink for a substance in the environment. Work by Struijs et al. (2010b) assumed a threshold of 0.3 mg/L as indicative of excess human nutrient inputs. An alternative approach has also differentiated response levels, but with impacts occurring at all existing P levels – i.e., without a threshold concentration (Payet 2006).

The time scales over which eutrophying emissions reach and cause impact in a receiving location vary by compound and the emission compartment. The time frame over which first-order impacts occur is generally rapid (i.e., proportional to nutrient uptake during the growing season), but second-order impacts are more complex and varied (e.g., remobilisation of previously sequestered nutrients). Atmospheric emissions of NO_x and NH_3 can be transported over a continental scale; however, they have atmospheric residence times on the order of hours to days (Galloway 2003). Once deposited to terrestrial systems, reactive nitrogen can persist for time scales ranging up to centuries in unmanaged forests (Galloway 2003).

For freshwater emissions, the hydrological cycle tends to move nutrients downstream relatively efficiently, with natural and man-made reservoirs delaying transport. For phosphorus, impacts happen throughout a river system; time scales can range up to years for areas that are upstream of large water bodies (Helmes et al. 2012). For nitrogen, impacts in the coastal zones are delayed while nitrogen is transported; this transport is largely tied to river transport, though wetlands may remove nitrogen prior to reaching marine systems (Galloway 2003).

There is often a time lag before substances emitted to groundwater may reach a down gradient fresh or marine water. In a model of European nutrients, Beusen et al. (1995) used a typical time scale of 50 years for nitrate emissions to groundwater. However, the majority of nitrate discharge was in groundwater for less than 5 years. Over longer time scales, one method to account for temporal variability is to include future emission scenarios (and corresponding environmental concentrations or sensitivities) (EC-JRC 2011).

5.2 Variability

As noted in Sect. 3, there are seasonal changes in eutrophication impacts. It is also possible for the limiting nutrient in a water body to change, both on sub-annual time scales or over a longer time frame (for example, if there is an increase in atmospheric N input). However, LCIA models have not accounted for this variability. In the former case, the added complexity would not necessarily contribute to enhanced life cycle impact modelling; in the latter, improved ecosystem models and loading data would be necessary.

At short time scales, eutrophication can exhibit strong variability. Areas at high latitudes may have substantial seasonal differences in available light, and thus times of year when microbial activity does and does not drive eutrophication. Many areas experience seasonal stratification (see Sect. 3.2), which hinders replenishment of dissolved oxygen in lake hypolimnion. The LCIA perspective does an adequate job of capturing overall eutrophication impact trends at longer time scales. However, if the limiting nutrient changes over time, this would be imperative to capture in a model. Early LCIA models did not attempt to account for intra-annual variability (Udo de Haes et al. 2002).

Limnological science is not yet robust enough to understand the recovery of ecosystems after cessation of inputs (Schindler 2006), so using varying lengths of impact windows is not currently possible. However, it is clear that there are seasonal variations in eutrophication, e.g., for freshwater and marine systems (Conroy et al. 2010; Obenour et al. 2012); these subtleties pull LCIA characterisation towards models with higher temporal and spatial resolution, but the only models that capture these trends are highly parameterised, highly tailored models for specific systems (e.g., Obenour et al. 2012). Global models and the requisite input data for such models are not yet available.

6 Midpoint Methodologies

Eutrophication is a category with relatively few substances affecting the area of protection. At the effect level, assuming the critical distinction between P and N-limited conditions has been made, there are not large differences between such substances: from a nutrient supply perspective, the distinction between ammonia and nitrate is largely one of stoichiometry. Therefore, correctly modelling the fate and transport of eutrophying substances is critical. Those models that account for BOD and COD are able to tie in P or N inputs via assumptions about nutrients driving primary production and, hence, oxygen depletion.

Table 10.1 summarises currently available characterisation methods for eutrophication impact. The simplest approach, that of the initial LCIA models, is to assume a standard fate of eutrophying emissions, which can range from an implicit transport without losses to a fraction of emissions that reach terrestrial, freshwater, or marine ecosystems. CML 2002 (Guinée et al. 2002) and EDIP97 (Hauschild and Wenzel 1998; Wenzel et al. 1997) take the former approach for terrestrial eutrophication. Using the Redfield ratio (Redfield 1934), which is based on typical aquatic biomass stoichiometry, N and P substances are converted into phosphate or nitrate equivalents, representing an overall potential for eutrophication.

To varying degrees, all midpoint models make assumptions about or explicitly model fate and transport of eutrophying substances. At limited spatial scales, there may be minimal variation in fate and transport of airborne or aquatic emissions. As discussed in Sect. 5.1, the importance of spatial considerations has been established for emissions to air (Potting et al. 1998a) and freshwater. Therefore, refining transport models will be an active work area in future.

Another midpoint approach, used for terrestrial eutrophication, is to model the transport of substances, linking their environmental fate to deposition in sensitive areas. The accumulated exceedance model of Seppälä et al. (2006) uses the EMEP model for transport and a LRTAP critical load database (Posch et al. 1995) to determine the assimilative capacity of the receiving area. EDIP 2003 expresses its eutrophication midpoint as an increase in unprotected area; this is a binary, on/off model, which maps airborne deposition predictions of the RAINS model (Alcamo et al. 1990) to the LRTAP critical load database (Potting and Hauschild 2005).

For aquatic eutrophication, methods may also employ fate models to calculate a nutrient enrichment. EDIP 2003 couples the RAINS model for deposition with CARMEN (Beusen et al. 1995; De Haan et al. 1996; Klepper et al. 1995) for aquatic fate; substances are converted to nitrate equivalents. The EUTREND model (van Jaarsveld 1995) is used in ReCiPe, which distinguishes freshwater as P-limited and marine systems as N-limited (Struijs et al. 2009). The method behind the ReCiPe factors also uses the CARMEN model, which incorporates soil, topography, and land use data at a grid scale of $1/6° \times 1/6°$, allowing calculation of factors for gross application of fertiliser to agricultural lands (Struijs et al. 2010a). The TRACI method uses a topological hydrological model (Fekete et al. 2002) to consider fate, at the U.S. state level, for emissions to air and water, converting all substances to phosphate or nitrogen equivalents, for freshwater and marine eutrophication, respectively (Norris 2003). TRACI also uses the Redfield ratio to relate N and P on a molar basis and to quantify a eutrophication midpoint (see Sect. 1).

An alternative approach is offered by the LIME method, which was developed specifically for marine waters in Japan (Itsubo and Inaba 2003). Nutrient inputs (N and P) and organic matter inputs, as BOD and COD, are linked to oxygen depletion in the hypolimnia of coastal bays.

Although there is still discussion regarding the treatment of organic matter, it is clear that making the distinction between freshwater and marine systems is essential. In future, as the supporting science grows stronger and models are more spatially and temporally explicit, there will also be considerations of multi-nutrient limitation paradigms. At present, however, differences in fate modelling assumptions are the main distinction at the midpoint level.

7 Endpoint Methodologies

Ecosystem quality can be expressed in a variety of ways that are currently challenging to measure (Curran et al. 2011). Energy, matter, and information flows are quantities for which measurement is, to some extent, possible (Goedkoop et al. 2009). Ecosystem information can be expressed at the ecosystem, species, or gene level. In LCIA, most endpoint methodologies use species as indicators of overall ecosystem quality. Most such approaches consider overall species, based on the observational record. It is challenging to estimate species distributions prior to human intervention.

Indeed, even the current distribution of species is complex, and modelling the extinction of species is challenging and would rarely realistically be attributable to the emissions associated with a single product or service, as would be modelled in LCIA. Therefore, it is more common to model the disappearance of species from a given area over a given time, assuming that removing the environmental stressor would allow the species to re-colonise an affected area (Goedkoop et al. 2009). See Goedkoop et al. (2009) for a discussion of the weighting of species across terrestrial, freshwater, and marine environments.

Table 10.1 summarises currently available eutrophication impact methods. Some endpoint methods have adapted information available from regional or country-specific studies of species occurrence. For example, the Dutch Nature Planner (Latour et al. 1997) relates acidifying and eutrophying substances to threatened terrestrial species. This study is used in Eco-Indicator 99 (Goedkoop and Spriensma 2000) to characterise impacts in terms of a potentially disappeared fraction over an area and time (PDF·m^2·year). A Swedish study on the fraction of endangered species related to terrestrial and aquatic eutrophication is used in the EPS2000 method (Steen 1999a, b), which applies a generic fate assumption to N and P, as well as oxygen-depleting substances. One limitation of such approaches is the difficulty in extrapolating to other or larger areas, from continents to the globe. A study of macrofauna in Dutch waters was used to develop an effect factor relating P concentration to species sensitivity in the ReCiPe method, which is coupled to the midpoint fate model to calculate impacts as PDF·m^3·year (Struijs et al. 2009, 2010a). This model assumes no impact at P concentrations below 100 μg/L. A study of French freshwater invertebrates (Tachet et al. 2000) was used to create a model of species effects due to increasing P concentrations (Payet 2006); this was added to the IMPACT World + method (www.impactworldplus.org).

Considering the flow of resources from ecosystems, the LIME method (Itsubo and Inaba 2003) translates the oxygen depletion in marine ecosystems to reductions in biomass in the benthos as well as fishery catch decreases. These impacts are expressed in monetary units. Having been developed for specific locations, it is difficult to apply the LIME endpoints to other regions.

As the science to support endpoint modelling is generally not as robust as midpoint modelling (e.g., endpoint species sensitivity vs. midpoint fate and transport), these endpoint methods all carry higher uncertainty.

8 Recent Developments and Research Needs

The arc of developments for eutrophication continues forward, often following the themes of improving modelling of transport and effect, with respect to refined geographic resolution and increased geographic coverage.

Geographic specificity of fate models, such as the CARMEN model used by Struijs et al. (2010a), allows for more precise conversion of inventory (e.g., nutrient application to an agricultural field) into emissions to freshwater. Accurately modelling the fate of such emissions is an area of continuing work.

On the effect side, existing species sensitivity distribution (SSD) approaches are likely limited in their taxonomic coverage and geographic transferability. Extrapolating from taxa is difficult, as responses to stressors between taxa is weak (Wolters et al. 2006), introducing uncertainty. Whether the underlying data source is Dutch (Struijs et al. 2010b) or French (Payet 2006), empirical relationships are constrained in application by the geographical specificity of their input data. SSD are not meant to be used in specific water bodies, nor can they be applied outside the region on which they were based without careful thought. These two impact methods are probably applicable in temperate climates, but use for (sub)tropical climates may be problematic. New work attempts to mine a richer dataset, adding analysis for freshwater eutrophication due to P at a global scale, comparing linear, average, and marginal effects, for autotrophs and heterotrophs (Azevedo et al. 2013b).

As understanding of ecological systems' responses to nutrient inputs improves, empirical and theoretical approaches to modelling eutrophication may change. Marine modelling of eutrophication impacts has yet to be formalised, and has been identified as a major challenge in this century (Cederwall and Elmgren 1990; Turner and Rabalais 2003).

LCIA has traditionally not dealt with combinatory (synergistic or antagonistic) effects of compounds (UNEP 2003), although work in ecotoxicity has explored this concept (van Zelm et al. 2007a). Future models should work to shed light on those situations where water bodies may not be not limited by single nutrients. Beyond nitrogen, iron may limit productivity, and silica can also play a role, via blooms of diatoms, in eutrophication (Schindler 2006; Sterner 2008); this should be addressed in future research. Such models will require understanding time scales of eutrophication; if a dynamic model were available, it could be used to create temporally variable impacts, as has been done for acidification (van Zelm et al. 2007b).

However, there is a need for a re-conceptualisation of underlying assumptions in current LCIA models for eutrophication. First, the choice of linear, average, or marginal effect model can significantly influence calculated effect factors (Azevedo et al. 2013a). Beyond the numerical values of effect factors, though, the meaning and goals of those factors should be considered: the marginal effect model discounts the importance of additional stresses to an already-degraded area, thus providing little impetus for remediation of degraded areas (Huijbregts et al. 2011). In contrast, an average effect factor based on a target level may serve societal goals for LCA, such as restoration or minimising further damage to eutrophied areas.

Secondly, the PDF category indicator for intrinsic ecosystem value assumes that species occurrence adequately represents biodiversity, which in turn is assumed to adequately represent a suite of ecosystem metrics (Struijs et al. 2009). There is a need to revisit assumptions of linearity of PDF with scale, and to consider the different implications of local, regional, and global loss of habitat. However, as Curran et al. (2011) point out, LCA captures composition of ecosystems, but there is a need to incorporate other metrics of biodiversity, including function and structure. For example, considering impacts on native, as opposed to all, species may be a useful area of investigation. Going beyond loss of habitat (which is implied with the PDF-area or PDF-volume approaches), LCIA could also consider a phylogenetic indicator as well as a metric that links species to ecosystems; the often-used but ill-defined term 'ecosystem services' may capture this idea (Curran et al. 2011).

References

Alcamo J, Shaw RW, Hordijk L (1990) The RAINS model of acidification: science and strategies in Europe. Kluwer Academic Publishers, Dordrecht

Anderson DM (1989) Toxic algal blooms and red tides: a global perspective. In: Okaichi T, Anderson DM, Nemoto T (eds) Red tides: biology, environmental science and toxicology. Elsevier, New York, pp 11–16

Azevedo LB, Henderson AD, van Zelm R, Jolliet O, Huijbregts MAJ (2013a) Assessing the importance of spatial variability versus model choices in life cycle impact assessment: the case of freshwater eutrophication in Europe. Environ Sci Technol 47:13565–13570

Azevedo LB, van Zelm R, Elshout PMF et al (2013b) Species richness–phosphorus relationships for lakes and streams worldwide. Glob Ecol Biogeogr 22:1304–1314. doi:10.1111/geb.12080

Bare JC, Norris GA, Pennington DW, McKone TE (2003) TRACI: the tool for the reduction and assessment of chemical and other environmental impacts. J Ind Ecol 6:49–78

Beusen AHW, Klepper O, Meinardi CR (1995) Modelling the flow of nitrogen and phosphorus in Europe: from loads to coastal seas. Water Sci Technol 31:141–145. doi:10.1016/0273-1223(95)00364-S

Boehrer B, Schultze M (2008) Stratification of lakes. Rev Geophys 46(2). doi:10.1029/2006RG000210

Bouwman AF, Beusen A, Billen G (2009) Human alteration of the global nitrogen and phosphorus soil balances for the period 1970–2050. Glob Biogeochem Cycles 23:GB0A04. doi:10.1029/2009GB003576

Carpenter SR, Caraco NF, Correll DL et al (1998) Nonpoint pollution of surface waters with phosphorus and nitrogen. Ecol Appl 8:559–568

Cederwall H, Elmgren R (1990) Biological effects of eutrophication in the Baltic Sea, particularly the coastal zone. Ambio Stockholm 19:109–112

Conroy JD, Boegman L, Zhang H et al (2010) "Dead Zone" dynamics in Lake Erie: the importance of weather and sampling intensity for calculated hypolimnetic oxygen depletion rates. Aquat Sci 73:289–304. doi:10.1007/s00027-010-0176-1

Curran MA, de Baan L, De Schryver AM et al (2011) Toward meaningful end points of biodiversity in life cycle assessment. Environ Sci Technol 45:70–79. doi:10.1021/es101444k

De Haan BJ, Klepper O, Sauter FJ et al (1996) The CARMEN status report 1995. RIVM Report 461501005

Diaz RJ (2001) Overview of hypoxia around the world. J Environ Qual 30:275. doi:10.2134/jeq2001.302275x

EC-JRC (2010a) Framework and requirements for life cycle impact assessment models and indicators. ILCD Handbook—International Reference Life Cycle Data System, European Union EUR24571EN. http://lct.jrc.ec.europa.eu/

EC-JRC (2010b) Analysis of existing environmental impact assessment methodologies for use in life cycle assessment. ILCD Handbook—International Reference Life CycleData System, European Union EUR24571EN. http://lct.jrc.ec.europa.eu/

EC-JRC (2011) Recommendations based on existing environmental impact assessment models and factors for life cycle assessment in European context. ILCD Handbook—International Reference Life Cycle Data System, European Union EUR24571EN. http://lct.jrc.ec.europa.eu/

EEA (2009) EMEP/EEA air pollutant emission inventory guidebook: technical guidance to prepare national emission inventories. European Communities, Luxembourg

Elser JJ, Fagan WF, Denno RF et al (2000) Nutritional constraints in terrestrial and freshwater food webs. Nature 408:578–580. doi:10.1038/35046058

Fekete BM, Vörösmarty CJ, Grabs W (2002) High-resolution fields of global runoff combining observed river discharge and simulated water balances. Glob Biogeochem Cycles 16(15):1

Finnveden G, Potting J (1999) Eutrophication as an impact category. Int J Life Cycle Assess 4:311–314. doi:10.1007/BF02978518

Frischknecht R, Steiner R, Jungbluth N (2009) The ecological scarcity method – eco-factors 2006: a method for impact assessment in LCA. Federal Office for the Environment FOEN, Bern
Galloway JN (2003) 8.12 – the global nitrogen cycle. In: Holland HD, Turekian KK (eds) Treatise on geochemistry. Pergamon, Oxford, pp 557–583
Goedkoop M, Spriensma R (2000) The Eco-indicator 99: a damage oriented method for life cycle assessment. PRé Consultants BV, Amersfoort
Goedkoop M, Heijungs R, Huijbregts MAJ et al (2009) ReCiPe 2008: a life cycle impact assessment method which comprises harmonised category indicators at the midpoint and the endpoint level; Report 1: Characterisation. VROM Den Haag 2012
Goldman CR (1988) Primary productivity, nutrients, and transparency during the early onset of eutrophication in ultra-oligotrophic lake Tahoe, California-Nevada. Limnol Oceanogr 33:1321–1333. doi:10.2307/2837293
Grouzet P, Leonard J, Nixon SW et al (2000) Nutrients in European ecosystems. European Environment Agency, Copenhagen
Guinée J, Gorrée M, Heijungs R et al (2002) Handbook on life cycle assessment: operational guide to the ISO standards. Kluwer Academic Publishers, Dordrecht
Hansson M, Axe P, Andersson L (2009) Extent of anoxia and hypoxia in the Baltic Sea, 1960–2009. Swedish Meteorological and Hydrological Institute (SMHI)
Hauschild MZ, Wenzel H (1998) Environmental assessment of products vol 2: scientific background. Chapman & Hall/Kluwer Academic Publishers, London/Hingham, 1997. ISBN 0-412-80810-2
Heijungs R, Guinée JB, Huppes G et al (1992) Environmental life cycle assessment of products: guide and background. Center of Environmental Science (CML), Leiden
Helmes RJK, Huijbregts MAJ, Henderson AD, Jolliet O (2012) Spatially explicit fate factors of phosphorous emissions to freshwater at the global scale. Int J Life Cycle Assess 17:646–654. doi:10.1007/s11367-012-0382-2
Hertwich EG, Hammitt JK, Pease WS (2000) A theoretical foundation for life-cycle assessment: recognizing the role of values in environmental decision making. J Ind Ecol 4:13–28. doi:10.1162/108819800569267
Hornung MW, Ineson P, Bull KR et al (1994) Impacts of nitrogen deposition in terrestrial ecosystems. United Kingdom Review Group, London
Huijbregts MAJ, Seppälä J (2000) Towards region-specific, European fate factors for airborne nitrogen compounds causing aquatic eutrophication. Int J Life Cycle Assess 5:65–67. doi:10.1007/BF02979719
Huijbregts MAJ, Seppälä J (2001) Life cycle impact assessment of pollutants causing aquatic eutrophication. Int J Life Cycle Assess 6:339–343, doi: 09483349
Huijbregts MAJ, Schöpp W, Verkuijlen E et al (2000) Spatially explicit characterisation of acidifying and eutrophying air pollution in life-cycle assessment. J Ind Ecol 4:75–92. doi:10.1162/108819800300106393
Huijbregts MAJ, Hellweg S, Hertwich E (2011) Do we need a paradigm shift in life cycle impact assessment? Environ Sci Technol 45:3833–3834. doi:10.1021/es200918b
Itsubo N, Inaba A (2003) A new LCIA method: LIME has been completed. Int J Life Cycle Assess 8:305–305. doi:10.1007/BF02978923
Jolliet O, Margni M, Charles R et al (2003) IMPACT 2002+: a new life cycle impact assessment methodology. Int J Life Cycle Assess 8:324–330
Jørgensen BB, Richardson K (1996) Eutrophication in coastal marine ecosystems. Coastal Estuarine Stud 52:1–273. doi:10.1029/CE052
Kemna R, van Elburg M, Li W, van Holsteijn R (2005) MEEuP: methodology report. 188
Klepper O, Beusen AH, Meinardi C (1995) Modelling the flow of nitrogen and phosphorus in Europe: from loads to coastal seas. National Institute of Public Health and Environmental Protection (RIVM), Bilthoven

Kristensen P, Hansen HO (1994) European rivers and lakes. Assessment of their environmental state. Environmental Monographs Copenhagen, Denmark: European Environment Agency. 122 pages. http://www.eea.europa.eu/publications/87-90198-01-8
Kros J, Reinds GJ, De Vries W et al (1995) Modelling of soil acidity and nitrogen availability in natural ecosystems in response to changes in acid deposition and hydrology. SC-DLO, report 95. Wageningen, The Netherlands: Winand Staring Centre for Integrated Land, Soil and Water Research. 90 pages. http://library.wur.nl/WebQuery/wurpubs/302176
Latour JB, Staritsky IG, Alkemade JRM, Wiertz J (1997) De natuurplanner, Decision support system natuur en mileu. Report Number 711901019. National Institute of Public Health and Environmental Protection (RIVM), Bilthoven
Levine SN, Stainton MP, Schindler DW (1986) A radiotracer study of phosphorus cycling in a eutrophic Canadian Shield Lake, Lake 227, Northwestern Ontario. Can J Fish Aquat Sci 43:366–378. doi:10.1139/f86-047
Lindfors L-G, Christiansen K, Hoffman L et al (1995) Nordic guidelines on life-cycle assessment. Nord 1995:20. Nordic Council of Ministers, Copenhagen, Denmark
Liu D, Keesing JK, He P et al (2013) The world's largest macroalgal bloom in the Yellow Sea, China: formation and implications. Estuar Coast Shelf Sci 129:2–10. doi:10.1016/j.ecss.2013.05.021
Margni M, Gloria T, Bare JC et al (2008) Guidance on how to move from current practice to recommended practice in life cycle impact assessment. UNEP/SETAC Life Cycle Initiative. Life Cycle Impact Assessment Programme. http://www.lifecycleinitiative.org/wp-content/uploads/2012/12/2008%20-%20Guidance%20to%20move%20to%20LCA.pdf
McGroddy ME, Daufresne T, Hedin LO (2004) Scaling of C:N:P stoichiometry in forests worldwide: implications of terrestrial redfield-type ratios. Ecology 85:2390–2401. doi:10.1890/03-0351
MEA (2005) Ecosystems and human well-being. Island Press, Washington, DC
Mortimer CH (1942) The exchange of dissolved substances between mud and water in lakes. J Ecol 30:147–201. doi:10.2307/2256691
NAS (1969) Eutrophication: causes, consequences, correctives; proceedings of a symposium. U.S. National Academy of Sciences, Washington, DC
Nilsson J, Grennfelt P (1988) Critical loads for sulphur and nitrogen. Report from a workshop at Skokloster, Sweden, 19–24 Mar 1988. Miljörapport 1998: 15. Nordic Council of Ministers, Copenhagen
Norris GA (2003) Impact characterisation in the tool for the reduction and assessment of chemical and other environmental impacts: methods for acidification, eutrophication, and ozone formation. J Ind Ecol 6:79–101
NRC (1992) Restoration of aquatic ecosystems: science, technology, and public policy. National Research Council, Washington, DC
Obenour DR, Michalak AM, Zhou Y, Scavia D (2012) Quantifying the impacts of stratification and nutrient loading on hypoxia in the Northern Gulf of Mexico. Environ Sci Technol 46:5489–5496. doi:10.1021/es204481a
OECD (1982) Eutrophication of waters: monitoring, assessment and control. Organization for Economic Cooperation and Development, Paris
Paerl HW, Fulton RS III, Moisander PH, Dyble J (2001) Harmful freshwater algal blooms, with an emphasis on cyanobacteria. Sci World J 1:76–113. doi:10.1100/tsw.2001.16
Payet J (2006) Report describing a method for the quantification of impacts on aquatic freshwater ecosystems resulting from different stressors (e.g., toxic substances, eutrophication, etc). Novel Methods for Integrated Risk Assessment of Cumulative Stressors in Europe (NOMIRACLE). Report Number 003956. École Polytechnique Fédérale de Lausanne. http://nomiracle.jrc.ec.europa.eu/webapp/ViewPublicDeliverables.aspx
Posch M, de Smet PAM, Hettelingh J-P, Downing RJ (1995) Calculation and mapping of critical thresholds in Europe. Status Report 1995. National Institute of Public Health and Environmental Protection (RIVM), Bilthoven

Posch M, Seppälä J, Hettelingh J-P et al (2008) The role of atmospheric dispersion models and ecosystem sensitivity in the determination of characterisation factors for acidifying and eutrophying emissions in LCIA. Int J Life Cycle Assess 13:477–486. doi:10.1007/s11367-008-0025-9

Potting J (2000) Spatial differentiation in life cycle impact assessment a framework, and site-dependent factors to assess acidification and human exposure. Int J Life Cycle Assess 5:77. doi:10.1007/BF02979725

Potting J, Blok K (1994) Spatial aspects of life-cycle impact assessment. In: Udo de Haes HA, Jensen AA, Klöpffer W, Lindfors L-G (eds) Integrating impact assessment into LCA. Society of Environmental Toxicology & Chemistry, Brussels, pp 91–98

Potting J, Hauschild MZ (2005) Background for spatial differentiation in LCA impact assessment – the EDIP2003 methodology. Environmental Project No. 996 2005. Danish Ministry of the Environment

Potting J, Schöpp W, Blok K, Hauschild M (1998a) Comparison of the acidifying impact from emissions with different regional origin in life-cycle assessment. J Hazard Mater 61:155–162, 16/S0304-3894(98)00119-8

Potting J, Schöpp W, Blok K, Hauschild MZ (1998b) Site-dependent life-cycle impact assessment of acidification. J Ind Ecol 2:63–87. doi:10.1162/jiec.1998.2.2.63

Redfield AC (1934) On the proportions of organic derivatives in sea water and their relation to the composition of plankton. In: Cole FJ (ed) James Johnstone memorial volume. University Press of Liverpool, Liverpool, pp 176–192

Scheffer M, Carpenter S, Foley JA et al (2001) Catastrophic shifts in ecosystems. Nature 413:591–596. http://dx.doi.org.www5.sph.uth.tmc.edu:2048/10.1038/35098000

Schindler DW (1977) Evolution of phosphorus limitation in lakes. Science 195:260–262

Schindler DW (2006) Recent advances in the understanding and management of eutrophication. Limnol Oceanogr 51:356–363

Seppälä J, Posch M, Johansson M, Hettelingh J-P (2006) Country-dependent characterisation factors for acidification and terrestrial eutrophication based on accumulated exceedance as an impact category indicator. Int J Life Cycle Assess 11:403–416. doi:10.1065/lca2005.06.215

Shannon JD, Clark TL (1985) User's guide for the advanced statistical trajectory regional air pollution (ASTRAP) model. Atmospheric Sciences Research Laboratory, Office of Research and Development, Environmental Protection Agency, Lemont

Smith VH (1983) Low nitrogen to phosphorus ratios favor dominance by blue-green algae in lake phytoplankton. Science 221:669–671. doi:10.1126/science.221.4611.669

Smith VH, Tilman GD, Nekola JC (1999) Eutrophication: impacts of excess nutrient inputs on freshwater, marine, and terrestrial ecosystems. Environ Pollut 100:179–196. doi:10.1016/S0269-7491(99)00091-3

Steen B (1999a) A systematic approach to environmental priority strategies in product development (EPS). Version 2000–General system characteristics. Chalmers University of Technology, Gothenburg

Steen B (1999b) A systematic approach to environmental priority strategies in product development (EPS). Version 2000–Models and data of the default method. Chalmers University of Technology, Gothenburg

Sterner RW (2008) On the phosphorus limitation paradigm for lakes. Int Rev Hydrobiol 93:433–445. doi:10.1002/iroh.200811068

Struijs J, Beusen AHW, van Jaarsveld HA, Huijbregts MAJ et al (2009) Aquatic eutrophication (chapter 6). In: Goedkoop M, Heijungs R, Huijbregts MAJ (eds) ReCiPe 2008: a life cycle impact assessment method which comprises harmonised category indicators at the midpoint and the endpoint level; Report 1: Characterisation. Ministry of Housing, Spatial Planning, and Environment (VROM), The Netherlands, p 132

Struijs J, Beusen A, Zwart D, Huijbregts M (2010a) Characterisation factors for inland water eutrophication at the damage level in life cycle impact assessment. Int J Life Cycle Assess 16:59–64. doi:10.1007/s11367-010-0232-z

Struijs J, De Zwart D, Posthuma L et al (2010b) Field sensitivity distribution of macroinvertebrates for phosphorus in inland waters. Integr Environ Assess Manag 7:280–286. doi:10.1002/ieam.141

Tachet H, Richoux P, Bournaud M, Usseglio-Polatera P (2000) Invertébrés d'eau douce: systématique, biologie, écologie. CNRS Editions, Paris

Tilman D (1999) Global environmental impacts of agricultural expansion: the need for sustainable and efficient practices. Proc Natl Acad Sci U S A 96:5995–6000

Toffoleto L, Bulle C, Godin J et al (2007) LUCAS – a new LCIA method used for a Canadian-specific context (10 pp). Int J Life Cycle Assess 12:93–102. doi:10.1065/lca2005.12.242

Turner RE, Rabalais NN (2003) Linking landscape and water quality in the Mississippi river basin for 200 years. BioScience 53:563–572. doi:10.1641/0006-3568(2003)053[0563:LLAWQI]2.0.CO;2

Udo de Haes HA, Finnveden G, Goedkoop M et al (2002) Life cycle impact assessment: striving towards best practice. SETAC Press, Pensacola

UNEP (2003) Evaluation of environmental impacts in life cycle assessment. United Nations Environment Program Division of Technology, Industry, and Economics, Rome

UNEP/SETAC (2005) Life cycle approaches: the road from analysis to practice. United Nations Environment Program, Division of Technology, Industry and Economics (DTIE), Paris

Van Jaarsveld JA (1995) Modelling the long-term atmospheric behaviour of pollutants on various spatial scales. Ph.D. thesis, Universiteit Utrecht

Van Zelm R, Huijbregts MAJ, Harbers JV et al (2007a) Uncertainty in msPAF-based ecotoxicological effect factors for freshwater ecosystems in life cycle impact assessment. Integr Environ Assess Manag 3:203–210. doi:10.1897/IEAM_2006-013.1

Van Zelm R, Huijbregts MAJ, van Jaarsveld HA et al (2007b) Time horizon dependent characterisation factors for acidification in life-cycle assessment based on forest plant species occurrence in Europe. Environ Sci Technol 41:922–927. doi:10.1021/es061433q

Vermeire TG, Jager DT, Bussian B et al (1997) European Union System for the Evaluation of Substances (EUSES). Principles and structure. Chemosphere 34:1823–1836. doi:10.1016/S0045-6535(97)00017-9

Vermeire T, Rikken M, Attias L et al (2007) European Union System for the evaluation of substances: the second version. Chemosphere 59:473–485. doi:10.1016/j.chemosphere.2005.01.062

Vitousek PM, Mooney HA, Lubchenco J, Melillo JM (1997) Human domination of earth's ecosystems. Science 277:494–499. doi:10.1126/science.277.5325.494

Vollenweider RA (1968) Scientific fundamentals of the eutrophication of lakes and flowing waters, with a particular reference to phosphorus and nitrogen as factor in eutrophication. Organization for Economic Cooperation and Development, Paris

Von Liebig J (1855) Principles of agricultural chemistry with special reference to the late researches made in England. In: Pomeroy LR (ed) Cycles of essential elements, Benchmark papers in ecology. Dowden, Hutchinson, and Ross, Stroudsburg, pp 11–28, United Kingdom

Weber CA (1907) Aufbau und Vegetation der Moore Norddeutschlands. Beibl Bot Jahrb 90:19–34

Wenzel H, Hauschild M, Alting L (1997) Environmental assessment of products vol 1: methodology, tools and case studies in product development. Chapman & Hall/Kluwer Academic Publishers, London/Hingham. ISBN 0 412 80800 5

Withers PJA, Jarvie HP (2008) Delivery and cycling of phosphorus in rivers: a review. Sci Total Environ 400:379–395. doi:10.1016/j.scitotenv.2008.08.002

Wolters V, Bengtsson J, Zaitsev AS (2006) Relationship among the species richness of different taxa. Ecology 87:1886–1895. doi:10.1890/0012-9658(2006)87[1886:RATSRO]2.0.CO;2

Chapter 11
Land Use

Llorenç Milà i Canals and Laura de Baan

Abstract Land use impacts are the effects caused by the use of land by humans, which range from changes in species composition and abundance to the disruption of ecosystem processes contributing to climate and water regulation. These impacts end up affecting key areas of protection such as human health, ecosystem quality, and natural resources. In life cycle assessment (LCA), land use impact assessment quantifies the difference between the land quality level of the studied system and a reference level over the duration and the area being used. For land transformations (change of land use), the time required for land to naturally regenerate is often considered as the duration when calculating impacts. Indicators to assess land use impacts on biodiversity in LCA focus on either species richness or on ecosystem metrics; whereas the indicators for land use impacts on ecosystem services are classed as pressure (describing land degradation processes) and state (describing overall quality) indicators. Many approaches to describe such impact indicators are presented in this chapter. The key areas for further research relate to (1) the description of the reference used to quantify the impacts; (2) aggregation of impacts on different ecosystem services; (3) the availability of inventory information to be able to inform the spatial differentiation level required for a more accurate impact assessment; (4) leveraging 'big data' processes to allow full utilisation of the data available on ecosystem quality; and (5) promote global consensus on impact indicators in order to facilitate comparison and stability of LCA results.

Keywords Biodiversity • Ecosystem services • Land use • Land use change • LCA • LCIA • Life cycle assessment • Life cycle impact assessment

---

L. Milà i Canals (✉)
Division of Technology, Industry and Economics, United Nations Environment Programme, 15 rue de Milan, Paris 75009, France
e-mail: llorenc.mila-i-canals@unep.org

L. de Baan
Institute of Environmental Engineering, ETH Zurich, John-von-Neumann-Weg 9, Zurich 8093, Switzerland

M.Z. Hauschild, M.A.J. Huijbregts (eds.), *Life Cycle Impact Assessment*, LCA Compendium – The Complete World of Life Cycle Assessment, DOI 10.1007/978-94-017-9744-3_11

1 Introduction

Land is a key resource for many human activities such as agriculture, forestry, mining, or housing. Land use refers to the use of land for any of such human purposes, and generally entails the modification of the conditions of the natural environment into a set of conditions suitable for its use by humans. Such modifications are linked to significant effects (usually damages) on the natural environment, which are often referred to as land use impacts. Land use and particularly land use change from natural habitats to cropland and other human land uses is one of the main drivers of biodiversity loss (Sala et al. 2000; Millennium Ecosystem Assessment 2005; Lenzen et al. 2009), and land competition is very likely to increase in the future (UNEP 2014).

Land use impacts have long been discussed in life cycle assessment (LCA) (see, e.g. Audsley et al. 1997; Cowell 1998; Lindeijer et al. 2002; Milà i Canals 2003; Milà i Canals et al. 2007a), but operational methods for their practical inclusion have only begun to be applied since the 2010s. Part of the reason for this slow uptake is the strong spatial dependency of land use impacts, which is difficult to capture in traditional LCA.

Within this chapter, the principles and fundamentals of land use impact assessment in LCA (Sect. 1.1) and considerations of temporal and spatial variability (Sect. 1.2) are introduced. Section 2 then describes the impact pathways affected by land use, while Sect. 3 provides the life cycle inventory (LCI) interventions linked to land use. Section 4 describes the methodologies suggested to date for the assessment of land use impacts on biodiversity (Sect. 4.1), ecosystem services (Sect. 4.2) and climate change (Sect. 4.3). Finally, Sect. 5 reviews some of the most recent developments in land use impact assessment in LCA and suggests key research needs.

1.1 Principles of Land Use Impact Characterisation Modelling

According to the land use impact assessment framework defined by the SETAC working group on Impact Assessment (Lindeijer et al. 2002), which was further refined in the UNEP/SETAC Life Cycle Initiative projects (Milà i Canals et al. 2007a; Koellner et al. 2013a), human land uses generate environmental impacts through land occupation (also known as land use, LU) and the transformation of land's properties to fit a new use (also known as land use change, LUC). As it can be seen in Fig. 11.1, occupation impacts (areas II; IV; and VI) refer to the maintenance of a quality difference between current quality and a suitable reference during the period of the occupation, whereas transformation impacts (areas I; III; V) arise from a change in quality that could be subsequently reverted during a regeneration time if land was left to recover after the transformation. In other

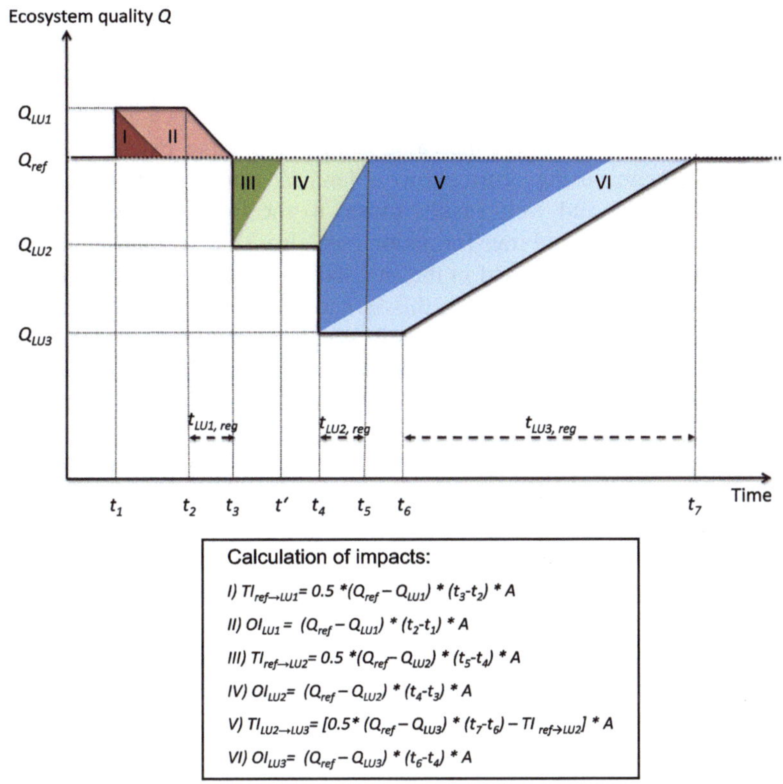

Fig. 11.1 Simplified illustration of transformation impact (*TI*) and occupation impact (*OI*) for three land use types with different regeneration rates (tLU1, reg, tLU2, reg and tLU3, reg). For simplicity, the area A of occupation or transformation, which would embrace the third dimension, is not shown in the graph, but in the equations (Source: Koellner et al. (2013a))

words, transformation impacts relate to the difference in land quality maintained during the regeneration time, whereas occupation impacts relate to the delay in this regeneration. The total impact would be calculated by multiplying the quality difference maintained by the studied land use with the occupation time or the regeneration time for transformation, and the area affected (which would be represented by a third axis on Fig. 11.1, not represented here for simplicity (see Koellner et al. 2013a)).

The key elements identified in this modelling are (1) the measure or indicator of land quality being used, which varies for each impact pathway affected; (2) the reference land use; and (3) the regeneration time. In terms of the quality indicators, or indicators of land use impacts, Sect. 4 reviews the main approaches suggested in the last 15 years.

As explained above and illustrated in Fig. 11.1, a reference land quality is required to quantify the impacts from land use. Most references in the literature (e.g. Milà i Canals et al. 2007a; Goedkoop et al. 2008; Koellner et al. 2013a) recommend to use the Potential Natural Vegetation (PNV) as a reference. This reference situation is mainly suggested for practicality (availability of data on PNV) and because its apparent logic makes it easy to communicate. However, as highlighted by the references above, the choice of reference is subjective and influences the land use impact indicator results to a large extent. Koellner and Scholz (2008) suggest using a current mix of land uses, and this option may actually be more meaningful and yield results closer to the impacts actually caused by current land occupation and transformation (see also Milà i Canals et al. 2013).

The regeneration time is used in the calculation of impacts from land transformation, or land use change (LUC); this is usually quite uncertain, and strongly depends on the impact pathway considered (Weidema and Lindeijer 2001). Regeneration time depends on the degree of impact on the land, but also on the type of ecosystem and bio-geographical region. For instance, regeneration tends to be quicker in warmer and more humid climates, and slower in colder and drier ones. Some estimates have been put forward and used within the LCA context (Dobben et al. 1998; Bastian and Schreiber 1999). Curran et al. (2014) performed a meta-analysis of recovery of biodiversity after land use change and identified relevant factors that influence the speed of recovery using generalized linear models. These models have been used by de Baan et al. (2013b) to derive ecosystem specific recovery times for LCA (accounting for the effects of latitude, altitude, taxonomic groups, etc.), and quantifying their uncertainties. The assumption that ecosystems can fully recover after land use is not in all cases realistic, and permanent, irreversible changes in systems can occur. De Baan et al. (2013b) have provided a first attempt, to quantify such permanent land use impacts for biodiversity by quantifying the potential loss of endemic species, which would indicate permanent biodiversity loss. Alternatively, Koellner et al. (2013a) suggest modelling land use impacts over a long enough period (500 years) in order to make the 'permanent' difference in quality apparent in the final results.

1.2 Scale, Spatial Variability, Temporal Variability

Scale and shape matter in ecosystem analysis and landscape ecology: it is not the same to transform 1 ha or 1 Mha, nor to transform the edge of a forest or its core: non-linear ecosystem responses appear as tipping points or thresholds are trespassed (Chaplin-Kramer et al. in press). However, this is something that needs to be addressed at the goal and scope level of the LCA study, and whether these effects are incorporated or not in the LCA will change the outcome of the study. For macro-scale decision support (e.g. when land use policy is being assessed, or when the goal is informing public policy related to bio-based strategies such as bio-fuels), not considering such elements may generate misleading results (see e.g. Searchinger et al. 2008). On the other hand, the approaches considered in this chapter are focused on the traditional LCA premise where effects of scale and shape are not considered, i.e. the focus is on small (marginal) changes, and how these can be assessed in a product's life cycle. Whereas the afore-mentioned considerations mainly relate to the life cycle inventory phase, the incorporation of scale in the land use impact assessment for LCA is in its infancy, see some further discussion in the context of biodiversity in Sect. 4.1.1.

In the case of spatial variability, the impacts caused by land use change vary enormously depending on the location of the land use; this is a challenge that has kept land use impact assessment out from LCA up to recently (see e.g. Geyer et al. 2010; Núñez et al. 2013a; de Baan et al. 2013a, b, among others). Current Geographic Information Systems (GIS) technology and databases allow for fast modelling of spatially relevant characterisation factors, which coupled with expanded information in LCI databases offers a promising prospect for more accurate impact assessment. This is discussed in further detail in the following sections.

Temporal dependency, on the other hand, has received much less attention in land use impact assessment. Considering this would imply differentiating the time of the year when land is used, and the implications this may have on processes that vary through the year (e.g. erosion affected by timing of heavy rains, and the vegetation cover as affected by human activities; impact on migratory species when land is used during migrations, etc.).

2 Impact Pathway, Affected Areas of Protection

All traditional areas of protection (AoP) are affected by land use activities (see Fig. 11.2). The nature of land as a resource (competition for land) is already mentioned in the chapter on abiotic resource use (see Chap. 13 of this volume). In addition, other aspects linked to this AoP include the aesthetic and cultural value of land. As shown in Fig. 11.2, land occupation and transformation may take the form of physical (compaction, erosion...); chemical (pH; salts composition; toxicity...); or biological (vegetation cover, species composition...) alterations of land, which are linked to several direct impacts. Such changes and impacts affect land-based processes (e.g. albedo; water cycle, etc.) which may lead to effects measured in the midpoint of the cause-effect change (e.g. biotic production), in turn linked to all areas of protection.

This chapter focuses on land use impacts on Ecosystem Quality (via effects on biodiversity, see Sect. 4.1, and on ecosystem services, see Sect. 4.2); other impacts affected by land use, including effects on human health from climate change are briefly introduced in Sect. 4.3.

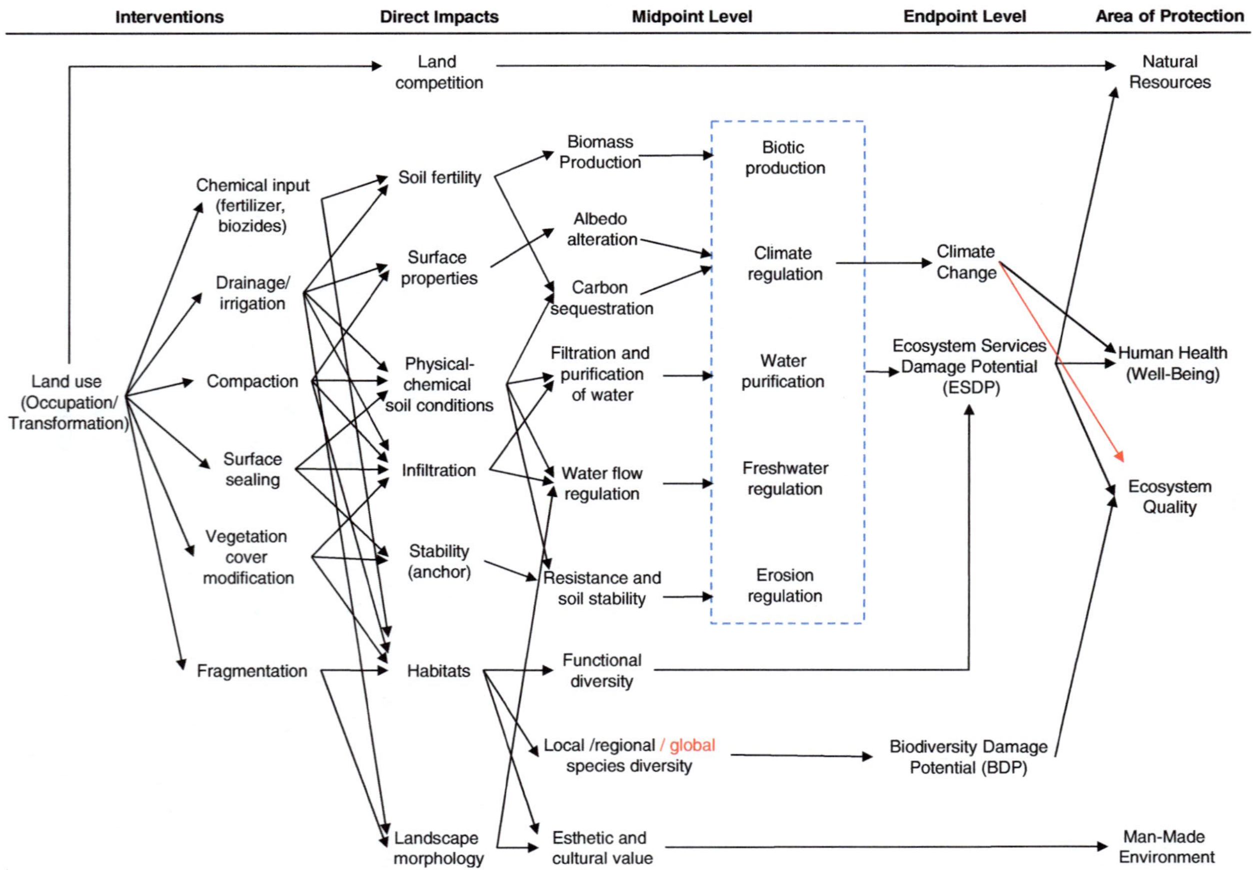

Fig. 11.2 Impact pathways from land use interventions to impacts on biodiversity and ecosystem services. Based on Koellner et al. (2013a), main modifications are marked in *red*. At the midpoint level most impacts should refer to land's 'capacity' (e.g. biomass productive capacity, see Koellner et al. 2013a), but the labels have been simplified for clarity

3 Contributing Life Cycle Interventions (Classification)

As explained in Sect. 1.1, the relevant life cycle interventions for land use impact assessment are of two main types: land occupation (measured in area × time units) and transformation (measured in area units). As developed in Koellner et al. (2013b), such interventions must specify the land use type and the region where the intervention occurs. Koellner et al. (2013b) provide a comprehensive and systematic classification of interventions at varying levels of detail, including 11 main land use types defined at four levels of detail and spatial differentiation at five levels of granularity from 'global' to the exact geo-referenced location. The first level of land use types proposed by Koellner et al. (2013b) encompasses:

1. Unspecified
2. Forest
3. Wetlands
4. Shrubland
5. Grassland
6. Agriculture
7. Agriculture, mosaic
8. Artificial areas
9. Bare area
10. Snow and ice
11. Water bodies

The above main types are then subdivided in categories depending on the needs (e.g. there are no more sub-classifications for shrub land or snow and ice, but 18 sub-types are offered for agriculture, distinguishing between arable and permanent crops at the second level and further specifying with, e.g. type of irrigation or input regime).

In terms of spatial differentiation, the five levels suggested by Koellner et al. (2013b) are:

1. Broad type of biome (terrestrial; freshwater; coastal water…)
2. Climatic region
3. Classification in 16 terrestrial and freshwater biomes and 3 marine ones
4. Olson terrestrial and freshwater ecoregions (n = 867) and Spalding coastal and shelf ecoregions (n = 232)
5. Geo-referenced information

Further work is still required in order to determine the optimum level of spatial differentiation, which will be a compromise between environmental relevance and practical feasibility and data availability. Milà i Canals et al. (2013) test some of the methods available mainly at the biome level (3), and suggest that the differentiation at this level is meaningful; on the other hand, they suggest it would be unlikely to obtain LCI information at finer levels (4 or 5) for global commodity supply chains. Characterisation factors are not provided at such levels of resolution for most impact pathways and indicators.

4 Methodologies

This section reviews the main approaches suggested to quantify the impacts on several endpoints (the indicators in the y-axis in Fig. 11.1): 'impacts on biodiversity' (Sect. 4.1); 'impacts on ecosystem services' (Sect. 4.2); and 'other impacts' (Sect. 4.3, which mainly focuses on effects from land use on greenhouse gas (GHG) emissions). A more simplified approach of simply quantifying the amount of land occupied or transformed has also been suggested (e.g. Goedkoop et al. 2008 recommend it as a midpoint approach in ReCiPe). The advantage of focusing the quantification on the inventory interventions is that it is straightforward, and it is certainly better to at least quantify how much land is being used or transformed by the product system than to not consider land at all. In addition, occupation may be a key driver of impacts from land use (e.g. Milà i Canals et al. 2013). However, such approaches are not further discussed in this chapter because they do not actually quantify the impacts arising from land use; besides, Mueller et al. (2014) find that a significant amount of impact is missed when focusing only on occupation processes, in contrast to what is reported in Milà i Canals et al. (2013). Tables 11.1 and 11.2 provide an overview of the main methods discussed for Biodiversity and Ecosystem Services, respectively, highlighting their key characteristics. Methods assessing the land use-derived GHG emissions are discussed in Sect. 4.3, but they are not further detailed in the tables as they are discussed more in detail under Climate change in Chap. 3 of this volume.

4.1 Methodologies Addressing Impacts on Biodiversity

Biodiversity is a complex and multi-faceted concept, which usually has a local value. Thus, it is very difficult to measure using a single metric that allows aggregation of impacts occurring across the product's life cycle and around the world, such as is required in LCA. The 'less is better' approach which is generally accepted for pollution does not always work for biodiversity; in general, less loss of species is preferred, but not all species are equivalent, and sometimes more species is not necessarily better (e.g. with invasive species). Several angles compromising between feasibility and relevance have been suggested in the last two decades to account for impacts on biodiversity in LCA; these have been grouped in two main types in the rest of this section: those focused on species richness as an indicator for biodiversity, and those attempting to capture other aspects of biodiversity such as ecosystem scarcity and vulnerability. Other approaches to assess biodiversity include a measure of 'naturalness' (Hemeroby, see Brentrup et al. 2002) and indicators based on exergy (Wagendorp et al. 2006), or emergy (Hau and Bakshi 2004); such approaches have received much less attention in recent years and are not considered further here.

Table 11.1 Overview of impact assessment methods for land use impacts on Biodiversity in LCA

Method	Indicator	Species richness	Ecosystem abundance	Rare/ threatened species	Ecosystem vulnerability	Absolute/ relative indicator	Reference situation	Relaxation (if considered: data source used)	Geographic coverage
Lindeijer (2000)	Vascular plant species diversity	X				Relative	Total regional species richness	Not considered	South America/ Europe (Global)
Weidema and Lindeijer (2001)	Vascular plant species richness, Ecosystem Scarcity and Vulneratability	X	X		X	Relative	Total regional species richness	Source: van Dobben et al. (1998)	Global
Toffoletto et al. (2007)	Based on Weidema and Lindeijer (2001)	X	X		X	Relative	Total regional species richness	Source: van Dobben et al. (1998)	Canada
Michelsen (2008)	Conditions for Maintained Biodiversity weighted with ecosystem vulnerability and scarcity		X	X (Percentage of invasive species)	X	Relative	Intact biodiversity	Not considered	Norway
Coelho and Michelsen (2014)	Hemeroby weighted with ecosystem vulnerability and scarcity		X		X	Relative	PNV	Not considered	New Zealand (Global)
Brentrup et al. (2002)	Hemeroby, Naturalness Degradation Potential (NDP)					Relative	PNV	Not considered	Europe
Koellner (2000)	Species-pool effect potential (SPEP), relative local and regional species loss	X			X	Relative	Regional species pool	Not considered	Switzerland/ Germany
Koellner and Scholz (2007, 2008)	Ecosystem Depletion Potential (EDP), relative local species loss	X		X (threatened species)		Relative	Regional average species richness	Source: Bastian and Schreiber (1999)	Switzerland/ Germany
Schmidt (2008)	Species richness weighted by ecosystem vulnerability	X			X	Absolute	PNV	Source: Bastian and Schreiber (1999), van Dobben et al. (1998)	Denmark/ Malaysia/ Indonesia
De Schryver et al. (2010)	Potentially disappeared fraction of species, PDF, local and regional effects	X			X (species-area relationship approach)	Relative	PNV	Not considered	United Kingdom

(continued)

Table 11.1 (continued)

Method	Indicator	Species richness	Ecosystem abundance	Rare/ threatened species	Ecosystem vulnerability	Absolute/ relative indicator	Reference situation	Relaxation (if considered: data source used)	Geographic coverage
Itsubo and Inaba (2012)	Increased extinction risk; Expected Increase in Number of Extinct Species (EINES)	X		X		Absolute	Natural extinction rate	Not considered	Japan
Geyer et al. (2010)	(1) hemeroby, (2) richness, (3) abundance, (4) evenness	X		X (species abundance)		Relative and absolute	Total regional species richness	Not considered	California
de Baan et al. (2013a)	Biodiversity Depletion Potential (BDP), local loss of species	X				Relative	PNV	Not considered	Global
de Baan et al. (2013b)	Regional and global species loss	X		X	X	Absolute	PNV	Source: Curran et al. (2014), irreversible impacts considered	Global
Mueller et al. (2014)	Local species loss weighted with species richness, endemism and ecosystem vulnerability	X		X	X	Relative	PNV	Source : own review and van Dobben et al. (1998)	South America and Europe
de Souza et al. (2013)	Functional species diversity	X (functional diversity)				Relative	PNV	Not considered	North and South America
Vogtländer et al. (2004)	1. Species richness indicator (SRI)	X	X	X		Relative	Regional average species richness	Not considered	Netherlands
	2. Botanical value								
Kyläkorpi et al. (2005)	Biotope method; changes in biotope size and quality		X (Biotope)	X		Absolute or relative change in area	Current situation	Not considered	Sweden
Burke et al. (2008)	Biotope method, changes in biotope size and quality		X (Biotope)	X		Absolute or relative change in area	Current situation	Not considered	Namibia
Jeanneret et al. (2009)	SALCA-Biodiversity; sensitivity of indicator species groups on agricultural management activities	X		X		Impacts assessed on a scale	Situation with out management	Not considered	Switzerland

Table 11.2 Overview of impact assessment methods for land use impacts on Ecosystem Services in LCA

Method	Description	Indicator	Absolute/ relative indicator	Reference situation	Relaxation	Coverage
Pressure indicators						
Cowell (1998) and Cowell and Clift (2000)	Soil quality assessed with Soil Erosion, loss of Organic Matter and Compaction	Soil lost tonnes/ha, characterised with its "Static Reserve Life"; OM Indicator (tonnes OM^{-1}); Soil Compaction Indicator (tonnes of machinery × hours of field work)	Absolute	Values at the beginning of the crop cycle	Not considered (impacts from land transformation not modelled)	Guidance provided for calculations, but no CF provided. Spatial differentiation suggested for LCI, not for LCIA
Núñez et al. (2013a)	Soil Erosion linked to effects on Soil resource Depletion (SRD), and effects on NPP leading to impacts on ecosystem quality (EQ)	SRD: MJ solar energy m^2 × year, combining soil loss through erosion and Solar Energy Factor of Soil	EQ: Relative to max NPP_0 in the world	Max soil depth (3 m)	Not considered (impacts from land transformation not modelled)	Guidance provided for calculation of CF in agricultural soils globally; application in other land use types is not explored yet. CFs are provided at 10 × 10 km^2
		Altered soil function effects on NPP affecting ecosystem quality and human health (NNP depletion leading to HH effects)	SRD: relative to max soil depth in the world	Max NPP_0 in the world		
		EQ: Loss in Net Primary Productivity (NPP_0), from soil loss through erosion				
Feitz and Lundie (2002)	Salinisation, considering the effect of irrigation water's SAR (Sodium Absorption Rate) and EC (Electrical Conductivity) on soil integrity	Salinisation potential, function of SAR and EC in soil and irrigation water	Relative to threshold	Threshold electrolyte concentration to maintain soil integrity	Not considered	Only relevant/calculated for irrigation water; values provided for a few Australian irrigation waters

(continued)

Table 11.2 (continued)

Method	Description	Indicator	Absolute/relative indicator	Reference situation	Relaxation	Coverage
Milà i Canals et al. (2009)	Freshwater cycling/regulation: Effect of land occupation on water availability through changes in evapo-transpiration (ET), Infiltration (I) and Runoff (R)	Lost precipitation (mm/m^2 year)	Absolute	PNV	Not modelled for transformation	2 types of land use (natural and artificial or sealed) and 2 classes of rainfall area (>600 mm $year^{-1}$ or <600 mm $year^{-1}$). Global CF provided for 22 land use types
Maes et al. (2009)	Freshwater cycling/regulation: Effect of land occupation and transformation on water availability through changes in evapo-transpiration (ET)	Change in water available for ecosystems (mm)	Absolute amount of water; CF Relative to ET of PNV	PNV	Not considered	2 main types of land use, depending on whether actual ET is $<$ or $>$ than ET of PNV
Núñez et al. (2013b)	Change in long-term freshwater availability downstream linked to a specific land use upstream	Delta green water (dGW) deprivation impacts: impacts of GW consumed by the system studied minus impacts of GW consumed by the reference system (WSI × dGW)	Relative to GW consumption of the reference system	Alternative contemporary land use or PNV	Not considered	CFs used are WSI by Pfister et al. (2009)
Beck et al. (2010) and Baitz (2002), Saad et al. (2011, 2013)	Freshwater regulation; Erosion resistance; Water purification	Groundwater recharge (mm $year^{-1}$); erosion resistance (Mg soil ha^{-1} $year^{-1}$); physic-chemical filtration (centimoles of cation fixed per kilogram of soil); mechanical filtration (cm day^{-1})	Absolute	PNV	Dobben et al. (1998)	Global, 7 land use types at terrestrial biome (14) and Holdridge life zones (38) levels)

State indicators						
Lindeijer (2000)	Life Support Functions	fNPP (free NPP) (tonnes biomass m^{-2} $year^{-1}$)	Absolute	fNPP of natural (pristine) vegetation in the region (initial NPP for transformation impacts)	Not included as part of the modelling	Reference NPP values provided globally for 8 classes of NPP, and initial evaluation provided for 7 land use types across the world
Weidema and Lindeijer (2001)	Life Support Functions through altered soil functions	Change in NPP (tonnes biomass m^{-2} $year^{-1}$)	Absolute	Climax vegetation	Dobben et al. (1998)	NPP values provided for the main global terrestrial ecosystems, as well as for key types of land uses
Milà i Canals (2003), Milà i Canals et al. (2007b), Brandão and Milà i Canals (2013), Brandão et al. (2011)	Soil quality	Soil Carbon Deficit (Mg C year)	Absolute	Natural relaxation towards PNV	Depending on land use type and location	Global, distinction of 8 land use types at climatic region level
Achten et al. (2008)	Indicator system based on Ecosystem Structural Quality (ESQ) and Ecosystem Functional Quality (EFQ). All indicators are quantified relative to a local reference, and equally weighted in a final Impact Score (%)	ESQ = f[soil fertility (cation exchange capacity, base saturation), biomass production (fNPP, total aboveground biomass) and species diversity (vascular plant species richness)]	Relative	PNV	Not described	Guidance provided to calculate CF, but actual values given
		EFQ = f[soil structure (soil organic matter, soil compaction), vegetation structure (leaf area index, vertical space distribution) and on-site water balance (evapotranspiration, soil cover)]				
Brandão (2012)	Impacts on ecosystem services and biodiversity	HAPECS – Human Appropriation of Ecosystem Carbon Stock (Mg C year)	Absolute	Natural relaxation towards PNV	Depending on land use type and location	Provided for 4 land use types in 21 countries across the world

4.1.1 Species Richness

Species richness has been often suggested as an operational indicator for biodiversity, given the ease of communication and availability of data. Species richness depends on the taxonomic groups considered, where insects are one of the most diverse groups (ca. 60 %), followed by fungi (ca. 10 %), plants (ca. 2 %) and vertebrates (ca. 0.4 %; including mammals, birds, etc.; Heywood and Watson 1995). To reduce the bias introduced by the choice of taxonomic group used as indicator for overall biodiversity, the initial concepts in LCA assessed reduction in species richness in relative instead of absolute terms, based on the unit potentially disappeared fraction of species (PDF). Due to relatively good data availability and the important role of plants as primary producers in ecosystems, most studies focused on the reduction of vascular plant species richness, mostly based on data from Europe (Müller-Wenk 1998; Koellner 2000; Vogtländer et al. 2004; Koellner and Scholz 2007, 2008; Schmidt 2008; De Schryver et al. 2010), but also from Asia (Schmidt 2008; Itsubo and Inaba 2012), or global data (Lindeijer 2000). Geyer et al. (2010) developed an approach based on habitat suitability models of Californian vertebrate species. Species richness is strongly scale dependent and increases in a non-linear way when larger areas are analysed. This is described as species-area relationship (Arrhenius 1921), expressed in the equation:

$$S = cA^z. \tag{11.1}$$

where S stands for species richness, A for the area of the ecosystem, c and z are constants and z is smaller than 1.

Due to the non-linear species-area relationship in Eq. 11.1, the conversion of the first or the last piece of an ecosystem leads to very different species losses. Converting the first piece of land of a completely undisturbed ecosystem to an agricultural field might lead to a local displacement of most species, but it might not lead to a reduction in regional species richness, because all species still find sufficient habitat in the remaining ecosystem. However, converting additional land in an ecosystem that is already heavily impacted by human land use activities might contribute to a large number of species getting regionally or even globally extinct, e.g., if the species are endemic to this region. Based on these considerations, suggestions have been made to assess local and regional species loss caused by land use separately in LCA (e.g., Koellner 2000; De Schryver et al. 2010; de Baan et al. 2013a, b). However, across impact categories, the same endpoint unit for biodiversity loss, PDF, has been used to assess relative species loss at different spatial scales, which can result in a misleading equal weighting of local and global species loss in LCA applications (Curran et al. 2011). The search for a better endpoint unit for biodiversity loss than PDF is therefore still ongoing, including proposals for assessing absolute instead of relative species loss (e.g., de Baan et al. 2013b). The study of De Baan et al. (2013b) calculated absolute, regional species loss based on the so-called matrix-calibrated species-area relationship models developed by Koh and Ghazoul (2010). For nearly all global ecoregions and for five taxonomic groups (plants, mammals, birds, amphibians, and reptiles), de Baan et al. (2013b) modelled regional losses of endemic and non-endemic species caused by all land use activities within each ecoregion and allocated this total damage to the different land use types that occur within each region. They distinguish between potentially reversible regional loss of non-endemic species, which are used to calculate regional characterisation factors for occupation and transformation, and irreversible, global loss of endemic species to calculate characterisation factors for permanent impacts. They also quantify the uncertainties of the characterisation factors using Monte Carlo Simulations.

Most land use Life Cycle Impact Assessment (LCIA) methods were developed based on regionally available data, and extrapolation of results to other regions is difficult, because the necessary data are not readily available. Only few globally applicable methods were developed. Lindeijer (2000) proposed to use global plant species richness maps as the reference state and combine it with plant species richness data for different land use types. Quantifying the impacts for specific land use situations remained a challenge and was illustrated only for specific cases in Europe and South America (Lindeijer 2000). De Baan et al. (2013a) used empirical species richness data from both human-modified and undisturbed land in the same region to calculate relative local species losses per type of land use. The underlying data were derived from a global literature review of biodiversity surveys (GLOBIO3, Alkemade et al. 2009) and national biodiversity monitoring data (BDM 2004) and encompass multiple taxonomic groups and world regions. Local characterisation factors for occupation for specific land use types in 9 of the world's 14 biomes were calculated. For a subset of data, characterisation factors were also calculated using four other biodiversity indicators (including abundance of species and similarity of species composition between used land and reference situation), which lead to significantly different results. De Souza et al. (2013) calculated characterisation factors across several regions in the Americas based on functional diversity. Functional diversity considers the role of species for ecosystem functioning and is highest, when each species present in an ecosystem covers a different function. De Souza et al. (2013) compared the results of functional diversity with species richness for multiple taxonomic groups and land use types and found significant differences in impacts between some, but not all, land use types and taxa.

4.1.2 Other Biodiversity Metrics

Cowell (1998) proposed three indicators for biodiversity in addition to species diversity: the relative area of a specific ecosystem type, the number of rare species and the net primary production (NPP) as an indicator for the number of individuals. Weidema and Lindeijer (2001) suggested evaluating the biodiversity value of land based on vascular plant species richness of an ecosystem, ecosystem vulnerability, and ecosystem scarcity. Ecosystem vulnerability should indicate the relative number of species affected by a change in the ecosystem area, as expressed by the species-area relationship. Ecosystem scarcity accounts for the natural (global) scarcity of an ecosystem type, expressed as the inverse of the total potential ecosystem area. Weidema and Lindeijer (2001) quantified all three factors per biome and multiplied the factors after normalisation with the relative reduction in land quality for different land use types. This relative reduction was quantified based on rough estimates and not on empirical data. This approach was implemented in the Canadian impact assessment method LUCAS (Toffoletto et al. 2007), recalculating the approach on the level of the 15 Canadian ecozones. Several authors have further developed the approach of Weidema and Lindeijer (2001). Michelsen (2008) calculated ecosystem vulnerability and scarcity per ecoregion and combined this with relative biodiversity reduction factors for forestry in Scandinavia. Instead of using reduction in species richness, Michelsen (2008) proposed to use 'Conditions for Maintained Biodiversity, CMB' as an index for distinguishing between different forestry management regimes. This index includes information about relevant factors that help to maintain biodiversity in boreal forests, such as the amount of decaying wood, areas set aside, and introduction of alien tree species. A similar approach was applied to Kiwi fruit production in New Zealand (Coelho and Michelsen 2014), using hemeroby values (Brentrup et al. 2002) as an indication of biodiversity value of different land use types. Mueller et al. (2014) calculate three factors for the biodiversity value of land for all global ecoregions: species richness (based on mammals, birds, amphibians and reptiles), ecosystem vulnerability (based on the Conservation Risk Index developed by Hoekstra et al. 2005) and ecosystem irreplaceability (based on the endemism of mammals, birds, amphibians and reptiles). They multiply these factors with the relative reduction in vascular plant species richness of different land use types compared to a natural reference.

The 'biotope method' is another ecosystem metric, developed by Kyläkorpi et al. (2005) for Scandinavia and later also tested for Namibia (Burke et al. 2008). This method considers the losses and gains in area of different habitat types, i.e. biotopes, in a particular region for a specific change in land use. The relative importance of the biotope types for the local context are evaluated based on indicators such as red-list species and biotopes are categorized as either critical, rare, general biotopes, or technotopes (i.e. not supporting any biodiversity).

4.2 Methodologies Addressing Impacts on Ecosystem Services

Ecosystems provide a variety of goods and processes known as ecosystem services. The Millennium Ecosystem Assessment classifies them as provisioning, regulating, cultural and supporting services (Millennium Ecosystem Assessment 2005):

- *Provisioning services* refer to the provision of ecosystem products: food, fibre, fuel, genetic resources, ornamental resources and fresh water.

- *Regulating services* such as air quality regulation, climate regulation, water regulation, erosion regulation, water purification and waste treatment, disease regulation, pest regulation, pollination, and natural hazard regulation.
- *Cultural services* refer to non-material aspects of ecosystems, such as cultural diversity, spiritual and religious values, knowledge systems, educational values, inspiration, aesthetic values, social relations, sense of place, cultural heritage values, and recreation and ecotourism.
- *Supporting services* include soil formation, photosynthesis, primary production, nutrient cycling, and water cycling, and as suggested by their name they underpin all other services.

It is difficult to integrate the impacts on all these complex service systems in a simple way within LCA. Cultural services are usually left out of environmental LCA due to the fact that they concern a social or economic issue. With regards to the others, in the Milennium Ecosystem Assessment classification, there is an overlap of some of these services, e.g. primary production is considered a supporting service, but when biomass is harvested then it is considered a provisioning service. In essence, most of the provisioning, regulating and supporting services have been often considered together in LCA in what has been called 'Life Support Functions' (Lindeijer 2000; Milà i Canals et al. 2007b).

Roughly speaking, LCIA authors have either chosen to model specific impact pathways (e.g. pressure on specific ecosystem services) or single indicators that could be considered representative of overall impact on ecosystem services, or the state of such services. The next two sub-sections provide an overview of what methods have been put forward to date in these two types of approaches.

4.2.1 Pressure Indicators

Given the diversity of ecosystem services to be protected, many authors have suggested modelling the land use effects on one or few of them. Cowell (1998) and Cowell and Clift (2000) offered some early recommendations on the incorporation of soil quality and quantity in LCA, including the quantification of soil erosion as an impact on abiotic resources. Núñez et al. (2013a) further develop the indicator for soil erosion through the RUSLE (Revised Universal Soil Loss Equation) and linking the lost soil to soil organic carbon as an indicator of soil quality (see below); the authors also provide the operational methods for LCA practitioners to calculate characterisation factors in a variety of situations with global coverage.

Salinisation has also been suggested as a relevant degradation process to be included in LCA, and a quantification method has been proposed (Feitz and Lundie 2002) based on levels of salinity in irrigation water. This can indeed be the most limiting process from a soil productivity point of view, although it is quite localised, and its quantification may be meaningless outside agricultural life cycle stages where irrigation is used within salinisation-prone areas.

Land is key in regulating freshwater, and the importance of this ecosystem service has granted it some attention by LCIA modellers. Heuvelmans et al. (2005) suggest using changes in the magnitude of infiltration and evapotranspiration flows alongside other indicators for land use impacts on water (namely change in surface runoff and precipitation surplus), in order to quantify the amount of rainwater available for ecosystems. Milà i Canals et al. (2009) go a bit further in the calculation of the 'net' green water, or land use effect on the availability of rainwater for ecosystems. Maes et al. (2009) suggest a similar yet slightly more sophisticated approach to assess this impact pathway, although data availability may hinder its global application. Núñez et al. (2013b) address changes in green water availability as affected by the studied system by calculating the differences between the green water consumed in the system and by the reference vegetation (using PNV as reference).

Some authors have attempted to model the impacts of land use on an array of ecosystem processes, providing multi-indicator assessment of ecosystem services. Beck et al. (2010) provide an operationalization of the approaches earlier suggested by Baitz (2002), and align them to the framework of the UNEP/SETAC Life Cycle Initiative (Milà i Canals et al. 2007a); the result is a set of indicators describing five ecosystem services: erosion resistance; mechanical filtration; physic-chemical filtration; groundwater replenishment and biotic production. Similarly, Saad et al. (2011, 2013) focus on three key ecosystem services, namely erosion regulation; freshwater regulation; and water purification, and provide characterisation factors with global coverage and spatial differentiation at the biome level for seven differentiated land use types. Achten et al. (2008) proposed the two endpoints Ecosystem Structural Quality (ESQ) and Ecosystem Functional Quality (EFQ). For each endpoint, three midpoints and indicators for quantifying them (in brackets) were defined. For ESQ, these midpoints are soil fertility (cation exchange capacity, base saturation), biomass production (fNPP, total aboveground biomass) and species diversity (vascular plant species richness). For EFQ, they propose soil structure (soil organic matter, soil compaction), vegetation structure (leaf area index, vertical space distribution) and on-site water balance (evapotranspiration, soil cover) as midpoints.

The difficulty with the above mentioned indicators and processes is that the focus on one specific degradation process or ecosystem service rends them less appropriate when applied on a life cycle scale, due to the fact that for many land uses they are not really relevant. For example, salinisation and erosion do not occur in a sealed surface, and yet the sealing of the surface has impeded or removed many ecosystem services. Providing more holistic assessments encompassing several ecosystem services (e.g. as advocated by Beck et al. (2010) or Saad et al. (2013)) partly deals with this problem, although a new challenge arises: that of increasing the number of indicators to be interpreted by the decision-maker in an already rich set of LCA results.

4.2.2 State Indicators

A different approach to representing the impacts on ecosystem services has been the definition of indicators describing the state of such services as a function of different types of land use. Brandão (2012) offers a comprehensive review of indicators for ecosystem services based on soil quality and ecosystem productivity.

Many authors have indeed focused on soil quality/fertility to define such a comprehensive indicator. Many have recommended soil organic carbon (SOC) as the best stand-alone indicator for soil quality (see, e.g. Reeves 1997). Within LCA there have been several suggestions to use SOC (or soil organic matter, SOM) as a single indicator for soil quality and ultimately as an indicator for ecosystem services: Cowell (1998), Mattsson et al. (1998), Cowell and Clift (2000), Schenck (2001), Milà i Canals (2003), Milà i Canals et al. (2007b), Brandão et al. (2011), Brandão and Milà i Canals (2013). The recent EC recommendation on the use of common methods to measure and communicate the life cycle environmental performance of products and organisations (EU 2013, following European Commission 2010) also suggests using SOC deficit as an indicator for land use impacts.

Other authors (Lindeijer 2000; Weidema and Lindeijer 2001) defined impact indicators related to the amount of biomass supported by the ecosystems, as a measure of the success of delivery of ecosystem services. Initial indicators looked at net primary productivity (NPP) and free NPP (fNPP). Haberl et al. (2007) calculated the Human Appropriation of NPP (HANPP) (roughly equivalent to fNPP) for global ecosystems, thus facilitating the use of this indicator in LCA studies. The trouble with productivity based indicators is that they represent the current growth as affected by climate and land management, but not the longer-term impact on land (see, e.g., Milà i Canals et al. 2007b). Brandão (2012) suggests a further elaboration of these indicators focusing on biomass or carbon stocks rather than flows (as with NPP and HANPP), suggesting that the Human Appropriation of Ecosystem Carbon Stock (HAPECS) provides a better approximation to the impacts on ecosystem services. After finding a strong correlation between ecosystem carbon stocks and species richness (Strassburg et al. 2010), Brandão (2012, p. 103) suggests that this indicator could encompass impacts on biodiversity as well.

In terms of defining endpoint indicators for ecosystem services, surplus energy to restore the functions of land could be considered as suggested by Stewart and Weidema (2005), e.g. considering hydroponics as a backup technology to produce biomass after land degradation. The limitation of such approaches of course is that land is multi-functional and it would be hard to envisage a backup technology capable of reproducing all its functions, including the emergent properties related to life and its diversity. Considering that ecosystem services are an essential support for Life, then an indicator for the amount of Life could provide a relevant endpoint indicator. As mentioned above, though, productivity-based indicators like NPP or HANPP indicate a flow which can vary significantly in the short term, and miss the system's resilience or lack of it. On the other hand a measure of stock (capital) might help capturing the shorter-term changes induced by land management, but also the longer-term resilience. It could hence be argued that HAPECS-like indicators might be suggested as endpoint indicators for ecosystem services. This is an area that requires further research and consensus-building.

4.3 Methodologies Addressing Other Impacts from Land Use

As suggested in Fig. 11.2, land use may also be linked to other endpoints than those covered by biodiversity and ecosystem services. Among these, climate effects due to the impacts on climate regulation caused by land use and land use change tend to be modelled separately from other ecosystem services, even though strictly speaking climate regulation is part of ecosystem services. This is due to the importance of climate change in many other impact pathways, and the relevance of land use and land use change in global warming (ca, 18 % of global GHG emissions arise from land use and land use change according to Bellarby et al. 2008). It is worth noting the approach to measuring the contribution to changes in atmospheric concentration of GHG caused by LUC (Müller-Wenk and Brandão 2010) or the support to estimating the changes in carbon stocks caused by LUC (e.g. Flynn et al. 2012). Land use also affects land's contribution to climate change through the albedo effect; this has been described by Muñoz et al. (2010), and the potential importance of such effect on an agricultural system's contribution to global warming raised, although this potentially significant contribution of land use and land use change to climate has not been taken up yet by common LCIA methods.

As discussed in Chap. 13 of this volume,[1] land use may also lead to its loss as a resource (when land is degraded e.g. through erosion). When not degraded through its use, land use may be linked to the impact categories addressed in this chapter, and also to land competition, which is not essentially an environmental impact but is of economic nature.

5 New Developments and Research Needs

The evolution of land use impact assessment in LCA has progressed rapidly in the last decade, usually going hand in hand with development of new databases and GIS tools supporting landscape ecology. Easy access to global databases for biodiversity, climate and several soil properties, coupled with increased computing capacity in GIS software has enabled increased ecological relevance and spatial resolution of impact characterisation for the pathways considered in this chapter. However, some of the basic challenges and subjective choices remain in terms of

[1] Abiotic resource use by Pilar Swart, Rodrigo A. F. Alvarenga, Jo Dewulf.

the modelling approaches and the practical implementation of land use impact assessment in LCA. The choice of a reference system based on "quasi-natural" states (e.g. Potential Natural Vegetation) or current states provides different information to decision-makers, assigning high land use impacts to regions with high past or high future impacts, respectively. In order to use current mixes of land use as a reference, consensus would be required on a "baseline or reference year", and data be provided with global coverage for its practical application. In the field of biodiversity impact assessment, the choice of absolute vs. relative indicators implies a value choice that biases the results in favour of either species-rich or species-poor ecosystems. For absolute impacts, value choices have to be made to aggregate impacts to different taxonomic groups, i.e., how detrimental we consider the extinction of an insect species compared to a mammal. The spatial scale of biodiversity loss modelled also implies a value choice. Do we only consider global extinction of species, or should we also assign damage to local or regional extinction? To finally allow integrating impacts on biodiversity from different drivers of biodiversity loss, such as land and water use, climate change, and pollution, endpoint units need to be better harmonised, including the spatial scale of biodiversity loss, integration of different taxonomic groups, and if impacts are assessed in relative or absolute terms.

As for the assessment of impacts on ecosystem services by land use the key area to address is the aggregation among the different impact pathways, possibly targeting the spatially dependent degradation process which is closest to a tipping point or threshold (Milà i Canals et al. 2006). In ecosystem services modelling there is also great scope for increasing the relevance of current indicators by leveraging global databases, and this has been explored e.g. with productivity and biomass-based maps, as well as global maps and databases for soil properties. There has been perhaps less attention on the development of an endpoint indicator for ecosystem services, although biomass stocks as affected by human processes may be a promising avenue (e.g., Human Appropriation of Ecosystem Carbon Stock, HAPECS, as suggested by Brandão 2012). There remains the question of whether impacts on Biodiversity could be quantified with the same indicator as ecosystem services (an endpoint indicator for ecosystem quality). Developing comparable endpoint indicators would help in determining the key processes contributing to impacts on ecosystem quality (hotspots) although such an aggregated indicator would lose an important amount of information e.g. on the emergent properties that come with diversity such as resilience.

From an implementation point of view, tracking supply chain information on sourcing locations remains a challenge for global supply chains sourcing commodities from around the world (Milà i Canals et al. 2013). Thus, the implementation of LCIA models and approaches that allow very fine spatial resolution may be impeded by access to LCI information with equivalent resolution. As suggested by Hellweg and Milà i Canals (2014) 'Big Data' efforts could be leveraged to combine increased information on ingredients' sourcing areas with databases and maps of biodiversity and ecosystem services in order to increase the accuracy of land use impact assessment in LCA.

As land use impact assessment in LCA becomes more spatially differentiated by leveraging GIS, the accuracy and environmental relevance will also increase. However, as these results become more widely available and spatially-dependent LCI databases become the norm (at least for land based products), it will become more apparent that some locations are simply not good to source certain products (e.g. Núñez et al. 2013b), even under better land management conditions. In this sense, the LCA results would be providing better information to support decision making, and it would be up to the user of such information to decide what to do with it: e.g. work with suppliers in sensitive areas to reduce their impacts (e.g. promoting compliance with good practice and certification) vs. shifting suppliers.

Disclaimer The views expressed in this article are those of the author and do not necessarily reflect those of UNEP.

References

Achten WMJ, Mathijs E, Muys B Proposing a life cycle land use impact calculation methodology. In: 6th international conference on LCA in the agri-food sector, 1 Nov 2008, Zurich

Alkemade R, van Oorschot M, Miles L, Nellemann C, Bakkenes M, ten Brink B (2009) GLOBIO3: a framework to investigate options for reducing global terrestrial biodiversity loss. Ecosystems 12(3):374–390

Arrhenius O (1921) Species and area. J Ecol 9(1):95–99

Audsley E (coord.), Alber S, Clift R, Cowell S, Crettaz P, Gaillard G, Hausheer J, Jolliet O, Kleijn R, Mortensen B, Pearce D, Roger E, Teulon H, Weidema B, Van Zeijts H (1997) Harmonisation of environmental life cycle assessment for agriculture. Final report. concerted action AIR3-CT94-2028. European Commission. DG VI Agriculture

Baitz M (2002) Die Bedeutung der funktionsbasierten Charakterisierung von Flächen-Inanspruchnahmen in industriellen Prozesskettenanalysen: Ein Beitrag zur ganzheitlichen Bilanzierung. Dissertation. Berichte aus der Umwelttechnik. Institut für Kunststoffprufung und Kunstsoffkunde, Universität Stuttgart, Shaker Verlag, Aachen

Bastian O, Schreiber K-F (1999) Analyse und ökologische Bewertung der Landschaft. Spektrum Akademischer Verlag, Heidelberg

BDM (2004) Biodiversity monitoring Switzerland. Indicator Z9: species diversirty in habitats. In: de Baan L, Alkemade R, Koellner T (2013a)

Beck T, Bos U, Wittstock B, Baitz M, Fischer M, Sedlbauer K (2010) LANCA. Land use indicator value calculation in life cycle assessment. Fraunhofer Verlag, Suttgart

Bellarby J, Foereid B, Hastings A, Smith P (2008) Cool farming: climate impacts of agriculture and mitigation potential. Greenpeace, Amsterdam, Retrieved from http://marktcheck.greenpeace.at/uploads/media/Cool_Farming_Report_Final_web_01.pdf

Brandão M (2012) Food, feed, fuel, timber or carbon sink? Towards sustainable land use – a consequential life-cycle approach. PhD thesis, University of Surrey, Guildford

Brandão M, Milà i Canals L (2013) Global characterisation factors to assess land use impacts on Biotic Production. Int J Life Cycle Assess 18(6):1243–1252

Brandão M, Milà i Canals L, Clift R (2011) Soil Organic Carbon changes in the cultivation of energy crops: implications for GHG balances and soil quality for use in LCA. Biomass Bioenergy 35:2323–2336

Brentrup F, Kusters J, Lammel J, Kuhlmann H (2002) Life cycle impact assessment of land use based on the Hemeroby concept. Int J Life Cycle Assess 7(6):339–348

Burke A, Kyläkorpi L, Rydgren B, Schneeweiss R (2008) Testing a Scandinavian biodiversity assessment tool in an African desert environment. Environ Manage 42(4):698–706
Chaplin-Kramer B, Sharp R, Mandle L, Sim S, Johnson J, Butnar I, Milà i Canals L, Eichelberg B, Ramler I, Mueller C, McLachlan N, Yousefi A, King H, Kareiva PM (in press) Where matters: understanding how spatial patterns of agricultural expansion impact biodiversity and carbon storage at a landscape level. Proc Natl Acad Sci USA
Coelho CRV, Michelsen O (2014) Land use impacts on biodiversity from kiwifruit production in New Zealand assessed with global and national datasets. Int J Life Cycle Assess 19:285–296
Cowell S (1998) Environmental life cycle assessment of agricultural systems: integration into decision-making. PhD thesis, University of Surrey, Guildford
Cowell SJ, Clift R (2000) A methodology for assessing soil quantity and quality in life cycle assessment. J Clean Prod 8(4):321–331
Curran M, de Baan L, De Schryver A, van Zelm R, Hellweg S, Koellner T, Sonnemann G, Huijbregts MAJ (2011) Toward meaningful end points of biodiversity in life cycle assessment. Environ Sci Technol 45(1):70–79
Curran M, Hellweg S, Beck J (2014) Is there any empirical support for biodiversity offset policy? Ecol Appl 24(4):617–632
de Baan L, Alkemade R, Koellner T (2013a) Land use impacts on biodiversity in LCA: a global approach. Int J Life Cycle Assess 18(6):1216–1230
de Baan L, Mutel CL, Curran M, Hellweg S, Koellner T (2013b) Land use in life cycle assessment: global characterization factors based on regional and global potential species extinctions. Environ Sci Technol 47(16):9281–9290
De Schryver AM, Goedkoop MJ, Leuven RSEW, Huijbregts MAJ (2010) Uncertainties in the application of the species area relationship for characterisation factors of land occupation in life cycle assessment. Int J Life Cycle Assess 15(7):682–691
de Souza DM, Flynn DFB, Declerk F, Rosenbaum RK, de Melo Lisboa H, Koellner T (2013) Land use impacts on biodiversity: proposal of characterization factors based on functional diversity. Int J Life Cycle Assess 18(6):1231–1242
EU (2013) 2013/179/EU Commission recommendation of 9 April 2013 on the use of common methods to measure and communicate the life cycle environmental performance of products and organisations. Off J Eur Union L 124:0001–0210
European Commission (2010) Recommendations based on existing environmental impact assessment models and factors for Life Cycle Assessment in a European context (draft). International Reference Life Cycle Data System (ILCD) handbook. Joint Research Centre, Institute for Environment and Sustainability, Ispra
Feitz AJ, Lundie S (2002) Soil salinisation: a local life cycle assessment impact category. Int J Life Cycle Assess 7(4):244–249
Flynn HC, Milà i Canals L, Keller E, King H, Sim S, Hastings A, Wang S, Smith P (2012) Quantifying global greenhouse gas emissions from land use change for crop production. Glob Change Biol 18(5):1622–1635
Geyer R, Lindner JP, Stoms DM, Davis FW, Wittstock B (2010) Coupling GIS and LCA for biodiversity assessments of land use: part 2: impact assessment. Int J Life Cycle Assess 15 (7):692–703
Goedkoop MJ, Heijungs R, Huijbregts M, De Schryver A, Struijs J, Van Zelm R (2008) ReCiPe 2008, a life cycle impact assessment method which comprises harmonised category indicators at the midpoint and the endpoint level, 1st edn Report I: Characterisation; 6 Jan 2009. http://www.lcia-recipe.net
Haberl H, Erb KH, Krausmann F, Gaube V, Bondeau A, Plutzar C, Gingrich S, Lucht W, Fischer-Kowalski M (2007) Quantifying and mapping the human appropriation of net primary production in earth's terrestrial ecosystems. Proc Natl Acad Sci U S A 104(31):12942–12947
Hau JL, Bakshi BR (2004) Promise and problems of emergy analysis. Ecol Model 178(1–2):215–225

Hellweg S, Milà i Canals L (2014) Emerging approaches, challenges and opportunities in life cycle assessment. Science 344(6188):1109–1113

Heuvelmans G, Muys B, Feyen J (2005) Extending the life cycle methodology to cover impacts of land use systems on the water balance. Int J Life Cycle Assess 10:113–119

Heywood VH, Watson RT (1995) Global biodiversity assessment. Cambridge University Press, Cambridge

Hoekstra JT, Boucher TM, Ricketts T (2005) Confronting a biome crisis: global disparities of habitat loss and protection. Ecology 8:23–29

Itsubo N, Inaba A (2012) LIME2. LCIA method based on endpoint modeling. Newsletter. Life-Cycle Assessment Society of Japan

Jeanneret P, Baumgartner DU, Freiermuth Knuchel R, Gaillard G (2009) Methode zur Beurteilung der Wirkung landwirtschaftlicher Aktivitäten auf die Biodiversität für Ökobilanzen (SALCA-Biodiversität). Agroscope Reckenholz-Tänikon ART, Zürich/Ettenhausen, 74 p

Koellner T (2000) Species-pool effect potentials (SPEP) as a yardstick to evaluate land-use impacts on biodiversity. J Clean Prod 8:293–311

Koellner T, Scholz R (2007) Assessment of land use impacts on the natural environment. Part 1: an analytical framework for pure land occupation and land use change. Int J Life Cycle Assess 12(1):16–23

Koellner T, Scholz R (2008) Assessment of land use impacts on the natural environment. Part 2: generic characterization factors for local species diversity in Central Europe. Int J Life Cycle Assess 13(1):32–48

Koellner T, de Baan L, Brandão M, Civit B, Margni M, Milà i Canals L, Saad R, Maia de Souza D, Beck T, Müller-Wenk R (2013a) UNEP-SETAC Guideline on global land use impact assessment on biodiversity and ecosystem services in LCA. Int J Life Cycle Assess 18(6):1188–1202

Koellner T, de Baan L, Beck T, Brandão M, Civit B, Goedkoop M, Margni M, Milà i Canals L, Müller-Wenk R, Weidema B, Wittstock B (2013b) Principles for life cycle inventories of land use on a global scale. Int J Life Cycle Assess 18(6):1203–1215

Koh LP, Ghazoul L (2010) A Matrix-calibrated species-area model for predicting biodiversity losses due to land-use change. Conserv Biol 24(4):994–1001

Kyläkorpi K, Rydgren B, Ellegård A, Miliander S (2005) The biotope method 2005: a method to assess the impact of land use on biodiversity. Vattenfall, Sweden

Lenzen M, Lane A, Widmer-Cooper A, Williams M (2009) Effects of land use on threatened species. Conserv Biol 23:294–306

Lindeijer E (2000) Biodiversity and life support impacts of land use in LCA. J Clean Prod 8:313–319

Lindeijer E, Müller-Wenk R, Steen B et al (2002) Resources and land use. In: Udo de Haes H, Jolliet O, Finnveden G, Goedkoop M, Hauschild M, Hertwich E (eds) Life-cycle impact assessment: striving towards best practice. SETAC Press, Pensacola

Maes WH, Heuvelmans G, Muys B (2009) Assessment of land use impact on water-related ecosystem services capturing the integrated terrestrial-aquatic system. Environ Sci Technol 2009(43):7324–7330

Mattsson B, Cederberg C, Ljung M (1998) Principles for environmental assessment of land use in agriculture. SIK Rapport (642). SIK, The Swedish Institute for Food and Biotechnology, Goteborg

Michelsen O (2008) Assessment of land use impact on biodiversity. Int J Life Cycle Assess 13 (1):22–31

Milà i Canals L (2003) Contributions to LCA methodology for agricultural systems. Site-dependency and soil degradation impact assessment. Ph.D thesis. Autonomous University of Barcelona, Barcelona. Available from http://www.tdx.cesca.es/TDX-1222103-154811/

Milà i Canals L, Clift R, Basson L, Hansen Y, Brandão M (2006) Expert Workshop on Land Use Impacts in Life Cycle Assessment (LCA). 12–13 June 2006 Guildford, Surrey (UK). Int J Life Cycle Assess 11(5):363–368

Milà i Canals L, Bauer C, Depestele J, Dubreuil A, Freiermuth Knuchel R, Gaillard G, Michelsen O, Müller-Wenk R, Rydgren B (2007a) Key elements in a framework for land use impact assessment in LCA. Int J Life Cycle Assess 12(1):5–15

Milà i Canals L, Romanyà J, Cowell SJ (2007b) Method for assessing impacts on life support functions (LSF) related to the use of 'fertile land' in Life Cycle Assessment (LCA). J Clean Prod 15:1426–1440

Milà i Canals L, Chenoweth J, Chapagain AK, Orr S, Antón A, Clift R (2009) Assessing freshwater use impacts in LCA part I: inventory modelling and characterisation factors for the main impact pathways. Int J Life Cycle Assess 14(1):28–42

Milà i Canals L, Rigarlsford G, Sim S (2013) Land use impact assessment of Margarine. Int J Life Cycle Assess 18(6):1265–1277

Millennium Ecosystem Assessment (2005) Ecosystems and human well-being: synthesis. Island Press, Washington, DC

Mueller C, de Baan L, Koellner T (2014) Comparing direct land use impacts on biodiversity of conventional and organic milk—based on a Swedish case study. Int J Life Cycle Assess 19 (1):52–68

Müller-Wenk R (1998) Land use – the main threat to species. How to Include land use in LCA. IWÖ – Diskussionsbeitrag No. 64. IWÖ, Universität St. Gallen, St. Gallen

Müller-Wenk R, Brandão M (2010) Climatic impact of land use in LCA—carbon transfers between vegetation/soil and air. Int J Life Cycle Assess 15:172–182

Muñoz I, Campra P, Fernández-Alba AR (2010) Including CO_2-emission equivalence of changes in land surface albedo in life cycle assessment. Methodology and case study on greenhouse agriculture. Int J Life Cycle Assess 15:672–681

Núñez M, Antón A, Muñoz P, Rieradevall J (2013a) Inclusion of soil erosion impacts in life cycle assessment on a global scale: application to energy crops in Spain. Int J Life Cycle Assess 18(4):755–767

Núñez M, Pfister S, Antón A, Muñoz P, Hellweg S, Koehler A, Rieradevall J (2013b) Assessing the environmental impacts of water consumption by energy crops grown in Spain. J Ind Ecol 17(1):90–102

Pfister S, Koehler A, Hellweg S (2009) Assessing the environmental impacts of freshwater consumption in LCA. Environ Sci Technol 43:4098–4104

Reeves DW (1997) The role of soil organic matter in maintaining soil quality in continuous cropping systems. Soil Tillage Res 43:131e67

Saad R, Margni M, Koellner T, Wittstock B, Deschênes L (2011) Assessment of land use impacts on soil ecological functions: development of spatially differentiated characterization factors within a Canadian context. Int J Life Cycle Assess 16:198–211

Saad R, Koellner T, Margni M (2013) Land use impacts on freshwater regulation, erosion regulation and water purification: a spatial approach for a global scale. Int J Life Cycle Assess 18:1253–1264

Sala OE, Chapin FS III, Armesto JJ, Berlow E, Bloomfield J, Dirzo R, Huber-Sanwald E, Huenneke LF, Jackson RB, Kinzig A, Leemans R, Lodge DM, Mooney HA, Oesterheld M, LeRoy Poff N, Sykes MT, Walker BH, Walker M, Wall DH (2000) Global biodiversity scenarios for the year 2100. Science 287:1770–1774

Schenck RC (2001) Land use and biodiversity indicators for life cycle impact assessment. Int J Life Cycle Assess 6(2):114–117

Schmidt JH (2008) Development of LCIA characterisation factors for land use impacts on biodiversity. J Clean Prod 16:1929–1942

Searchinger T, Heimlich R, Houghton RA, Dong F, Elobeid A, Fabiosa J, Tokgoz S, Hayes D, Yu T-H (2008) Use of U.S. Croplands for biofuels increases greenhouse gases through emissions from land-use change. Science 319(5867):1238–1240

Stewart M, Weidema B (2005) A consistent framework for assessing the impacts from resource use. A focus on resource functionality. Int J Life Cycle Assess 10(4):240–247

Strassburg BBN, Kelly A, Balmford A, Davies RG, Gibbs HK, Lovett A, Miles L, Orme CDL, Price J, Turner RK, Rodrigues ASL (2010) Global congruence of carbon storage and biodiversity in terrestrial ecosystems. Conserv Lett 3(2):98

Toffoletto L, Bulle C, Godin J, Reid C, Deschenes L (2007) LUCAS – a new LCIA method used for a Canadian-specific context. Int J Life Cycle Assess 12(2):93–102

UNEP (2014) Assessing global land use: balancing consumption with sustainable supply. A Report of the Working Group on Land and Soils of the International Resource Panel. Bringezu S, Schütz H, Pengue W, O'Brien M, Garcia F, Sims R, Howarth R, Kauppi L, Swilling M, and Herrick J. http://www.unep.org/resourcepanel/Publications

van Dobben H, Schouwenberg E, Nabuurs G, Prins A (1998) Biodiversity and productivity parameters as a basis for evaluating land use changes in LCA. In: Lindeijer E, van Kampen M, Fraanje P et al (eds) Biodiversity and life support indicators for land use impacts in LCA, vol (Publication series raw materials 1998/07). Dienst Wegen Waterbouwkunde, Delft

Vogtländer J, Lindeijer E, Witte J, Hendriks C (2004) Characterizing the change of land-use based on flora: application for EIA and LCA. J Clean Prod 12:47–57

Wagendorp T, Gulinck H, Coppin P, Muys B (2006) Land use impact evaluation in life cycle assessment based on ecosystem thermodynamics. Energy 31(1):112–125

Weidema BP, Lindeijer E (2001) Physical impacts of land use in product life cycle assessment. Final report of the EURENVIRONLCAGAPS sub-project on land use. Technical University of Denmark, Lyngby

Chapter 12
Water Use

Stephan Pfister

Abstract Water use impacts have two different dimensions: pollution (degradative use) and consumption (consumptive use). Degradative use is mainly tackled by impact assessment of pollutant emissions. This chapter focuses on water deprivation due to consumption. The impacts considered here address the case of ecosystems and human users being deprived of water, but also the depletion of stock resources, potentially depriving future users of water.

The short-term water cycle is dominated by evaporation from sea, precipitation on land and runoff in rivers. Groundwater and lakes play a longer-term role but are crucial for the assessment of effects of water use competition. While the global water cycle is not heavily influenced by human activities, the impacts can be significant in specific regions, and therefore regional water cycles are relevant. The temporal variability of the hydrological processes is also important for considerations of water use effects on environment.

The existing methods consider the main features of the hydrological cycle but still many improvements in the global, regionalised data are required for proper integration of relevant aspects in life cycle impact assessment (LCIA) of water use. Moreover, the available life cycle inventories have generally low data quality for the relevant flows for most processes and often lack global coverage.

The impact assessment methods can be grouped into four main categories: water scarcity as midpoint, impacts on human health, impacts on ecosystem quality, and resource depletion. They cover a variety of impact pathways, while still many important issues such as in-stream use of dams are currently missing.

Keywords Evaporation • Consumptive use • LCA • LCIA • Life cycle assessment • Life cycle impact assessment • Water depletion • Water footprint • Water scarcity • Water stress • Withdrawal

---

S. Pfister (✉)
Institute of Environmental Engineering (IfU), ETH Zurich, HPZ E 33,
John-von-Neumann-Weg 9, Zurich, CH-8093, Switzerland
e-mail: pfister@ifu.baug.ethz.ch

M.Z. Hauschild, M.A.J. Huijbregts (eds.), *Life Cycle Impact Assessment*,
LCA Compendium – The Complete World of Life Cycle Assessment,
DOI 10.1007/978-94-017-9744-3_12

1 Introduction

1.1 Water Use and LCA

Water is an abiotic resource and an environmental compartment receiving emissions. Therefore two principal dimensions of water need to be considered in LCA: resource use and pollution of water bodies. From a resource perspective we can further distinguish resources into stocks and flows as previously suggested (Guinée 2001). Stocks are classical resources like deposits and funds and are addressed as depletion of an abiotic resource. For water, this mainly concerns use of fossil groundwater as well as overuse of large lakes or groundwater bodies beyond their annual renewability (recharge rate) such as the depletion of the Aral Sea. Resource flows are resources where mainly competition for resource access is causing impacts, such as for water, or in the cases of wind and solar radiation, for instance.

Water is the most extracted resource (Zekster and Everett 2004) and has a special role since it is required for life. It is therefore ultimately needed for human activities as well as the natural environment. As such, its role in LCA can be compared to the one of land use. However, water is a flowing resource and consequently impacts not only occur at the point of use, but also downstream or within the area that is hydrologically connected (e.g. through groundwater level drops).

Water use in LCA typically only accounts for freshwater use and impacts of the use of sea or brackish water are disregarded so far. In LCA, the framework presented in Pfister et al. (2009, see Fig. 12.1) and updated in Bayart et al. (2010) is largely accepted. It distinguishes between degradative and consumptive use of withdrawals as well as in-stream water use (e.g. run-of-the-river hydroelectricity or shipping). Degradative water use relates to pollution of water during its use, while consumptive use refers to water that is not released back to the original watershed and therefore not available for downstream users. Consumptive water use is consequently set equal to water consumption and includes water that is evaporated, transpired, incorporated into products, released to a different watershed or directly to saline water.

Degradative water use is mainly covered in the chapters of the respective emissions. Some aspects of degradative water use have also been integrated in a few impact assessment methods concerning water resources, as explained below. Thermal emissions to water, addressed by Verones et al. (2010), also belong to water degradation (pollution with heat).

In addition, the framework distinguishes in-stream water use and storage which is addressed in land use (water surface). While shipping and recreation mainly refer to the area of water bodies used, the built environment affects the river flow regime largely through channeling river stretches or storing large water amounts in dams and reservoirs. These aspects are highly important but so far not properly addressed in LCA. Increased evaporation from reservoirs due to storage is also accounted for by consumptive water use.

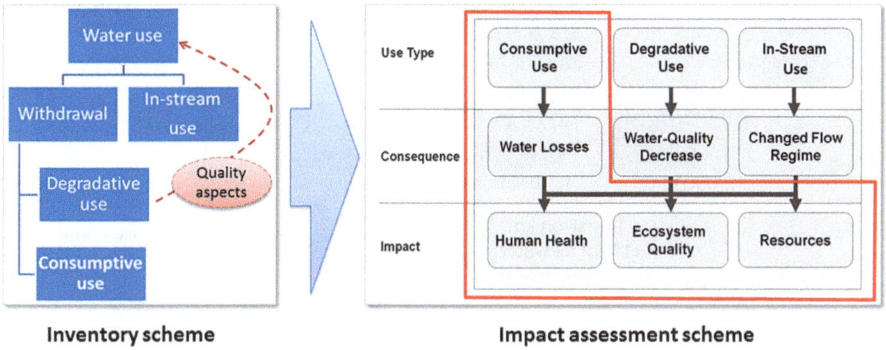

Fig. 12.1 Framework for integrating water-use related environmental impacts into LCA and water footprinting. *Left part*: Life cycle inventory (LCI) data addresses different types of water use: In-stream use, consumptive use and degradative use. Quality aspects, which are important for water-reuse options, are covered by degradative use (emissions of pollutants) of freshwater. *Right part*: While the focus of this chapter is on consumptive water use (*red frame*), all use types can have impacts on 'human health', 'ecosystem quality' and 'resources', depending on the regional conditions. The graph has been adjusted from Pfister et al. (2009)

1.2 Global Water Cycle (Water Availability)

Water is a flowing resource and every water flow is part of the water cycle which depends on the interaction of the static water storage between atmosphere, water bodies, aquifers, ice and to a minor extent between soils and living organisms (Fig. 12.2). As summarised by Shiklomanov (1999), the total amount of water in the hydrosphere "consists of the free water in liquid, solid, or gaseous states in the atmosphere, on the earth's surface, and in the crust down to the depth of 2,000 m. By approximate estimates (Korzoun 1974; Korzoun et al. 1978), the earth's hydrosphere contains a huge amount of water, about 1,386 million cubic kilometers (km^3). However, 97.5 % of this amount is saline water, and only 2.5 % is fresh water. The greater portion of the fresh water (68.7 %) is in the form of ice and permanent snow cover in the Antarctic, the Arctic, and mountainous regions. Fresh groundwater comprises 30 % of fresh water resources." Finally, only 0.3 % of the total amount of fresh water on the earth is 'visible' in lakes and river systems, which are most accessible for humans and are very important for ecosystems (Shiklomanov 1999).

While the freshwater stocks are dominated by ice and groundwater, the largest freshwater flows occur in rivers due to their high dynamics. These flows are also highly dependent on short time intervals: the average residence time of water in a river is ~2 weeks, while for lakes it is 17 years and for ice and deep groundwater more than 1,000 years (Shiklomanov and Rodda 2003). These different temporal dimensions are crucial aspects for the assessment of water uses, especially when considering interactions of ground- and surface water or snow/ice melt. Furthermore the numbers presented above are not constant and vary among seasons and

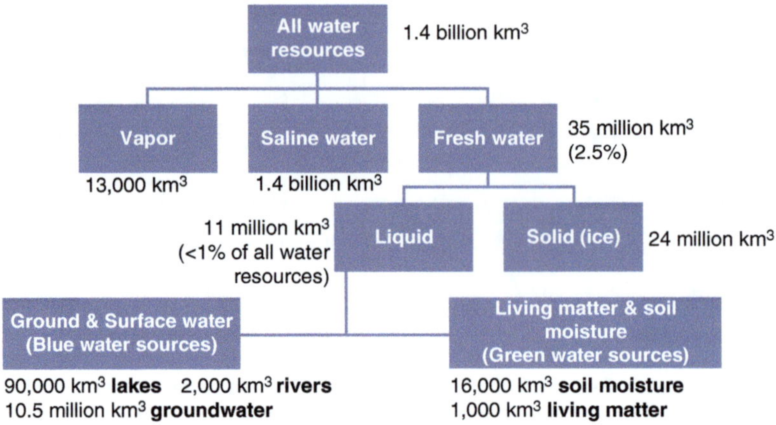

Fig. 12.2 Water cycle and the hydrosphere. Water volumes of water resources in the hydrosphere are based on Shiklomanov and Rodda (2003). Blue water is the freshwater part flowing through water bodies and aquifers which are accessible to human users while Green water refers to the water that is stored as soil moisture and is only available to plants or unproductive evaporation

Table 12.1 Global water flows

Flow	Flow [km^3/year]	Main reference
Precipitation on land	~100,000	Mitchell and Jones (2005)
Net available freshwater	77,000	Alcamo et al. (2003)
Transpiration (all plants)	~40,000	Rost et al. (2008)
Crops	~6,000	Rost et al. (2008)
Runoff	~35,000	Rost et al. (2008)
Human water withdrawal	3,600	Alcamo et al. (2003)
Irrigation water consumption	~1,000–2,000	Pfister et al. (2011b)

shorter time periods and consequently need additional attention, especially for water 'storages' with small residence times (rivers, water stored in biomass and in the atmosphere) in order to properly assess the impacts caused by consumption of renewable resources.

While water stocks are in the focus of classical resource impact assessment for future generations, water flows analysis is required for the assessment of current impacts of water consumption (Bayart et al. 2010). The typical flow rates (renewability) on a global level are presented in Table 12.1. These numbers indicate the low share of water used by humans on a global scale. Rockström et al. (2009) therefore suggested a safe operation limit of 4,000 km^3/year, which has not yet been reached. However, water use and consumption have led to severe environmental damages and human suffering since there is a severe distribution problem of water. Therefore the scale at which the water cycle is considered is crucial when assessing the environmental consequences. Usually, water scarcity is based on a watershed level (e.g. Alcamo et al. 2003; Pfister et al. 2009; Mekonnen and Hoekstra 2011) or higher spatial resolution (Oki and Kanae 2006; Fekete et al. 2002). Effects of recirculation of consumed water (e.g. increased precipitation after evaporation) within a relevant time and geographical scale are generally neglected due to lack of data, although these might be important in specific cases (Berger and Finkbeiner 2011). Water is required in massive amounts and has severe transport limitations and is therefore not a tradable good. Only a small fraction of drinking water is shipped around the world (mainly mineral waters), and the large water demand for agriculture and power plant cooling cannot be economically traded. Some large water transfer projects diverting rivers from one to another watershed exist but are heavily criticized such as the Chinese South–north water transfer that could divert 40 km^3/year from the Yangtse Basin to the Northern China Plain, a region with 300–325 million people (Berkoff 2013). The possibility of such a transfer is a controversial issue, since it would lead to additional water consumption and impacts in the Southern watershed. Moreover, even if an overall benefit were observed concerning water scarcity (Lin et al. 2012), negative social and other environmental consequences of such a large project might prevail. Therefore, water scarcity is typically identified as a ratio between water use and availability within a hydrological unit. Availability is not a consistent term since it can include both ground and surface water flows and it can refer to naturally or currently available amounts. Some studies also deduct environmental water requirements from availability (Smakhtin et al. 2004). A typical threshold for water scarcity is based on withdrawal-to-availability ratio of 20 and 60 % for extreme water scarcity (e.g. Alcamo et al. 2000).

The human influence on the global water cycle is mainly driven by agriculture which is responsible for ~85 % of water use (Shiklomanov 1999) and even more for water consumption. Irrigation water consumption is in the range of 1,000–2,000 km^3/year. Additionally, water consumption of agriculture often occurs in areas of high water scarcity (Pfister et al. 2011b) since arid zones with high irrigation water demand are often already water stressed. What is also very important is water consumption of dams and reservoirs (additional evaporation due to storage) with estimated flows of >65 km^3/year (Pfister et al. 2011a) to 222 km^3/year (Shiklomanov 1999).

1.3 Water Footprinting

Water use impacts are also addressed by the recently emerging water footprint approach initially suggested by Hoekstra and Hung (2002), which has been institutionalised in the water footprint network and by a related standard (Hoekstra et al. 2011). From a carbon and LCA perspective, a different definition has been proposed for the water footprint (Pfister and Hellweg 2009). The Water Footprint Network defines the water footprint as the total amount of water consumed from water withdrawals (blue water) and rain water plus soil moisture (green water), combined with an emission-based, calculated dilution water volume required to meet legal standards (grey water), and with an optional sustainability assessment

(Hoekstra et al. 2011). In accordance with carbon footprints, the common definition from the LCA community accounts only for the human induced water consumption (i.e. blue water) and requires a weighting with a spatially-explicit characterisation factor in order to account for the water scarcity and environmental vulnerability in each location (Ridoutt and Pfister 2010). Green water consumption is already covered as part of land use impacts, as is grey water by assessing the pollutants in an LCA framework. The ISO 14046 standard on water footprint has a public draft and is committed to follow LCA principles (ISO 2013). A water footprint approach aiming at fulfilling these ISO requirements has been presented by Ridoutt and Pfister (2013). The methods reported below are therefore also relevant for water footprinting. The two presented definitions, according to the WFN and LCA, are not principally divergent but rather suffer from lack of proper communication (Pfister and Ridoutt 2013). The suggested water footprint standard by WFN follows the procedure of ISO LCA standards (McGlade et al. 2012): a full water footprint assessment always includes an accounting (inventory), impact assessment and interpretation part, accounting for the local conditions.

2 Impact Pathways, Affected Areas of Protection

Impacts related to water use and consumption can affect all Areas of Protection (Pfister et al. 2009; Bayart et al. 2010; Kounina et al. 2013). A framework is presented in Fig. 12.3. For all impact pathways the location of the water use is a relevant parameter.

2.1 Ecosystem Quality

Water use and consumption disrupts ecosystems by temporarily reducing water availability and has been a major driver for biodiversity loss (MA 2005). Ecosystems affected include surface water bodies and aquifers but also groundwater dependent ecosystems. Therefore it is relevant to know from what kind of water body water is withdrawn. The complex hydrogeological conditions and interactions between surface and groundwater make it a very complicated task to follow cause-effect chains with global coverage. However, only global coverage allows accounting for all impact pathways.

2.2 Human Health

Human health impacts can be caused by water use and consumption but the cause-effect relation is difficult to assess due to socioeconomic mechanisms and high

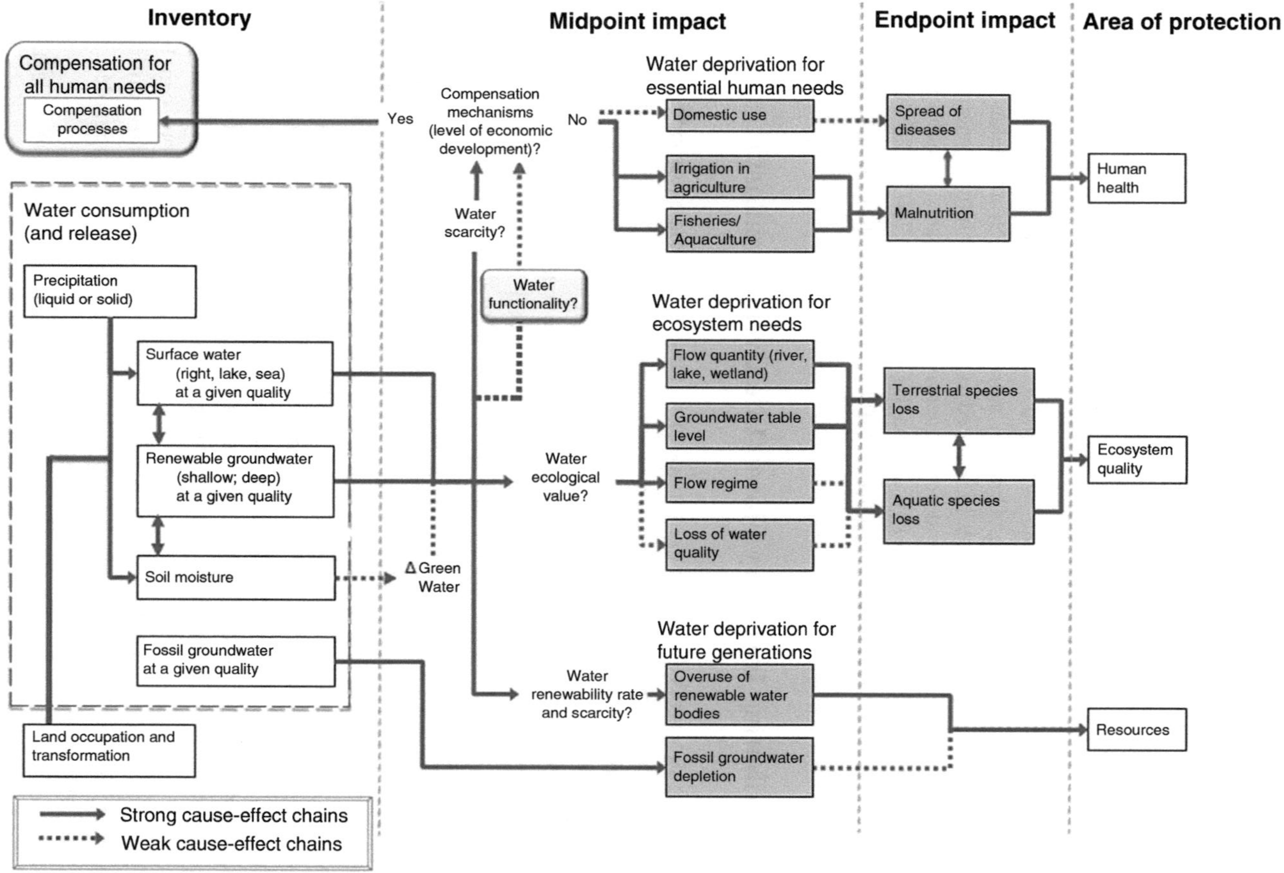

Fig. 12.3 Cause-effect chains leading from the inventory to the areas of protection human health, ecosystem quality and resources (Adapted from Kounina et al. (2013))

spatial variability. Most health problems related to lack of access to safe water are not mainly caused by water deprivation but by lack of infrastructure and by water pollution (Rijsberman 2006). Water scarcity effects and effective impact pathways for human health are therefore heavily dependent on socio-economic conditions which need to be accounted for (UN 2013; WHO 2008).

2.3 Resources

Resource depletion caused by water use and consumption primarily refers to overuse of groundwater stocks (e.g. Ogallala aquifer in the US) and lakes (such as the Aral Sea). Such depletion leads to reduced availability for future generations. In addition to impacts on stocks, and consequently future impacts, contemporary effects are addressed under the exergy approach (Bosch et al. 2007) and through external, consequential effects of water pollution by Boulay et al. (2011b).

3 Contributing Inventory Flows (Life Cycle Inventory)

Contributing inventory flows are exchanges of water between the environment and the studied processes. In order to account for water consumption the full water balance of each process is necessary and all inputs and outputs need to be recorded. The relevant flows are described in Fig. 12.4, with water flows from and to the environment, and flows of water within the technosphere (water content of products as well as tap and waste water flows). For agricultural processes, the process environment (agricultural land) needs to be included for detailed analysis of the 'green water' use (Falkenmark and Rockström 2004) and consequent net soil water use (Δ Soil moisture) compared to the reference land cover. The calculation of the total green water is part of the impact assessment step (by adding reference evapotranspiration calculated from land use information). However, current inventory databases do not account for this issue and it is not considered the most relevant in most cases. Δ Precipitation flows are relevant in rainwater collection systems (in industry or agriculture). The Δ Evapotranspiration flow is man-induced evapotranspiration, i.e. from water inputs to the technosphere and is equivalent to 'blue water' consumption (Falkenmark and Rockström 2004). This flow is also the main factor for water consumption. Salt water is usually not considered but can be necessary to balance the flows over a process, especially if seawater cooling systems are involved. Water consumption of seawater is generally disregarded in water use impact assessment.

Until recently, the two largest LCI databases ecoinvent v2.2 (ecoinvent Centre 2010) and GaBi databases 2006 (PE International 2012) accounted for water use, but the data quality and consistency did not match the level of other substances. In the years 2012 and 2013, updates of ecoinvent (v3) and GaBi include updated data

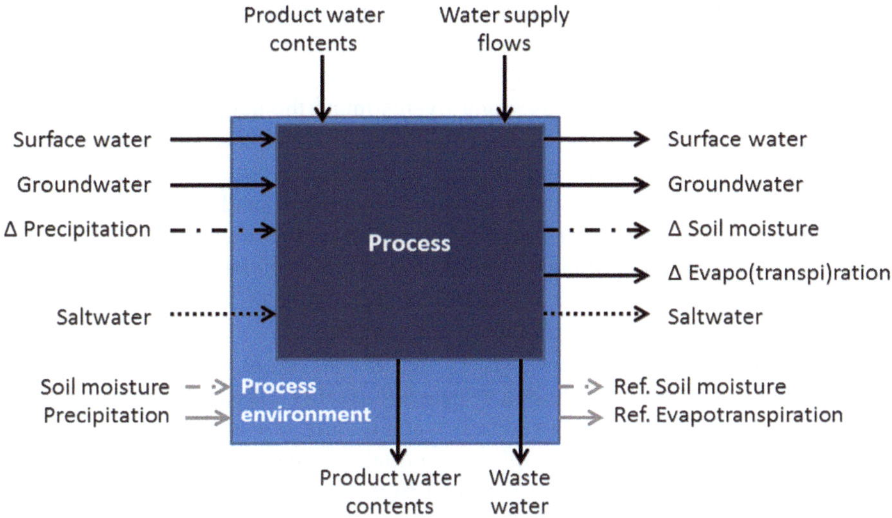

Fig. 12.4 Inventory flows relevant to the assessment of water use impacts. The horizontal flows are exchanges with the environment and the vertical flows are technosphere flows. The *Process box* shows the system boundaries for the process considered. For agricultural processes the process environment is usually included in water management (i.e. the water flows of natural water supply from soil and precipitation) and there is no clear border in LCA. These can be considered as 'neutral' flows as they present flows of the natural or reference environment (ref. flows), and are therefore considered to occur outside the system and not entering the technosphere

and a consistent structure that allows inclusion of water consumption as well as degradative use. However, water data has still been collected with limited efforts compared to other flows and consumptive water use numbers for most processes are based on estimates. ecoinvent v3 has very limited user experience to date and the GaBi database features limited transparency, because the inventory results are only available in aggregated form. Therefore, the robustness of the databases is still not fully tested. Additionally, both of these databases lack spatial differentiation beyond country level. Another LCI data source for a limited set of processes (main focus on crops) is the blue water accounting data provided by the Water Footprint Network (Hoekstra et al. 2011). However, careful consideration is required to avoid unintended inclusion of green water (soil moisture) consumption and 'grey water' (characterised degradative water use), which does not qualify for the purposes of LCA. Global coverage of water consumption data on country level, including uncertainty estimates, is also available for power production (Pfister et al. 2011a) and for agricultural products (Pfister et al. 2011a, b). These databases have been used to update parts of the GaBi database and in the Quantis water database (Vionnet et al. 2012), which has been the basis for evoinvent v3 water use inventory.

An important and controversial point is the inclusion of 'green water use' i.e. the water input from precipitation and soil moisture. By definition, green water is natural water supply received through land use. According to Pfister et al. (2009), only the difference in consumption of such water compared to natural situation should be included, whereas Milà i Canals et al. (2009) add surface water runoff from impervious areas as consumptive use, since it is not returned to the natural system in a proper manner. This might overestimate the negative impact of rapid runoffs, but accounts for the temporal aspects of water use and release. These approaches agree on the fact that precipitation and soil moisture use are indirect forms of water consumption which need to be linked to land use inventories. For this purpose, Núñez et al. (2013) derived regionalised factors for the assessment of green water consumption of the reference land use, with global coverage. Net green water consumption can then be assessed by analysing the difference of green water consumption between current and reference land use.

In addition to consumptive water use, Bayart et al. (2010) and Boulay et al. (2011a) suggest to differentiate specific quality classes in order to assess degradative use. This is useful to address water quality as a 'sum-parameter' from a resource perspective, but in principle, water quality impacts are better addressed by accounting for all substance flows related to water use and a consequent impact assessment.

3.1 Regionalisation

Spatial distinction of water sources is crucial since there is no global water scarcity (Rockström et al. 2009) and regional differences are crucial in any impact assessment method regarding consumptive water use. Regionalisation becomes more and more standard in LCIA and is prominent in land use impact assessment, which also requires regionalisation on a higher spatial resolution than country level (Pfister et al. 2010; de Baan et al. 2013). However, standard software tools feature very limited spatial differentiation. If water flows are transferred from one watershed to another, this has to be considered in the assessment, since it results in consumptive use in one watershed and negative consumptive use in the watershed of release.

3.2 Temporal Resolution

Temporal distinction is usually neglected in LCA. In water use, environmental effects often depend on the period within a year during which the water use takes place. Some water scarcity indicators take monthly time steps into account (Hoekstra et al. 2011; Pfister and Bayer 2013) and therefore inventories should include this information too, at least for foreground processes and especially for agricultural production, where large variability exists (Pfister and Bayer 2013). Such temporal resolution has also been applied for thermal pollution where timing is a key issue, too (Verones et al. 2010).

4 Impact Assessment Methodologies

Three main types of impact assessment methodologies are presented in Table 13.2 and further described below. Detailed analysis of the methods and a critical evaluation is presented in Kounina et al. (2013) for most methods. Newer methods not addressed in the aforementioned reference include Verones et al. (2013a, b), Loubet et al. (2013) and Tendall et al. (2014).

4.1 Methodologies Addressing Water Scarcity as Midpoint

Impact assessment methods addressing water scarcity at the midpoint level are mainly based on water use to availability ratios and result in a dimensionless characterisation factor (CF). The older methods base the CF on annual withdrawal-to-availability (WTA), as presented in Table 12.2.

The Swiss Ecoscarcity method (UBP06, Frischknecht et al. 2009) defines a distance-to-target approach with a target threshold of 20 % water used compared to availability. The function of the CF is an exponentially increasing one and is therefore sensitive to high WTA. In their original form the CFs were documented as country-specific values (Switzerland and other OECD countries), an average value for OECD countries (to be used when source region is unknown), and values for six different scarcity scenarios (low, moderate, medium, high, very high and extreme), spanning over an order of magnitude (Frischknecht et al. 2009). CF can also be calculated on watershed level.

The water stress index (WSI, Pfister et al. 2009) is based on the same watershed level data (>11,000 units; Alcamo et al. 2003) and is scaled between 0.01 and 1 with global coverage. The scaling was proposed to avoid extreme values of water stress which could dominate LCA results, and to account for the fact that a stress level is already extreme before using all available water. The WSI scaling is a logistic function with a value of 0.5 for a WTA ratio of 40 %, which is assumed to be the threshold between moderate and severe water scarcity (Alcamo et al. 2000). In addition, the CF includes a factor for monthly and inter-annual variability to account for higher stress in regions with highly variable water flows. The CF is interpreted as the share of water consumption that downstream users are deprived of. In a recent update, these factors have been adjusted to monthly CF accounting for temporal aspects of water consumption (Pfister and Bayer 2013).

Also based on WTA is the freshwater ecosystem impacts scheme, accounting for environmental water requirements in rivers (EWR) as suggested by Smakhtin et al. (2004) and which is available for a wide range of watersheds but without global coverage (WSI_{EWR}; Milà i Canals et al. 2009). It assumes that EWR needs to be deducted from availability. EWR is set between 20 and 50 % of the total available water. WSI_{EWR} is not further scaled.

Table 12.2 Overview of impact assessment methods and impact pathways addressed. The methods are grouped into three main types (vertical) and main mechanisms (horizontal grouping)

	Scarcity assessment		Human health assessment		Ecosystem quality impact		Other features		
	Withdrawal-based	Consumption-based	Agricultural use	Domestic use	Terrestrial impacts	Aquatic impacts	Endpoint level	Monthly resolution	Global coverage
Water scarcity indicators									
WSI (Pfister et al. 2009)	X								X
WII (Bayart et al. 2014)	X								X
UBP06 (Frischknecht et al. 2009)	X						(X)[a]		X
WSI_{bg} (Ridoutt and Pfister 2010)	X								X
EWS (Milà i Canals et al. 2009)	X								
BWSI (Mekonnen and Hoekstra 2011)		X						X	
(Boulay et al. 2011b)		X							X
$WSI_{monthly}$(Pfister and Bayer 2013)	X							X	X
Human health impacts									
Pfister et al. (2009)			X				X		X
Motoshita et al. (2010a)			X				X		X
Boulay et al. (2011b)			X				X		X
Motoshita et al. (2010b)				X			X		X
Boulay et al. (2011b)				X			X		X

Ecosystem quality impacts									
Pfister et al. (2009)					X		X		X
van Zelm et al. (2011)					X		X		
Verones et al. (2012)					X		X		
Verones et al. (2013a, b)							X		(X)[b]
Hanafiah et al. (2011)						X	X		
Tendall et al. (2014)						X	X		
Resource depletion									
Pfister et al. (2009)							X		X
Milà i Canals et al. (2009)									
Bosch et al. (2007)									(X)[c]

[a]It is arguable whether Swiss Ecoscarcity method is an endpoint method, since it does not address classical areas of protection. However, it leads to a common unit which can be aggregated with other impact categories
[b]While the coverage is principally global, there is only about 20 % of all wetlands included (those classified under the Ramsar convention)
[c]This method is not regionalised

Some methods also account for quality issues such as the approach suggested in Ridoutt and Pfister (2010). This method is not intended to be included in full LCA, since the emissions are characterised by the WSI, in addition to a distance-to-target-method for water quality (grey water; WSI_{bg}), leading to double counting with other impact categories such as freshwater ecotoxicity or eutrophication.

A slightly different approach including quality aspects is described in Bayart et al. (2014): the water impact index (WII), which applies the WSI of Pfister et al. (2009) to water quality classes, attributes lower CFs to consumption of polluted water quality classes. The issue of double counting arises in this approach too, but here the authors aim at covering an additional impact pathway on top of the pollution effect on ecosystem quality and human health: consuming high quality water and releasing lower quality water deprives further users of high quality water access, which is an effect not entirely covered by current impact assessment methods on emissions.

Along the same line of argumentation, Boulay et al. (2011b) define specific water quality class scarcities (α) to assess the impacts of withdrawal and release of specified water qualities. The classification requires >100 pollutants and therefore needs very detailed inventories and knowledge of environmental conditions. Also when applying these CFs, it is important to be aware of potential double counting of pollution impacts but it allows for the inclusion of external effects of water quality on the availability of downstream users that might not be covered by other impact pathways assessed in LCA. In addition, this method is based on consumption to availability ratios (CTA) and therefore permits the inclusion of purely consumptive effects on water scarcity.

Also based on CTA is the Blue water scarcity indicator (BWSI; Mekonnen and Hoekstra 2011). The BWSI extends the approach by Milà i Canals et al. (2009) to a monthly level, it is based on updated data but still does not have global coverage. While it is not designed as an LCA method its CF can be used for monthly water use impact assessment.

The uncertainties arising from all these methods have not been published, but it is known that data quality of water use and availability is generally low and in many regions only estimated and/or modelled. Pfister and Hellweg (2011) examined the uncertainties related to hydrological models on WTA and the consequent uncertainties in the total WSI model. The results showed high uncertainties although not all aspects of the CF model could be included: on average a dispersion factor for the 95 % interval of $k = 2.8$ was derived (based on Slob 1994). Moreover, uncertainties due to aggregating CFs from watershed to country level were analyzed and demonstrated the high relevance of spatial resolution in LCIA of water use (in most cases aggregation leads to a dispersion factor of $k > 2.0$).

In principle, all the methods presented above target the same impact of water stress due to water use or consumption, and are therefore not meant to be used in parallel. The main conceptual difference is whether water quality should be included in water scarcity assessment or not. Toxic, acidifying or eutrophying effects of pollution must be separated from water scarcity effects in order to avoid double counting. Further research is required in order to provide proper recommendations for the use of the different available approaches.

4.2 Methodologies Addressing Impacts on Human Health

The different methods presented in this section use the same principal approach to quantify impacts from water consumption as a function of water scarcity and socio-economic conditions. Impacts of water use concerning human health account for the lack of water for agricultural and domestic use as well as for fisheries. The two main impact pathways are lack of water for agriculture with a consequent lack of food leading to nutritional deficiencies, and lack of water for drinking, cooking and hygiene purposes with a resulting spread of diseases.

Pfister et al. (2009) and Milà i Canals et al. (2009) do not suggest to model impacts of lack of water for domestic use and spread of diseases due to the fact that these impact are mainly related to infrastructure and management problems. Pfister et al. (2009) provide CFs for lack of water for agriculture on a watershed level (>11,000 units) based on the WSI (see above) as a starting point for the assessment of water deprivation. In a second step the lack of water in agriculture is calculated and the nutritional consequences of lack of water for food are assessed via regression analysis including socio-economic conditions of countries (and subnational regions) together with per-capita water use requirements. Consequently, the damages are quantified in the standard unit of disability adjusted life year (DALY) lost based on the relationship between DALY from malnutrition and malnourished people. Finally, the factors are aggregated on a watershed level in order to include downstream effects, such as water consumption in the southern USA leading to impacts in Mexico.

The method of Boulay et al. (2011b) adopts a similar approach but on a different spatial scale, using larger aggregated watersheds and country borders. It also uses quality classes for both water availabilities and usability classification. Based on economic conditions, the method further distinguishes whether there is a human health effect due to lack of water for agriculture, fisheries and/or domestic use, or if the damage is compensated by technological measures (e.g. water treatment). For the latter case, the authors propose to add the LCIA of the resulting required actions for the prevention of human health damage (e.g. from water reclamation or desalination). There is a risk of double counting with toxicological effects from other LCIA methods for pollutants, but this method permits the integration of external effects that might be excluded otherwise, at least at the current availability of impact assessment methods. To assess impacts from water deprivation in this method, the water quality and scarcity as well as the distribution of human water users in the region are considered, resulting in DALY per m^3 of water consumed or released in an unsuitable quality for the required activity. The method does not account for downstream effects between countries.

Another similar method is the method of Motoshita et al. (2010a) which follows the same main principles but considers the problem purely from a country perspective. For economic conditions it considers a 'ripple' effect, which is a consequential approach looking at the global food market where countries of low per-capita income suffer the most from reduced food availability on the market. Therefore the effects are not only local but also influenced by the markets and global trade patterns. This method therefore adds an additional aspect that might be relevant, especially in consequential LCA settings. As a result, developed countries can also lead to human health impacts abroad through changes in market prices.

As in the method by Boulay et al. (2011b), Motoshita et al. (2010b) also suggest CFs for water use based on relationships between lack of access to safe water and water-borne diseases. While this is an important issue for global human health, it needs to be further investigated to which extent and under which circumstances water consumption and use cause such effects.

Pfister and Hellweg (2011) examined the uncertainties of the CF from Pfister et al. (2009) including all the steps in the cause-effect chains. The result shows very uncertain CFs (uncertainty intervals spanning orders of magnitudes), which reveals the need for better models and data to improve the assessment. The k-value is a dispersion factor denoting the 95 % confidence interval if the median of the value in question is divided (lower bound) and multiplied (upper bound) by k. For the WSI an average k-value of 2.76 was found, whereas for impacts addressing human health the average was $k = 18.1$. This means that the 95 % interval for characterising human health might span in many cases over four orders of magnitude, while for the midpoint it is in most cases within one order of magnitude. Aggregation from watershed to country level adds to the uncertainty and it is therefore concluded that spatial aggregation of impact factors should be carefully considered, as country-average uncertainties of CFs are considerably larger than those of watershed-level CFs.

4.3 Methodologies Addressing Impacts on Ecosystem Quality

The methods for ecosystem quality impacts of water use and consumption can be distinguished into different types and spatial coverage. The only method with global coverage and therefore applicable for full LCA is the method that was suggested by Pfister et al. (2009). It converts water consumption into land use equivalent of potentially disappeared fraction (PDF) of species on an area during a time. This is done by conceptually assuming all water being of equal importance whether it is used on land or in a river. Therefore the land-time requirement is calculated as the inverse of precipitation. The PDF is assumed proportional to the local limitation of plant growth by water availability. All this is analyzed using high spatial resolution data (0.5 arc minute) with global coverage, and subsequently aggregated on watershed level.

An approach specifically examining aquatic ecosystems provides CFs by relating river water deprivation to consequences on fish species richness (Hanafiah et al. 2011). It is principally a global approach but due to data limitation only provides CFs for a limited region between 42 arc degrees latitude and for watersheds which are not heavily disturbed. The method relates fish species richness to river flows at the river mouth for the effect factor (EF), and for the fate factor

(FF) assumes that the river flow is reduced by the volume of water consumed during an average residence time of water in the river.

Tendall et al. (2014) extended this approach on aquatic ecosystem damage by including more taxonomic groups in the study, as well as more detailed assessment of watershed characteristics and improved regression analyses of species richness and river flows. The method focused on specific areas in Switzerland and Europe and is not applicable on a global scale.

The assessment of specific impacts related to ground and surface water use on ecosystems has so far been limited to local approaches. van Zelm et al. (2011) presented an approach modelling groundwater drawdowns due to water extraction and effects on plant species richness for The Netherlands. The FF is derived using detailed groundwater models of The Netherlands, which quantify the average drawdown due to the altered water extraction rate. The effects are assessed based on extensive studies of the vegetation's occurrence of species at specific ground water depths. This method only represents the conditions in The Netherlands and can therefore not be applied for LCA in general. Other flat areas might have similar cause-effect chains but most regions in the world have different hydrogeological conditions.

An even larger geographical limitation applies to the CF developed by Verones et al. (2012). They modeled impacts on wetlands due to water extraction for a case study wetland in Peru. A focus was set on the use and release of surface and ground water which have completely different CFs for the studied wetland. However, extrapolation from this work to other areas is not directly possible.

Verones et al. (2013a, b) therefore further developed the local approach to make it generally applicable to global wetlands, and calculated CFs for groundwater and surface water use considering ~20 % of all wetlands globally. In this method the coverage of wetlands of international importance is global, but the rest of the wetlands, amounting to ~80 % of the total, are missing. Nevertheless, it is a large step forward towards a global assessment of ecosystem impacts on wetlands and other water dependent ecosystems.

The uncertainties arising in the aforementioned methods are generally not quantified. The most that is done is some basic sensitivity analysis. With increased level of detail in the global approaches uncertainties should also be addressed. Uncertainties will dramatically increase if local approaches are applied to global levels, since data tend to be of much lower quality on such a large scale.

The methods presented in this section do not address the same impact pathways but overlaps might still be possible. In particular, the method of Pfister et al. (2009), which addresses the impacts as land use equivalents, would lead to double counting when combined with more specific impacts addressed in the other presented methods. This limitation needs to be kept in mind, especially for the future when other impact assessment methods with global coverage might become available. For now, a main issue to avoid is applying methods developed for geographically limited regions to global studies without further uncertainty assessment.

4.4 Methodologies Addressing Impacts on Resource Depletion

The most obvious endpoint of an abiotic resource is typically resource depletion. As explained above, water as a resource has an absolutely vital function for humans and ecosystems and therefore more focus has been placed on the endpoints involving these two areas of protection.

In line with the abiotic depletion potential (ADP) used in CML (Guinée 2001), Milà i Canals et al. (2009) suggested to evaluate freshwater depletion with a corrected extraction rate to take into account regeneration for water fund resources. By doing so, it only accounts for the overuse of the resource.

Milà i Canals et al. (2009) remarked that extraction from underexploited groundwater bodies would result in negative ADP values and should not be assessed because this extraction does not lead to depletion of the water body. The problem of groundwater resources is that even deposits might not be static and slowly flow out to the sea. CFs are so far only calculated for a few cases and the authors suggested to determine whether overexploitation or extraction of fossil water occur and, if so, to obtain data specific to the water body in question. The lack of appropriate groundwater data is the main reason preventing global coverage with CFs.

Pfister et al. (2009) developed a method for resource depletion based on WTA (freshwater scarcity), relying on the backup-technology concept (Stewart and Weidema 2005), as done in other endpoint indicators, e.g. EI99 (Goedkoop and Spriensma 2001). The CF is calculated by multiplying the fraction of water depleted ($f_{depletion}$) within a watershed by the energy requirement for the backup technology. $f_{depletion}$is based on WTA and might therefore overestimate the fraction that is depleted as withdrawal might be returned to the watershed. On the other hand, local groundwater depletion might also occur in areas with low scarcity on a watershed level (e.g. in city areas). The energy factor is the energy required for seawater desalination (11 MJ/m^3), since this is the backup-technology considered by the authors.

Spatially-differentiated CFs at country and major watershed levels are provided. However, the energy requirement is considered the same for all regions, even though seawater desalination is not possible everywhere (for long distances, transportation of water may have higher impacts than desalination itself as shown by Pfister et al. (2009)).

A completely different approach is provided based on the Cumulative Exergy Demand (CExD) method by Bosch et al. (2007). Water resource is considered by its chemical exergy and its potential exergy (potential energy of water in hydropower generation) compared to seawater (reference state). It does not consider further aspects of resource scarcity or quality. While regionalisation is theoretically possible, there are no CFs available. This method is not well suited for the assessment of water use outside an CExD context.

5 New Developments and Research Needs

Ongoing methodological developments concerning missing specific impact pathways are necessary to make water use impact assessment more robust and consistent and allow standard integration into LCA.

One overarching issue is the inclusion of uncertainties, which is especially high in regionalised assessments due to the typically low data sample size available for the CF development (i.e. the more a sample is split, the less measurements per spatial unit are available). To judge the significance of LCA results, uncertainty information of the input parameters as well as the model uncertainty of the water impact assessment methods need to be consistently provided and analyzed. This also includes spatial uncertainty due to accuracy limitations, especially relevant along borders of geospatial units (Mutel et al. 2011). There is a trade-off between increased precision of regionalised data and accuracy of its application. Relative accuracy in spatial differentiation decreases with higher spatial resolution, as the probability of choosing the wrong geographical unit becomes higher. As an example, it is generally more accurate to define the country of production than the corresponding watershed, even though CFs on country level have significantly higher uncertainties, especially in locations close to the borders of the spatial units.

Improvement of the underlying hydrological data and models is crucial to, in turn, improve the impact assessment of water use for all impact pathways. Recent methodological developments are promising (e.g. Voss et al. 2008) but there is still room for improvement, especially in arid regions (e.g. the Nile watershed). In particular, the global assessment of groundwater resources and the distinct effect of water withdrawals from different sources (groundwater, surface water), as well as temporal aspects, all require better hydrological models. Although such distinction is made on country level by Boulay et al. (2011b), the quality of the underlying data is not of high robustness (Doll and Fiedler 2008; Wada et al. 2010; Siebert et al. 2010).

The assessment of impacts from degradative water use (quality change), in addition to the environmental consequences currently covered under other impact categories, such as ecotoxicity or eutrophication needs also further research to clearly define additions and overlaps. Moreover, the definition of freshwater is not entirely clear, and some basic quality levels (e.g. brackish, brine or produced water) should be integrated and better analysed.

One major area of research needed is the assessment of in-stream water use impacts, for instance those caused by dams. While some consider such impacts to constitute land use of water surface, the impacts of dams and other human infrastructure on aquatic ecosystem need to be clearly assessed in terms of water resource impacts. A first approach has been developed and was presented at international conferences (Humbert and Maendly 2008). However, this work needs further elaboration and operationalisation, to cover an additional important environmental impact pathway. For other constructions such as channeling of rivers and water supply systems, no method is available. However, the effects on the hydrological system and environmental consequences should be addressed in the future.

An additional aspect, which is not due to water use but a consequence of climate change, is altered precipitation and evaporation regimes which changes water availability and irrigation demand in agriculture. Therefore, the impacts on water-dependent ecosystems and the lack of water for humans due to emissions of greenhouse gases need to be further evaluated. The uncertainties related to predictions of changed precipitation patterns under climate change scenarios are extremely high and need to be carefully addressed. They might not be as relevant for the global picture (Pfister et al. 2011b) but can change the water stress in specific regions (Schlenker and Lobell 2010).

Finally, implementation of regionalisation into standard LCA is hampered by the absence of proper software solutions and good quality data. So far, regionalised assessment needs to be done manually by applying regionalised values retrieved from tools such as Google Earth or other geographic information systems (GIS). For inventory data on water use there is also a strong demand for spatially explicit values, since, especially in agriculture, water consumption and use is highly climate dependent.

Acknowledgement This chapter is partially based on the report on water use impact assessment methods delivered in the PROSUITE project (Funded by the European Commission under the 7th Framework Programme). I appreciate helpful comments by Mark Huijbregts and provision of graphical material by Anna Kounina, as well as input on resource depletion by Pilar Swart, Rodrigo A. F. Alvarenga and Jo Dewulf. Language editing was done by Catherine Raptis.

References

Alcamo J, Henrichs T, Rösch T (2000) World water in 2025: global modeling and scenario analysis. In: Rijsberman F (ed) World water scenarios. Earthscan Publications, London, pp 243–281

Alcamo J, Doll P, Henrichs T, Kaspar F, Lehner B, Rosch T, Siebert S (2003) Development and testing of the WaterGAP 2 global model of water use and availability. Hydrolog Sci J 48:317–337

Bayart JB, Bulle C, Deschenes L, Margni M, Pfister S, Vince F, Koehler A (2010) A framework for assessing off-stream freshwater use in LCA. Int J Life Cycle Assess 15:439–453. doi:10.1007/s11367-010-0172-7

Bayart J-B, Worbe S, Grimaud J, Aoustin E (2014) The Water Impact Index: a simplified single-indicator approach for water footprinting. Int J Life Cycle Assess 19:1336–1344. doi:10.1007/s11367-014-0732-3

Berger M, Finkbeiner M (2011) Correlation analysis of life cycle impact assessment indicators measuring resource use. Int J Life Cycle Assess 16:74–81

Berkoff J (2013) China: the south–north water transfer project – is it justified? Water Policy 5:1–28

Bosch ME, Hellweg S, Huijbregts MAJ, Frischknecht R (2007) Applying cumulative exergy demand (cexd) indicators to the ecoinvent database. Int J Life Cycle Assess 12:181–190

Boulay A-M, Bouchard C, Bulle C, Deschênes L, Margni M (2011a) Categorizing water for LCA inventory. Int J Life Cycle Assess 16:639–651

Boulay A-M, Bulle C, Bayart J-B, Deschênes L, Margni M (2011b) Regional characterisation of freshwater use in LCA: modelling direct impacts on human health. Environ Sci Technol 45:8948–8957

de Baan L, Mutel C, Curran M, Hellweg S, Koellner T (2013) Land use in life cycle assessment: global characterisation factors based on regional and global potential species extinctions. Environ Sci Technol 7:9281–9290

Doll P, Fiedler K (2008) Global-scale modelling of groundwater recharge. Hydrol Earth Syst Sci 12:863–885

ecoinvent Centre (2010) ecoinvent data v2.2. http://www.ecoinvent.org. Accessed 20 Nov 2013

Falkenmark M, Rockström J (2004) Balancing water for humans and nature: the new approach in ecohydrology. Earthscan, London

Fekete BM, Vörösmarty CJ, Grabs W (2002) High-resolution fields of global runoff combining observed river discharge and simulated water balances. Global Biogeochem Cycles 16:15-11-15-10

Frischknecht R, Steiner R, Jungbluth N (2009) The ecological scarcity method – eco-factors 2006. A method for impact assessment in LCA. Bundesamt für Umwelt (BAFU), Bern

Goedkoop M, Spriensma R (2001) The eco-indicator 99: a damage oriented method for life cycle impact assessment: methodology report. Ministerie van Volkshiusvesting, Ruimtelij-keOrdening en Milieubeheer, Den Haag

Guinée JB (2001) Life cycle assessment: an operational guide to the ISO standards; operational annex to guide. Centre for Environmental Science, Leiden University, Leiden

Hanafiah MM, Xenopoulos MA, Pfister S, Leuven RSEW, Huijbregts MAJ (2011) Characterization factors for water consumption and greenhouse gas emissions based on freshwater fish species extinction. Environ Sci Technol 45:5272–5278. doi:10.1021/es1039634

Hoekstra AY, Hung PQ (2002) Virtual water trade: a quantification of virtual water flows between nations in relation to international crop trade. Value of water research report series no 11. UNESCO-IHE, Delft

Hoekstra AY, Chapagain AK, Aldaya MM, Mekonnen MM (2011) The water footprint assessment manual: setting the global standard. Earthscan, London

Humbert S, Maendly R (2008) Characterisation factors for damage to aquatic biodiversity caused by water use especially from dams used for hydropower. Paper presented at the 35th LCA discussion forum, Zurich, June 5. http://www.lcainfo.ch/df/DF35/DF35_09_Humbert%20-%20CF%20for%20aquatic%20biodiv%20of%20dams.pdf. Accessed 10 Nov 2013

PE International (2012) GaBi 6. http://www.gabi-software.com/international/index/. Accessed June 2013

ISO (2013) ISO/DIS 14046 water footprint – principles, requirements and guidelines. http://www.iso.org/iso/catalogue_detail?csnumber=43263. Accessed 25 Oct 2013

Korzoun VI (1974) World water balance and water resources of the earth. Hydrometeoizdat, Leningrad

Korzoun VI, Budyko MI, Sokolov AA, Voskresensky KP, Konoplyantsev AA, Kalinin G et al (1978) Atlas of world water balance. USSR National Committee for the International Hydrological Decade. English translation. UNESCO, Paris

Kounina A et al (2013) Review of methods addressing freshwater use in life cycle inventory and impact assessment. Int J Life Cycle Assess 18:707–721. doi:10.1007/s11367-012-0519-3

Lin C, Suh S, Pfister S (2012) Does south-to-north water transfer reduce the environmental impact of water consumption in China? J Ind Ecol 16:647–654. doi:10.1111/j.1530-9290.2012.00500.x

Loubet P, Roux P, Núñez M, Belaud G, Bellon-Maurel V (2013) Assessing water deprivation at the sub-river basin scale in LCA integrating downstream cascade effects. Environ Sci Technol 47:14242–14249

McGlade J, Werner B, Young M et al (2012) Measuring water use in a green economy, a report of the working group on water efficiency to the international resource panel, UNEP (United Nations Environment Programme). http://www.unep.org/resourcepanel/Portals/24102/Measuring_Water.pdf. Accessed 12 Sept 2013

Mekonnen MM, Hoekstra AY (2011) Global water scarcity: monthly blue water footprint compared to blue water availability for the world's major river basins. Value of Water Research Report Series No 53. UNESCO-IHE. Delft

Milà i Canals L, Chenoweth J, Chapagain A, Orr S, Anton A, Clift R (2009) Assessing freshwater use impacts in LCA: part i-inventory modelling and characterisation factors for the main impact pathways. Int J Life Cycle Assess 14:28–42

Millennium Ecosystem Assessment (2005) Ecosystems and human well-being: biodiversity synthesis. World Resource Institute, Washington, DC

Mitchell TD, Jones PD (2005) An improved method of constructing a database of monthly climate observations and associated high-resolution grids. Int J Climatol 25:693–712

Motoshita M, Itsubo N, Inaba A (2010a) Damage assessment of water scarcity for agricultural use. In: Proceedings of 9th international conference on EcoBalance. Tokyo, 9–12 Nov 2010

Motoshita M, Itsubo N, Inaba A (2010b) Development of impact factors on damage to health by infectious diseases caused by domestic water scarcity. Int J Life Cycle Assess 16:65–73

Mutel CL, Pfister S, Hellweg S (2011) GIS-based regionalized life cycle assessment: how big is small enough? Methodology and case study of electricity generation. Environ Sci Technol 46:1096–1103. doi:10.1021/es203117z

Núñez M, Pfister S, Roux P, Anton A (2013) Estimating water consumption of potential natural vegetation on global dry lands: building an LCA framework for green water flows. Environ Sci Technol 47:12258–12265

Oki T, Kanae S (2006) Global hydrological cycles and world water resources. Science 313:1068–1072

Pfister S, Bayer P (2013) Monthly water stress: spatially and temporally explicit consumptive water footprint of global crop production. J Clean Prod 73:52–62. doi:10.1016/j.jclepro.2013.11.031

Pfister S, Hellweg S (2009) The water 'Shoesize' vs. footprint of bioenergy. Proc Natl Acad Sci U S A 106:E93–E94

Pfister S, Hellweg S (2011) Surface water use – human health impacts. Report of the LC-IMPACT project (EC: FP7). http://www.ifu.ethz.ch/ESD/downloads/Uncertainty_water_LCIA.pdf. Accessed 14 Feb 2014

Pfister S, Ridoutt BG (2013) Water footprint: pitfalls on common ground. Environ Sci Technol 48:4–4

Pfister S, Koehler A, Hellweg S (2009) Assessing the environmental impacts of freshwater consumption in LCA. Environ Sci Technol 43:4098–4104. doi:10.1021/es802423e

Pfister S, Curran M, Koehler A, Hellweg S (2010) Trade-offs between land and water use: regionalised impacts of energy crops. 7th international conference on LCA in the agri-food sector, Bari. http://www.ifu.ethz.ch/ESD/data/LCAfood2010_pfister.pdf. Accessed 14 Feb 2014

Pfister S, Saner D, Koehler A (2011a) The environmental relevance of freshwater consumption in global power production. Int J Life Cycle Assess 16:580–591

Pfister S, Bayer P, Koehler A, Hellweg S (2011b) Environmental impacts of water use in global crop production: hotspots and trade-offs with land use. Environ Sci Technol 45:5761–5768

Ridoutt B, Pfister S (2010) A revised approach to water footprinting to make transparent the impacts of consumption and production on global freshwater scarcity. Global Environ Chang 20:113–120

Ridoutt B, Pfister S (2013) A new water footprint calculation method integrating consumptive and degradative water use into a single stand-alone weighted indicator. Int J Life Cycle Assess 18:204–207

Rijsberman FR (2006) Water scarcity: fact or fiction? Agric Water Manag 80:5–22

Rockström J, Steffen W, Noone K et al (2009) A safe operating space for humanity. Nature 461:472–475

Rost S, Gerten D, Bondeau A, Lucht W, Rohwer J, Schaphoff S (2008) Agricultural green and blue water consumption and its influence on the global water system. Water Resour Res 44: W09405. doi:10.1029/2007wr006331

Schlenker W, Lobell DB (2010) Robust negative impacts of climate change on African agriculture. Environ Res Lett. doi:10.1088/1748-9326/5/1/014010
Shiklomanov IA (1999) World water resources at the beginning of the 21st century International Hydrological Programme. State Hydrological Institute (SHI)/UNESCO, St. Petersburg
Shiklomanov IA, Rodda JC (2003) World water resources at the beginning of the 21st century, International hydrology series. Cambridge University Press, Cambridge
Siebert S, Burke J, Faures JM, Frenken K, Hoogeveen J, Doll P, Portmann FT (2010) Groundwater use for irrigation – a global inventory. Hydrol Earth Syst Sci 14:1863–1880
Slob W (1994) Uncertainty analysis in multiplicative models. Risk Anal 14:571–576
Smakhtin V, Revenga C, Doll P (2004) A pilot global assessment of environmental water requirements and scarcity. Water Int 29:307–317
Stewart M, Weidema B (2005) A consistent framework for assessing the impacts from resource use – A focus on resource functionality. Int J Life Cycle Assess 10:240–247
Tendall DM, Hellweg S, Pfister S, Huijbregts MAJ, Gaillard G (2014) Impacts of river water consumption on aquatic biodiversity in life cycle assessment – a proposed method, and a case study for Europe. Environ Sci Technol. doi:10.1021/es4048686
UN (2013) The millennium development goals report. UN Department of Economic and Social Affairs, New York
van Zelm R, Schipper AM, Rombouts M, Snepvangers J, Huijbregts MAJ (2011) Implementing groundwater extraction in life cycle impact assessment: Characterisation factors based on plant species richness for the netherlands. Environ Sci Technol 45:629–635
Verones F, Hanafiah MM, Pfister S, Huijbregts MAJ, Pelletier GJ, Koehler A (2010) Characterization factors for thermal pollution in freshwater aquatic environments. Environ Sci Technol 44:9364–9369. doi:10.1021/es102260c
Verones F, Bartl K, Pfister S, Jiménez Vílchez R, Hellweg S (2012) Modelling the local biodiversity impacts of agricultural water use: case study of a wetland in the coastal arid area of Peru. Environ Sci Technol 46:4966–4974
Verones F, Pfister S, Hellweg S (2013a) Quantifying area changes of internationally important wetlands due to water consumption in LCA. Environ Sci Technol 47:9799–9807
Verones F, Saner D, Pfister S, Baisero D, Rondinini C, Hellweg S (2013b) Effects of consumptive water use on biodiversity in wetlands of international importance. Environ Sci Technol 47:12248–12257
Vionnet S, Lessard L, Offutt A, Levova T, Humbert S (2012) Quantis water database – technical report. Quantis International Lausanne, Switzerland. Available via Quantis International: http://www.quantis-intl.com/waterdatabase.php. Accessed 2 Feb 2012
Voss F, Alcamo J, Arnell N, Haddeland I, Hagemann S, Lammers R, Oki T, Hanasaki N, Kim H (2008) Technical report no 1. First results from intercomparison of surface water availability modules. EU sixth framework programme (2007–2011). Integrated Project Water and Global Change (WATCH). http://www.eu-watch.org/. Accessed 21 Dec 2011
Wada Y, van Beek LPH, van Kempen CM, Reckman JWTM, Vasak S, Bierkens MFP (2010) Global depletion of groundwater resources. Geophys Res Lett 37, L20402
WHO (2008) Death and DALY estimates for 2002 by cause for WHO Member States. http://www.who.int/healthinfo/bodestimates/en/index.html. Accessed 22 Feb 2008
Zekster IS, Everett LG (2004) Groundwater resources of the world and their use. UNESCO series on groundwater. UNESCO, Hannover

Chapter 13
Abiotic Resource Use

Pilar Swart, Rodrigo A.F. Alvarenga, and Jo Dewulf

Abstract Abiotic resource use in life cycle assessment (LCA) deals with the environmental concerns due to the use of resources such as metals, minerals, fossil energy, nuclear energy, atmospheric resources (e.g. argon), and flow energy resources (e.g. wind energy). Land and water may also be considered as abiotic resources, but these are dealt with elsewhere in the book series in dedicated chapters (Chap. 11 Land use by Llorenç Milà i Canals and Laura de Baan and Chap. 12 Water use by Stephan Pfister). Methods that evaluate 'abiotic resource use' in LCA were divided in three categories: (1) Resource accounting methods, which are methods that account for the overall natural resource use along the life cycle of a product; (2) Resource depletion methods at the midpoint level, which are methods that address the scarcity of resources (and therefore damage to the area of protection Resources), but at a midpoint level; and (3) Resource depletion methods at the endpoint level, which are methods that address the scarcity of resources at an endpoint level. Numerous methods are presented in this chapter, with different concepts and approaches. However, several gaps still exist in the evaluation of abiotic resource use in LCA, and more research is needed.

Keywords Area of Protection • Evaluation of abiotic resources • Impact assessment • LCA • LCIA • Life cycle assessment • Life cycle impact assessment • Resource accounting methods • Resource depletion • Resources • Typologies

1 Introduction

This chapter deals with the category 'abiotic resource use'. Resources may be defined as those elements that are extractable for human use and that have a functional value for society (Udo de Haes et al. 1999). This general definition

P. Swart • R.A.F. Alvarenga • J. Dewulf (✉)
Department of Sustainable Organic Chemistry and Technology, Faculty of Bioscience Engineering, Ghent University, Coupure Links 653, Blok B, Ghent 9000, Belgium
e-mail: jo.dewulf@ugent.be

M.Z. Hauschild, M.A.J. Huijbregts (eds.), *Life Cycle Impact Assessment*,
LCA Compendium – The Complete World of Life Cycle Assessment,
DOI 10.1007/978-94-017-9744-3_13

will be considered in this chapter. Resources may be classified according to different typologies:

- **Renewable and non-renewable**: According to the EPA (2014), renewable resources are natural resources that can be replenished at approximately the same rate at which they are used (e.g. wind and solar energy). On the other hand, nonrenewable resources cannot be produced (or re-grown) at the same rate at which they are consumed (e.g., coal and natural gas).
- **Biotic and abiotic**: Biotic resources are materials derived from presently living organisms. In addition to the resource value, they typically have an important role in maintaining ecosystem services and also intrinsic value (examples are tropical hardwood and ivory), while abiotic resources are the product of past biological processes (coal, oil and gas) or physical/chemical processes (deposits of metal ores) (Müller-Wenk 1998; Guinée 1995).
- **Funds, flows and stocks**: In the case of stocks extraction inevitably leads to the depletion of the resource, i.e. reduction of the available amounts in nature, whereas funds may be depleted but also have a renewal rate which is high enough to allow the resource to recover. Usually biotic resources are categorized as funds, but also groundwater can be considered as fund resource. Flow resources though cannot be depleted. Their availability per unit time however is limited, and thus their extraction is marked by competition (e.g. wind energy) (Heijungs et al. 1997; Lindeijer et al. 2002).

In contrast to many other impact categories discussed in the other chapters of this volume on life cycle impact assessment, 'abiotic resource use' is actually not an environmental impact category that is directly related to the Area of Protections (AoPs) 'natural environment' and 'human health', as climate change or human toxicity. The title of this chapter suggests that it deals with the environmental concerns due to the use of abiotic resources. The use of resources may cause environmental impacts to several AoPs, as shown in Fig. 13.1 in Sect. 2, and therefore can be considered an important issue when performing a Life Cycle Assessment (LCA). Some of these environmental impacts at the AoP Human Health and Natural Environment (ISO 2006) were already considered in previous chapters, and this discussion will not be repeated here. However, there are still damages to the AoP Resources to be discussed. According to Jolliet et al. (2003b), damages to this AoP consist in the reduced availability of the corresponding type of resource in the future, which is mainly known as 'resource depletion' (or 'resource scarcity'). On the other hand, we can also find several methods that are able to give answers to questions of environmental sustainability, by considering the abiotic resource use in a life-cycle perspective (Stewart and Weidema 2005). These methods do not evaluate 'resource depletion', but they are able to consistently account for the resource use of a product.

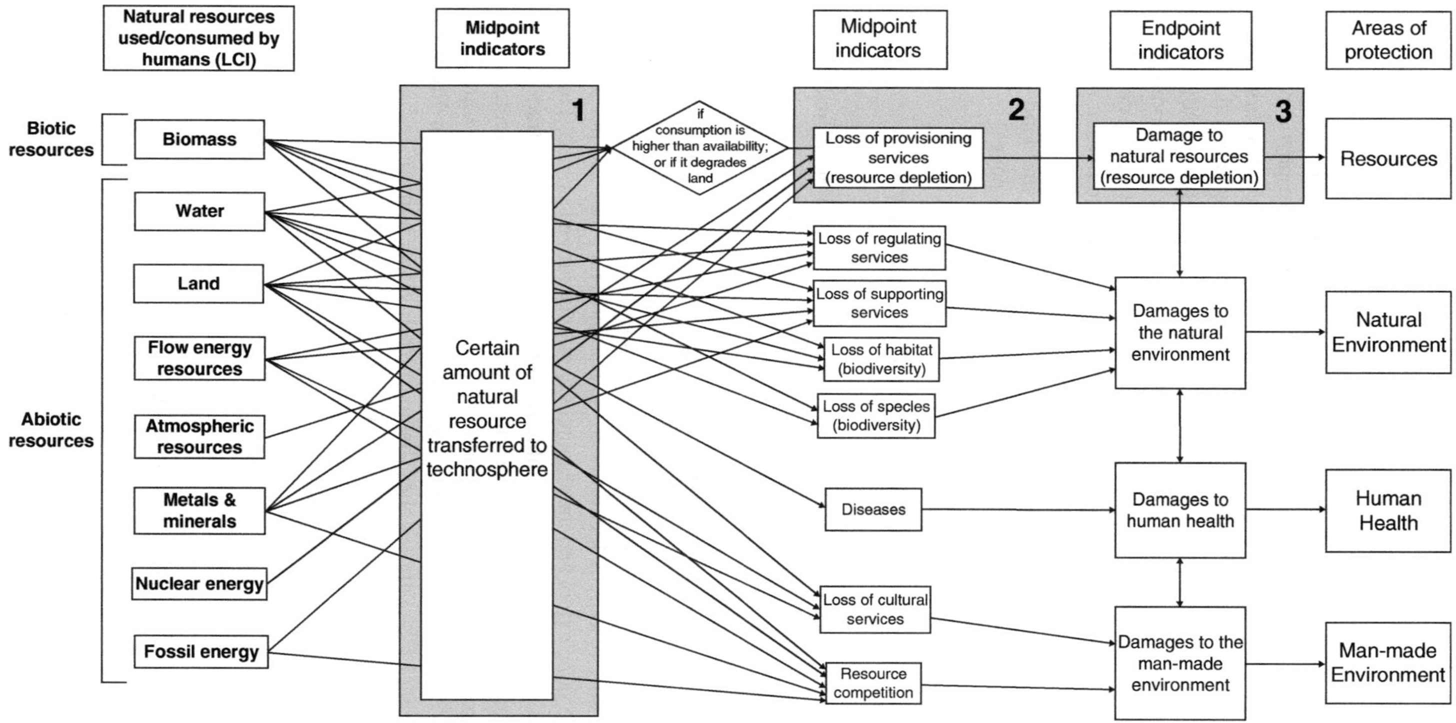

Fig. 13.1 Simplified impact pathway for the category abiotic resource use

2 Impact Pathways

Figure 13.1 shows the impact pathways for the category 'abiotic resource use'. At the left side there are types of resources, grouped as biomass, water, land, flow energy resources (e.g. wind and hydropower from dammed water), atmospheric resources (e.g. argon), metals and minerals, nuclear energy, and fossil energy. The first step in the impact pathway can be evaluated by the resource accounting method (RAM), which can affect several AoP (category #1). The next step in the impact pathway is a group of approaches that evaluate the scarcity of resources at a midpoint level (category #2), and affect solely the AoP Resources. Finally, the last step in the impact pathway is a group of approaches that evaluate the scarcity of resources at endpoint level (category #3), also affecting solely the AoP Resources. For reasons of completeness Fig. 13.1 also includes biomass (i.e. biotic resources), water and land (the two latter are dealt in dedicated chapters of this book, see the Contents page).

3 Scale, Spatial Variability, Temporal Variability

The scale of the impacts on the AoP Resources depends somewhat on the resource in question. For metals, nuclear and fossil energy carriers and atmospheric resources impacts can be considered to be on the global scale. Metals, fossil energy carrier and nuclear energy carrier are typically traded globally (except natural gas, which can be traded regionally as well) and atmospheric resources are in essence distributed equally over the planet (Hoekstra and Hung 2002). Some low value minerals like gravel are usually supplied from local sources. For this type of resources impacts can differ based on location. For flow resources location can be relevant, i.e., they are not available everywhere to the same extent and the resources cannot be moved as such. The energy content needs to be converted into a suitable form first, before it can be traded.

Short term temporal variability (diurnal, seasonal) can be neglected when determining impacts on the AoP Resources. Longer term temporal variability, however, can be deemed an important source of uncertainty for the life cycle impact assessment (LCIA) approaches discussed below. As humans influence the availability of resources, e.g. by their consumption pattern, but also by discovering new resources, and developing new technologies for resource extraction, e.g. fracking, impacts on the AoP Resources can be considered to be changing with time.

4 Methods for Abiotic Resource Use

LCIA methods for 'abiotic resource use' can be classified in different categories, considering some common characteristics. Finnveden (1996), Lindeijer et al. (2002) and Steen (2006) classified the approaches in four categories:

(1) Approaches based on energy or mass; (2) Approaches based on ratio of use to deposits; (3) Approaches based on future consequences of current resource extractions; and (4) Approaches based on exergy consumption or entropy production. The ILCD Handbook (European Commission 2011) classified the methods for abiotic resource use in four other categories: (1) Category 1 includes methods that use an inherent property of the material as basis for the characterisation; (2) Category 2 includes methods that address the scarcity of resource; (3) Category 3 includes methods focused on water depletion; and (4) Category 4 includes methods that evaluate the depletion of resources at an endpoint level.

The available amount of material can be evaluated using different definitions, e.g. '(economic) reserve' and 'reserve base' for metals and minerals from the United States Geological Survey (2010). The former are the resources that can currently be economically extracted and the latter includes also additional resources which meet certain criteria that are relevant for mining and production practice (e.g. depth of deposits). Another important definition for resource depletion used in LCIA methods is the 'ore grade' of a substance, which is the concentration of the targeted substance in the ore, and is typically represented in mass percentages.

4.1 Resource Accounting Methods (RAMs)

From the impact pathway in Fig. 13.1, we can see that the 'abiotic resource use' can be analysed through RAMs before reaching the environmental damages at specific areas of protection or even the environmental impacts at specific midpoint categories. These methods are far from giving a direct quantitative value for environmental damages, but they are still able to provide results on the environmental sustainability of a product due to the philosophy of 'less is better'.

These RAMs generally sum up all the resources consumed/used in the life cycle of a product. In order to provide results in single indicators, the resources are usually represented in common units (e.g. energy), otherwise the same information as given by the Life Cycle Inventory (LCI) would be obtained.

4.1.1 Mass

In Material Flow Analysis (MFA) resources are also typically aggregated based on mass. There are different MFA approaches. One of them is the Material Intensity Per Unit Service (MIPS) method (Ritthoff et al. 2002; Spangenberg et al. 1999). Though it is not usually classified as an LCIA method, it is similar to LCA in that it models at the system level (Finnveden and Moberg 2005). The MIPS method, pioneered by Schmidt-Bleek (Schmidt-Bleek et al. 1998), distinguishes between five resource categories: abiotic raw materials, biotic raw materials, movement of soil (agriculture and forestry, incl. soil erosion), water and air. These categories can be further divided into subcategories. A general guide to MFA was published by the OECD (2008a, b, c).

4.1.2 Energy

Accounting for energy use is a concept that was introduced in the 1970s (Boustead and Hancock 1979; Pimentel et al. 1973), and standardised by VDI (1997). Energy-based RAMs account for the energy extracted from the natural environment (i.e. the cradle) to support the technosphere system. They account not only for energy flows but also for material flows, by quantifying their energy content.

These methods have been operationalised for LCA, for instance as the Cumulative Energy Demand (CED) for the ecoinvent database (ecoinvent 2010; Hischier et al. 2009) and as the Primary Energy Demand (PED) for the GaBi database (PE International 2012). In principle, CED and PED are the same, only differing in names and compatibility to databases and/or software. The results are generated in a unit easily comprehended by stakeholders (e.g. MJ). However, since some materials have low energy value (e.g. water, metals and minerals), energy-based RAMs do not have a desirable completeness for abiotic resource use accounting. Also, biotic resources are accounted through their energy content and, to avoid double-counting, the use of land is neglected. This methodological approach results in two weaknesses: (1) agricultural systems with higher yields do not show better results although they have lower land use; (2) because these methods account for the energy content of the biomass harvested, the system boundaries may be inconsistent with the difference between natural environment and technosphere. The biomass harvested is basically an output from agricultural production, rather than a natural resource that should be accounted for. In this sense, the biomass harvested may be considered as still at the supply chain level (see Liao et al. (2012a)). In spite of these limitations, energy-based RAMs are applicable for LCA, but they should be complemented by methods accounting for the use of other resources (e.g. water, metals and minerals), for reasons of completeness.

Fossil energy consumption (one category of the CED and the PED) can be a useful screening tool (Huijbregts et al. 2010; Huijbregts et al. 2006) and is able to provide consistent results when LCA studies are interested in information solely regarding the consumption of fossil fuels during the product's life cycle. It is also common to find energy-based RAMs in some other traditional midpoint LCIA methods, i.e., the energy content of fossil fuels is used as characterisation factor (CF). For instance in the category 'Fossil depletion' of the method ReCiPe Midpoint, the mass of oil equivalent (kg oil eq.) is used as unit; this type of method will be discussed further later on.

4.1.3 Exergy

By definition, the exergy of a resource or a system is the maximum amount of useful work that can be obtained from it (Dewulf et al. 2008). Exergy analysis is usually used in industry to assess the efficiencies of processes. For natural resources that are exploited to convert their energy content into work (or heat) the idea that amounts of those energy carriers can be expressed in exergy terms is quite straightforward. Even though most metal and mineral resources are not extracted from nature with the aim to directly produce work from them, they still contain exergy. This is because these resources differ from the reference environment with respect to their chemical composition and their concentration. For example, the copper in a copper deposit is much more concentrated and occurs in another chemical form (e.g. $CuFeS_2$) than copper dissolved in seawater (the reference species for copper). These differences can in principle be used to produce work.

The cumulative exergy consumption (CExC), introduced by Szargut et al. (1988), is the exergy of the overall natural resources consumed in the life cycle of a product. Exergy-based RAMs have been operationalised to LCA through different LCI modelling approaches. For the process-based ecoinvent database, the Cumulative Exergy Demand (CExD) is operationalised in Bösch et al. (2007) and the Cumulative Exergy Extraction from the Natural Environment (CEENE) is operationalised in Dewulf et al. (2007). The two operational methods have some differences, including the approach to account for metals and minerals, but also the approach to account for biotic resources: While the exergy of the biomass is accounted in the CExD, in the CEENE the exergy deprived from nature due to land use is accounted. For the economic input-output U.S. 1997 database, the Industrial Cumulative Exergy Consumption (ICEC) is operationalised in Zhang et al. (2010).

The CExD and the CEENE are able to account for several resources, even though the results are generated through units not easily comprehended (e.g. MJ_{ex}). The system boundary of the CExD is similar to the CED, leading to similar limitations, i.e., land use is neglected and its system boundary does not correspond to the interface between natural environment and technosphere for biomass. The CEENE method accounts for land in exergy terms through the accounting for the quantity of photosynthetic solar exergy deprived from nature due to land use. This procedure allows accounting not only for land use for biotic resources, but also for other purposes (e.g. built-up land). However, by choosing to account for land, biotic resources from natural systems which had no land occupation during its production (e.g. wood from natural forests) are not accounted through CEENE. Nevertheless, the CEENE method was recommended as the most appropriate thermodynamic indicator for resource use accounting, in Liao et al. (2012b). Trying to tackle some limitations of CEENE and CExD, Alvarenga et al. (2013) proposed a new approach to account for land resources (i.e., land and biotic resources) by classifying the system studied as natural or human-made beforehand. On top of that, they provided spatially differentiated CFs for land occupation, through the potential natural net primary production.

4.1.4 Emergy and Similar Methods

Introduced by Odum (1996), emergy accounts for the total available energy that has been used to make a product (including human labor). It is sometimes referred to as energy memory, since it is supposed to be a record of the previously used-up available energy to make a product. In contrast to other RAMs (e.g. exergy-based), which usually set the natural environment as 'cradle', emergy has a different system boundary. The natural environment is part of the system, and the 'cradle' is considered to be the energy forces outside of the geobiosphere, e.g. the sun (Liao et al. 2012a) (Fig. 13.2). Emergy considers tidal, geothermal and solar energies as main energy sources that rule life on Earth, and the latter is taken as reference for its unit (Joules of solar energy – J_{se}). Emergy has received several criticisms from the scientific community, as combining disparate time scales, allocation problems, and the lack of uncertainty quantification on the numbers used to calculate transformities (Hau and Bakshi 2004a, b). Some propositions to overcome them, together with challenges to implement emergy into LCA, have been suggested in literature (Rugani and Benetto 2012; Ingwersen 2011).

Due to the limited acceptance in the scientific community, Hau and Bakshi (2004a) developed the ecological cumulative exergy consumption (ECEC). According to the authors, it overcomes some weaknesses of emergy, and if identical system boundaries, allocation and quantification methods are used, emergy and ECEC should produce equivalent results. The ECEC has been operationalised to LCA for the economic input-output U.S. 1997 database in Zhang et al. (2010). It accounts for several ecosystem services as well and is commonly used in complementation to the industrial ICEC (Zhang et al. 2010; Urban and Bakshi 2009; Baral and Bakshi 2010; Baral et al. 2012). However, it is questioned if certain emergy-based RAMs, as the ECEC, may be applicable to 'abiotic resource use' in LCA, since some regulating and supporting services considered in this methodology

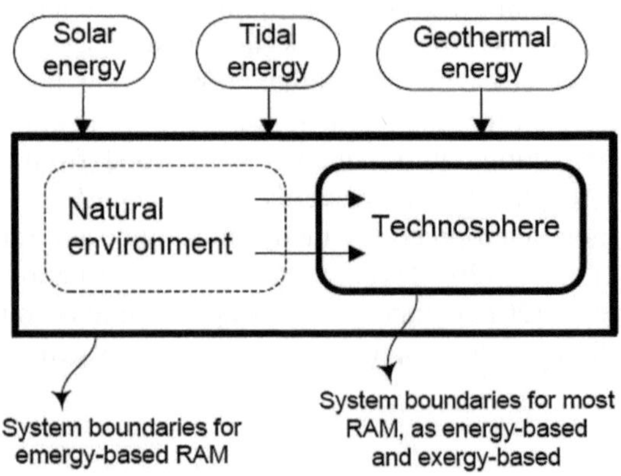

Fig. 13.2 Simplified scheme representing different system boundaries considered in RAM

(e.g. air quality regulation) appear not to be resources according to the definition from Udo de Haes et al. (1999).

Using emergy as starting point, the Solar Energy Demand (SED) accounts for the amount of solar energy needed to produce a certain product. It has been operationalised to the ecoinvent database (Rugani et al. 2011), and according to these authors, it shares the same conceptual rationale as emergy, although it does not use the same approach for allocation. Unlike emergy, the SED does not account for human labor and most of the ecosystem services: it accounts for provisioning services only.

4.1.5 Ecological Footprint

Developed by Wackernagel and Rees (1996) and further enhanced by the Global Footprint Network (2009) and Ewing et al. (2010), the Ecological Footprint is defined as the ecological surface area needed to sustain a certain system. When applied to products, the requirement of area to produce the raw materials and to absorb CO_2 emissions is calculated (m^2). It has been operationalised to LCA through the ecoinvent database (Huijbregts et al. 2008), where nuclear energy is also (indirectly) accounted for. In this methodology, solely land use, nuclear energy, and fossil energy (indirectly through the fossil CO_2 emissions) are accounted. Water, metals, minerals and other resources are not accounted; therefore the method does not provide satisfactory completeness for ‘abiotic resource use’. Nevertheless, it has a strong appeal to society, since it can directly be compared with the Earth’s carrying capacity, and as a consequence has a strong communication ability for dialogue with stakeholders (by its immediately understandable unit).

The Ecological Footprint is able to provide other information (e.g. the ecological deficit of nations) than solely resource accounting (see Global Footprint Network – www.footprintnetwork.org), but since this chapter is focused on ‘abiotic resource use’ in LCA, only its potential use as RAM is discussed.

4.1.6 Ecological Scarcity

Dating back to the 1990s, the ecological scarcity method (Ahbe et al. 1990; Brand et al. 1998) is a distance-to-target methodology developed based on the Swiss context and encompassing both resource use and emission-related environmental impacts. Its most recent implementation is described in Frischknecht et al. (2009). The resource types treated in the method and relevant for this chapter are flow energy resources, metals and minerals, fossil energy and nuclear energy.

The so-called ecofactors are derived on the basis of political targets (critical flow) and actual resource flows (current flow, normalisation flow) and expressed in units of eco-points. The targets are set for the year 2030. For the resources discussed in this chapter the derivation of the ecofactors does not involve an environmental impact pathway. The factors published in Frischknecht et al. (2009) are specific for Switzerland, but in principle the methodology can also be adopted to other regions, e.g. Baumann and Rydberg (1994) and Miyazaki (2006). Rather than a characterisation method, the ecological scarcity method can be seen as incorporating both normalisation and weighting resulting in Ecofactors, which are to be multiplied with the appropriate inventory flow to arrive at final scores which can be aggregated. Ecofactors are calculated according to (Eq. 13.1):

$$\text{Ecofactor} = K \times \frac{1 \times EP}{F_n} \times \left(\frac{F}{F_k}\right)^2 \times c \qquad (13.1)$$

With

K the CF ($\text{kg} \cdot \text{kg}^{-1}$ or MJ-equivalent/MJ), EP the unit eco-point, F the current flow ($\text{kg} \cdot \text{year}^{-1}$ or $\text{MJ} \cdot \text{year}^{-1}$), F_n the normalisation flow ($\text{kg} \cdot \text{year}^{-1}$ or $\text{MJ-equivalent} \cdot \text{year}^{-1}$), F_k the critical flow ($\text{kg} \cdot \text{year}^{-1}$ or $\text{MJ} \cdot \text{year}^{-1}$) and c the constant ($10^{12} \cdot \text{year}^{-1}$). The resulting unit is then $\text{EP} \cdot \text{MJ}^{-1}$ or $\text{EP} \cdot \text{kg}^{-1}$.

For energy resources flows are accounted in units of MJ. The only distinction made is between renewable and nonrenewable energy resources. Whereas the CF for nonrenewable energy is one MJ-equivalent/MJ, it is one third MJ-equivalent/MJ for renewable energy resources, due to political targets. For energy resources no environmental mechanism is involved.

The only other abiotic resource covered by the ecological scarcity method and relevant for this chapter is natural gravel. Due to its low market value on a mass basis resulting in low economic viability of extensive transport, gravel availability can be considered as a regional issue. In the calculations the critical flow is assumed to be equal to the current flow, because the life time of economic gravel reserves in Switzerland has been stable for quite some time, though eventually the resource is considered to be finite.

The methodology is interesting where the LCA target is an assessment with respect to country specific policy targets. As policy targets are the benchmark the environmental relevance of the method is limited to the relevance of these policy targets.

4.2 Resource Depletion at Midpoint Level

As illustrated in Fig. 13.1, the second category of resource impact assessment methods evaluates the scarcity of resources, but still at a midpoint level, i.e., they do not evaluate the actual damages to the AoP Resources.

4.2.1 EDIP 97 and EDIP 2003

The EDIP methodology has its baseline literature in Hauschild and Wenzel (1998). The approach for 'abiotic resource use' in EDIP 2003 and EDIP 97 is the same, except that the values used in the calculations in EDIP 2003 are updated for the year 2004, while in EDIP 97 the data is from 1990 to 1991. Based on the economic reserves, the 'abiotic resource use' is evaluated by the scarcity of resources naturally available, which means that even though metals may not disappear after their use (unlike fossil energy), they will no longer be available in their natural deposits, but in other places (e.g. landfill).

To calculate the CF, the authors divided the procedure in two steps. In the first, called normalisation, the global production of a substance (i) for a specific year (2004 in EDIP 2003) is considered, and this value is divided by the world population from that year. In the second step, called weighting, the economic reserve of the substance (i) is divided by the global production from the same substance (i) in a particular year (2004 in EDIP 2003), providing the supply horizon of the substance, in years. Finally, the CF for the substance (i) is calculated by the reciprocal of the product between the normalisation and the weighting factors (Eq. 13.2). Taking a look at the equation, we can notice that the 'global production' can be erased from the equation, and effectively the CF are based solely on the economic reserves, but normalized for the World population from the year 2004, expressing the resource use in so-called person-reserves – the available economic reserve per person in the world in 2004. We can conclude that this approach goes in accordance with 'option 2a' from Lindeijer et al. (2002).

$$\mathrm{CF_i} = \left[\frac{1}{\left(\frac{\text{Global prod.}_{\text{i.2004}}}{\text{World pop.}_{2004}}\right) \times \left(\frac{\text{Economic reserves}_\text{i}}{\text{Global prod.}_{\text{i.2004}}}\right)}\right] = \left[\frac{1}{\left(\frac{\text{Economic reserves}_\text{i}}{\text{World pop.}_{2004}}\right)}\right] \tag{13.2}$$

This LCIA method calculated CFs for several nonrenewable resources, including fossil energy, nuclear energy (uranium), metals and some minerals. In Hauschild and Wenzel (1998) the method is also elaborated for renewable resources for which a distinction is made according to whether the annual extraction exceeds the regeneration capacity. If this is not the case, the CF is zero (the resource use is sustainable) but if it is the case, the CF is calculated using Eq. 13.2, replacing the annual production by the annual exceedance of the regeneration capacity. CFs for water and wood on a global scale are presented, but they are often not considered due to the low significance of non-spatially differentiated CFs for these renewable resources.

4.2.2 Abiotic Depletion Potential

Guinée (1995) developed the abiotic depletion potential (ADP) as an approach applicable for 'abiotic resource use'. Later on the approach was implemented in the CML (Institute of Environmental Sciences Leiden) LCIA method by Guinée et al. (2002) and further updated by van Oers et al. (2002). These updates were implemented in the CML method in 2009 and 2010. The latest implementation of the CML method can be found on the CML website (http://www.cml.leiden.edu/software/data-cmlia.html). The CML method implements the ADP for metals and minerals, fossil energy, atmospheric resources and, since version 4.1, also for nuclear energy. In van Oers et al. (2002) CFs can also be found for some mineral configurations (e.g. fluorspar and bauxite). Due to relatively good data availability (for the reference years) a large number of substances are covered.

From a conceptual point of view the approach is similar to the approach used for resources in the EDIP methodology, as it is based on use-to-resource ratios. In contrast to the EDIP methodology the amount of remaining resource is squared in order to take into account that extracting 1 kg from a larger resource is not equivalent to extracting 1 kg from a small resource, even if the use-to-resource ratio is the same. In this sense, we can conclude that this approach goes in accordance with 'option 2c' from Lindeijer et al. (2002). Equation 13.3 represents the generic calculation of the ADP of substance i expressed in kg of reference substance (Guinée et al. 2002):

$$ADP_i = \frac{DR_i}{R_i^2} \times \frac{R_{ref}^2}{DR_{ref}} \tag{13.3}$$

With ADP_i the abiotic depletion potential of substance i (kg reference substance $\cdot$ kg^{-1}), R_i the ultimate reserve of substance i (kg), DR_i the extraction rate of substance i (kg $\cdot$ $year^{-1}$), R_{ref} the ultimate reserve of the reference substance (kg reference substance) and DR_i the extraction rate of the reference substance (kg reference substance $\cdot$ $year^{-1}$).

The ultimate reserve is the total amount (mass for elements and minerals, energy for fossil fuels) of the substance available on Earth, be it in the Earth's crust, in the oceans or the atmosphere. The reference substance is antimony for elemental species, and originally also for fossil energy. By now fossil energy is treated separately from elemental species in the CML implementation, with the total of fossil energy as reference. While Guinée (1995) had calculated ADPs for fossil energy based on their individual annual production and ultimate reserves, full substitutability based on energy content was assumed in Guinée et al. (2002), i.e. one MJ of oil is equal to one MJ of coal. As a consequence of these changes the ADPs of fossil energy are equal to their lower heating values when applied to elementary flows (ISO 2006) in units of kg, respectively m^3 for natural gas. Therefore for fossil energy the current CML implementation of the ADP is comparable to methods accounting only for the energy content of resources, as in the CED method discussed in Sect. 4.1.2. In the CML methodology the ADP is used as CF, while normalisation is done per impact category.

The ADP approach was criticised by Müller-Wenk (1998), because the amount of ultimate reserves could satisfy human consumption for 'millions of years', which would imply that there is no scarcity issue. Moreover, the approach lacks consideration of quality aspects of the resource. Already earlier Guinée (1995) had remarked that what was relevant were the reserves which can eventually be extracted, called 'ultimately extractable reserves', likely to be very different from the ultimate reserves. Guinée (1995) implicitly assumed 'the ratio between the ultimately extractable and ultimate reserve to be equal for all resource types.' With the updates to the ADP approach, alternative CFs are available which use economic reserves or the reserve base (United States Geological Survey 2010) in the reference year 1999 instead of ultimate reserves. For resource depletion the International Reference Life Cycle Data System (ILCD) (European Commission 2011) recommends the use of the CML method at midpoint, in particular the alternative CFs using reserve base. In addition, it is advised to perform a sensitivity analysis using economic reserves and ultimate reserves.

4.3 Resource Depletion at Endpoint Level

The approaches that evaluate the damages to the AoP Resources, for instance extra economic costs, are considered to be resource depletion methods at endpoint level (see Fig. 13.1), and are discussed below.

4.3.1 Eco-Indicator 99

The approach for 'abiotic resource use' in Eco-indicator 99 (EI99) (Goedkoop and Spriensma 2000) focused on the evaluation of the depletion of resources (minerals and fossil energy), which is based mainly on Müller-Wenk (1998). The decrease in resource concentration due to extraction is modelled and evaluated by the concept of *surplus energy*, i.e. the difference between the energy needed to extract a resource now and at some point in the future ('option 3' from the set of depletion approaches from Lindeijer et al. (2002)). The depletion of nuclear energy is not assessed in this methodology.

For metals and minerals, the authors considered geostatistical models in order to evaluate the relationship between availability and quality. As an approximation at higher ore grades, it was assumed that the logarithm of cumulative amount of minerals mined increases linearly with the decline in the logarithm of the ore grade. The slope of the curves between the cumulative amount of minerals mined and ore grade are of key interest. In Fig. 13.3 we can see that a metal with a high slope (e.g. metal A) has a small change in ore grade when a certain amount is mined, while another metal with a lower slope (e.g. metal B) has a higher change in ore grade when this same amount is mined. Since the authors considered that the energy requirement needed to extract, grind and purify an ore goes up as the grade goes down, the CFs were calculated by considering the slope. The calculation of the additional energy requirement is based on energy requirements per kg of ore treated that are fixed for each metal. In the EI99 methodology report (Goedkoop and Spriensma 2000) they argue that in line with the modelling of other impacts technological development, which might lead to efficiency increases, is also not

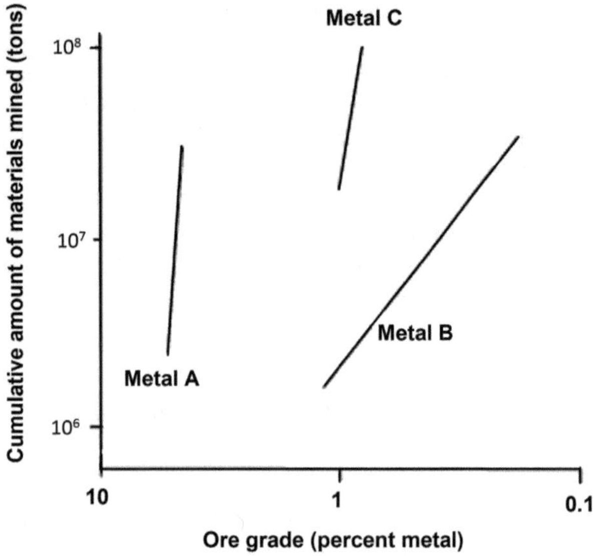

Fig. 13.3 Slope of the availability against the grade (Based on Chapman and Roberts (1983))

considered for the surplus energy method. This argumentation might be debatable as technology and grade are not fully independent, because mining has to be economic.

For fossil energy, the resource analysis had to be done differently from metals and minerals, mainly for two reasons: (1) The geological processes involved in the generation of fossils in nature are quite different from the processes that have caused the lognormal distribution of mineral ore grades in the Earth's crust; and (2) the effort to exploit a resource does not increase gradually when the resource is extracted (as for minerals), but rather more abruptly when the marginal production of fossil fuels changes from conventional to unconventional sources (e.g. natural gas from tight reservoirs). In this sense, the *surplus energy* (i.e., the CF) is calculated by the difference between energy requirements for current and future sources.

In the EI99 methodology, different cultural perspectives are used according to the cultural theory of risk (Individualist, Hierarchist and Egalitarian), and they play an important role in the resource analysis for fossil fuels (while for minerals the evaluation is the same independent of the cultural perspective). For the Individualist perspective, fossil fuels are considered not to be a problem, and since the long term perspective is not relevant with this cultural perspective, fossil fuels would not switch to unconventional sources, leading to null values of *surplus energy*. In the Hierarchist perspective, it is assumed that a certain fossil fuel (e.g. oil) would not be substituted by another fossil fuel (e.g. coal), but by unconventional sources of the same fossil fuel (e.g. oil shale). Finally, in the Egalitarian perspective, substitution between fossil fuels is assumed, for instance coal-shale mix is considered to substitute conventional natural gas in the future.

4.3.2 Impact 2002+

The LCIA methodology Impact 2002+ (Jolliet et al. 2003a) considers three types of resources in their approach for 'abiotic resource use': Metals and minerals, fossil energy and nuclear energy. In the former, the authors use the same approach as the methodology EI99 (Goedkoop and Spriensma 2000), i.e., the *surplus energy* concept. However, they assume an infinite time horizon for fossil and nuclear energy instead of assessing the surplus energy required for future unconventional technologies (as in EI99). This implies that the total energy content of the fossil energy and nuclear energy are lost due to their consumption. Therefore, CFs for fossil energy and nuclear energy are based on their energy content, making the approach to be classified as an energy-based RAM.

4.3.3 ReCiPe

ReCiPe is a Dutch LCIA method created in 2008, which combines the scientific efforts of several institutes. The main information can be found in their report (Goedkoop et al. 2009) and the updated CFs can be found in http://www.lcia-recipe.net/. This method provides indicators at two levels: midpoint and endpoint. The midpoint indicators for 'abiotic resource use' from ReCiPe will be discussed together with the endpoint indicators in this subsection.

The approach of ReCiPe for metals and minerals uses focuses on the depletion of deposits, instead of individual commodities. According to the authors, it allowed them to take into account the actual geological distribution of metals and to cover more commodities, especially those that are always mined as co-products (but the CFs are provided for the elements, e.g. 'copper, in ground'). ReCiPe provides midpoint and endpoint indicators for minerals. The damage to the AoP Resources due to the extraction of a certain mineral (resource depletion at endpoint level) is evaluated by the additional costs society has to pay due to this extraction, and it is expressed in US dollars, $ (present value in 2000). The CFs for the midpoint indicators are calculated by an equation similar to the equation from the endpoint indicator (except by the exclusion of some factors) and then normalized for the value obtained for iron, with iron equivalents as unit. Cultural perspectives are not considered for minerals, therefore the CF are the same for the Individualist, Egalitarian and Hierarchist versions. Nuclear energy (uranium) is considered together with other metals; therefore, the indicator 'Metal depletion' evaluates the depletion of metals and minerals and nuclear energy.

The approach for evaluating damage to AoP Resources from fossil energy use is also based on the marginal cost increase. However, as in the method EI99, the marginal increase is not related to the decrease of the grade of the metal, but to the shift from conventional to unconventional sources (Fig. 13.4). The method also provides CFs for midpoint and endpoint indicators. However, the characterisation model for the midpoint indicator is basically energy-based RAM, since the CFs are based on the energy content of the fossil fuels. The authors divided the fuel energy

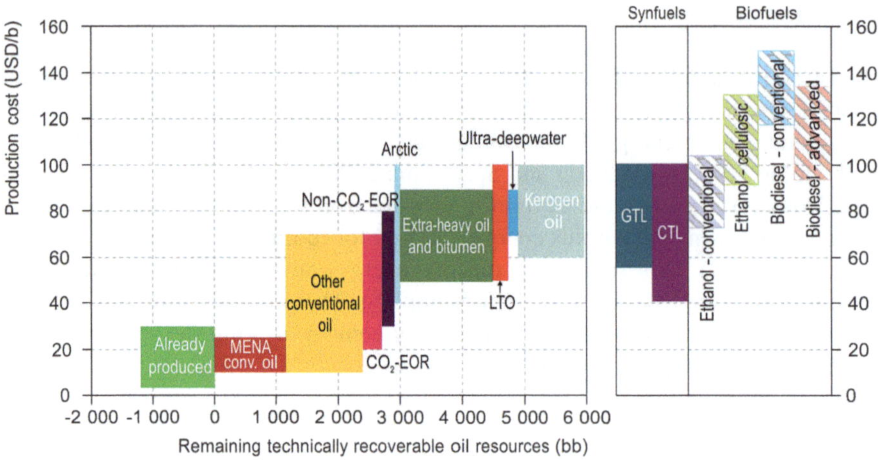

Notes: unless otherwise indicated, all material in figures and tables derives from IEA data and analysis. CO_2 = carbon dioxide; MENA = Middle East and North Africa. "Other conventional oil" includes deepwater. No carbon pricing included. Synfuel resources are difficult to assess due to competition with other natural gas and coal uses. Biofuels are renewable and, in theory, not resource constrained. Biofuels production costs have been credited with a "refiner's margin", using the ratio of gasoline and diesel spot prices in the United States compared to the West Texas Intermediate crude oil price. The ratio was, on average, 1.3 for gasoline and 1.35 for diesel between 2007 and 2012.

Fig. 13.4 Oil production costs for various resource categories (Resources to Reserves 2013© OECD/IEA (2013), fig. 8.3, p. 228)

content by the energy content of a specific type of crude oil (42 MJ/kg), generating values in mass of oil equivalents (kg oil-eq.). For the endpoint indicator, the authors calculated the marginal cost increase for oil when changing to unconventional sources (e.g. oil from tar/bituminous sands). Different cultural perspectives were considered, and the marginal cost increase for oil was considered to be lower in the Individualist perspective, making the CF of the Hierarchist and Egalitarian perspectives (which were the same) to be approximately three times higher. After the CF for oil was calculated for the three cultural perspectives, the CF of fossil fuel *i* (e.g. natural gas) was calculated by multiplying the CF of oil by the ratio between the energy content of fossil fuel *i* and the energy content of oil (42 MJ/kg).

4.3.4 Sustainable Process Approach (EPS2000)

The sustainable process approach is an endpoint method implemented in the Environmental Priority Strategies for product development (EPS2000) (Steen 1999a, b). Information on this method is also available on the website of the CPM (Swedish Life Cycle Center) database (http://cpmdatabase.cpm.chalmers.se/). In the EPS2000 methodology the approach is implemented for metals and minerals, fossil energy, nuclear energy and atmospheric resources. Alternative factors for some metals are available in Steen and Borg (2002).

As for the other impact categories in EPS2000, the idea is to quantify the willingness to pay (WTP) for restoring damage done to the safe-guard subject. In the case of abiotic stock resources, not only the present generation but also future generations are included; therefore the WTP is calculated based on hypothetical sustainable processes which could produce resources like those extracted today once these are depleted. Thus, the method can be classified as one based on future consequences of current abiotic resource extraction (Steen 2006; Lindeijer et al. 2002). The calculated costs include direct production costs and external costs due to emissions and resource use.

The sustainable processes are assumed to produce resources from average rock (most elements and gravel), from seawater or air. Fossil fuel alternatives are vegetable oil, charcoal and biogas, respectively. In this context it should be noted that it is acknowledged that the biomass products cannot fully substitute fossil use for energy production. The sustainable processes are further optimised, by using electricity from solar energy and wood as an energy source, among others.

The sustainable process approach has been criticised for its rather long time horizon and the many assumptions associated with it (Müller-Wenk 1998; European Commission 2011).

4.4 Summary of the Methods

A summary of the methods presented in this chapter can be found in Table 13.1, where they are classified into the three 'abiotic resource use' categories mentioned in the beginning of the chapter, their main literature reference are given, and the types of resource they evaluate are marked. We also included some well-known methods for water and land (dealt with in dedicated chapters of this volume[1]) for completeness.

5 New Developments and Research Needs

There are still many gaps in the way 'abiotic resource use' is evaluated in LCA. Especially for midpoint methods that are based on use-to-resource ratios, it is debatable how relevant the different resource definitions are for resource availability for future generations. Moreover, the dynamic nature of the concept of resources, which is affected by technology and economics to a varying extents depending on which resource definition is employed, and the estimate of the actual amount and quality of available stocks remain pending issues. At the same time, the needs of the future generations are not easy to define. For the future generations, needs may move to resources that are not in frequent use today. Overall pressure on

[1] See Chap. 12, Water use, by Stephan Pfister and Chap. 11, Land use, by Llorenç Milà i Canals and Laura de Baan.

Table 13.1 Summary of the methods for abiotic resource use, showing the type of resources they consider and to which category they belong (e.g. resource depletion at midpoint level)

Abiotic resource use category	Operational method	Water	Land	Flow energy resources[a]	Atmosph. resources	Metals and minerals	Fossil energy	Nuclear energy	Literature reference
Resource accounting methods (RAM)	CED/PED			X			X	X	Hischier et al. (2009), PE International (2012)
	CE_xD	X		X	X	X	X	X	Bösch et al. (2007)
	ICEC/ECEC	X		X	X	X	X	X	Zhang et al. (2010)
	CEENE	X	X	X	X	X	X	X	Dewulf et al. (2007)
	SED	X	X	X	X	X	X	X	Rugani et al. (2011)
	Ecological Footprint		X				X[b]	X[c]	Global Footprint Network (2009)
	Water Footprint	X							Hoekstra et al. (2011)
	Ecological Scarcity			X		Partial[d]	X	X	Frischknecht et al. (2009)
	Abiotic depletion potential (fossil fuels)[e]						X		van Oers (2012)
	Impact 2002+ ('non-renewable energy')						X	X	Jolliet et al. (2003a)
	ReCiPe Midpoint (fossil and water depletion)	X					X		Goedkoop et al. (2009)
Resource depletion at midpoint level	EDIP 97/2003			Partial[f]		X	X	X	Hauschild and Wenzel (1998)
	Abiotic depletion potential					X		X	Guinée et al. (2002), van Oers et al. (2002)
	Biotic Production Potential (BPP)		X						Brandão and Milài Canals (2012)
	Ecological Scarcity (water)	X							Frischknecht et al. (2009)
	Abiotic depletion potential (water)	X							Milà i Canals et al. (2009)

Resource depletion at endpoint level	EI99					X	X		Goedkoop and Spriensma (2000)
	Impact 2002+ ('mineral extraction')					X			Jolliet et al. (2003a)
	ReCiPe Endpoint					X	X	X	Goedkoop et al. (2009)
	EPS2000					X	X	X	Steen (1999a, b)
	Freshwater depletion	X							Pfister et al. (2009)[g]

[a]Examples of flow energy resources are wind energy and hydropower energy
[b]Indirectly accounted through the area needed to absorb the fossil CO_2 emissions
[c]Available in (Huijbregts et al. 2008), it is indirectly accounted through the equivalent average fossil fuel emission intensity of CO_2
[d]Solely gravel is accounted for this type of resource, in the Ecological Scarcity method
[e]Since version 3.6, fossil energy is rather an energy-based RAM (as CED)
[f]Method developed for renewable resources but factors only provided for wood and freshwater at a global level
[g]This method can also be used at midpoint level by not considering the backup-technology term ($E_{desalination}$)

the resource base is likely to increase as a consequence of increases in population numbers and material standard of living, and urban mining could become much more important. Overall, data availability for quantifying impacts at the AoP Resources is a considerable concern.

Recent research is being done to tackle some of these issues, for instance in the project LC-Impact (http://www.lc-impact.eu/). One example is the method from Vieira et al. (2012), which evaluates the resource depletion of metals based on the global ore grade information. Therefore, it would belong to the category 'resource depletion – midpoint', but considering future consequences. It provides CFs for three different types of copper deposits.

Other methods try to include the stocks of metals which have accumulated in the technosphere, e.g. Schneider et al. (2011), reflecting the idea of urban mining. Furthermore, the inclusion of competition and regional availability could be aspects to be included when assessing abiotic resource use in LCA (Yellishetty et al. 2011). By some it is even suggested to model process changes due to current resource dissipation, like increased energy needs, in the inventory phase of an LCA, rather than in the LCIA phase (Finnveden 2005; Weidema et al. 2005).

Abiotic resource availability is an important issue for society, yet there is still no clear consensus method for LCIA (European Commission 2011). The existing methods are quite diverse. Furthermore, uncertainties and data availability remain a challenge in the modelling of impacts of abiotic resource use as considerable parts of the planet are not fully explored and well documented.

References

Ahbe S, Braunschweig A, Müller-Wenk R (1990) Methodik für Ökobilanzen auf der Basis ökologischer Optimierung. Bundesamt für Umwelt, Wald und Landschaft (BUWAL), Bern

Alvarenga RAF, Dewulf J, Langenhove HV, Huijbregts MAJ (2013) Exergy-based accounting for land as a natural resource in life cycle assessment. Int J Life Cycle Assess 18:939–947

Baral A, Bakshi BR (2010) Thermodynamic metrics for aggregation of natural resources in life cycle analysis: insight via application to some transportation fuels. Environ Sci Technol 44:800–807

Baral A, Bakshi BR, Smith RL (2012) Assessing resource intensity and renewability of cellulosic ethanol technologies using eco-LCA. Environ Sci Technol 46:2436–2444

Baumann H, Rydberg T (1994) Life cycle assessment: a comparison of three methods for impact analysis and evaluation. J Clean Prod 2:13–20

Bösch M, Hellweg S, Huijbregts M, Frischknecht R (2007) Applying cumulative exergy demand (CExD) indicators to the ecoinvent database. Int J Life Cycle Assess 12:181–190

Boustead I, Hancock GF (1979) Handbook of industrial energy analysis. Wiley, New York

Brand G, Scheidegger A, Schwank O, Braunschweig A (1998) Bewertung in Ökobilanzen mit der Methode der ökologischen Knappheit – Ökofaktoren 1997. Bundesamt für Umwelt, Wald und Landschaft (BUWAL), Bern

Brandão M, Milài Canals L (2012) Global characterisation factors to assess landuse impacts on biotic production. Int J Life Cycle Assess 18(6):1243–1252

Chapman PF, Roberts F (1983) Metal resources and energy. Butterworths, Kent

Dewulf J, Bosch ME, Meester BD, Van der Vorst GV, Van Langenhove H, Hellweg S, Huijbregts MAJ (2007) Cumulative exergy extraction from the natural environment (CEENE): a comprehensive life cycle impact assessment method for resource accounting. Environ Sci Technol 41:8477–8483

Dewulf J, Van Langenhove H, Muys B, Bruers S, Bakshi BR, Grubb GF, Paulus DM, Sciubba E (2008) Exergy: its potential and limitations in environmental science and technology. Environ Sci Technol 42:2221–2232

ecoinvent (2010) ecoinvent data v2.2. ecoinvent reports No.1-25. Swiss Centre for Life Cycle Inventories, Dübendorf

EPA (2014) United States Environmental Protection Agency. Natural resources. Teacher fact sheet. http://www.epa.gov/osw/education/quest/pdfs/unit1/chap1/u1_natresources.pdf

European Commission (2011) International Reference Life Cycle Data System (ILCD) handbook-recommendations for life cycle impact assessment in the European context, 1st edn. European Commission – Joint Research Centre – Institute for Environment and Sustainability, Luxemburg

Ewing B, Reed A, Galli A, Kitzes J, Wachernagel M (2010) Calculation methodology for the national footprint accounts. Global Footprint Network, Oakland

Finnveden G (1996) Resources and related impact categories – part II. In: Udo de Haes HA (ed) Towards a methodology for life cycle impact assessment. SETAC-Europe, Brussels

Finnveden G (2005) The resource debate needs to continue. Int J Life Cycle Assess 10:372

Finnveden G, Moberg Å (2005) Environmental systems analysis tools: an overview. J Clean Prod 13(12):1165–1173

Frischknecht R, Steiner R, Jungbluth N (2009) The ecological scarcity method – eco-factors 2006. A method for impact assessment in LCA. Bundesamt für Umwelt (BAFU), Bern

Global Footprint Network (2009) Ecological footprint standards 2009. Global Footprint Network, Oakland

Goedkoop M, Heijungs R, Huijbregts M, de Schryver A, Struijs J, van Zelm R (2009) ReCiPe 2008: a life cycle impact assessment method which comprises harmonized category indicators at the midpoint and the endpoint level, 1st edn. Report I: characterisation

Goedkoop M, Spriensma R (2000) The eco-indicator 99 – a damage oriented method for life cycle impact assessment: methodology report. PRe Consultants, Amersfoort

Guinée J (1995) Development of a methodology for the environmental life-cycle assessment of products. Leiden University, Leiden

Guinée JB, Gorrée M, Heijungs R, Huppes G, Kleijn R, de Koning A, van Oers L, Wegener Sleeswijk A, Suh S, Udo de Haes HA, de Bruijn H, van Duin R, Huijbregts MAJ, Lindeijer E, Roorda AAH, van der Ven BL, Weidema BP (2002) Handbook on life cycle assessment: an operation guide to the ISO standards. Kluwer Academic Publishers, Dordrecht

Hau JL, Bakshi BR (2004a) Expanding exergy analysis to account for ecosystem products and services. Environ Sci Technol 38(13):3768–3777

Hau JL, Bakshi BR (2004b) Promise and problems of emergy analysis. Ecol Model 178 (1–2):215–225

Hauschild M, Wenzel H (1998) Environmental assessment of products vol 2: scientific background. Chapman & Hall/Kluwer Academic Publishers, London/Hingham, 1997. ISBN 0-412-80810-2

Heijungs R, Guinée J, Huppes G (1997) Impact categories for natural resource and land use. Section Substances & Products. CML Report. CML, Leiden University, Leiden

Hischier R, Weidema B, Althaus H-J, Doka G, Dones R, Frischknecht R, Hellweg S, Humbert S, Jungbluth N, Loerincik Y, Margni M, Nemecek T, Simons A (2009) Implementation of life cycle impact assessment methods: final report ecoinvent v2.1, vol 3. Swiss Centre for Life Cycle Inventories, St. Gallen

Hoekstra AY, Chapagain AK, Aldaya MM, Mekonnen MM (2011) The water footprint assessment manual: setting the global standard. Water Footprint Network, London

Hoekstra AY, Hung PQ (2002) Virtual water trade: a quantification of virtual water flows between nations in relation to international crop trade. Research report series no 11. IHE Delft, Delft
Huijbregts MAJ, Hellweg S, Frischknecht R, Hendriks HWM, Hungerbuhler K, Hendriks AJ (2010) Cumulative energy demand as predictor for the environmental burden of commodity production. Environ Sci Technol 44:2189–2196
Huijbregts MAJ, Hellweg S, Frischknecht R, Hungerbuhler K, Hendriks AJ (2008) Ecological footprint accounting in the life cycle assessment of products. Ecol Econ 64:798–807
Huijbregts MAJ, Rombouts LJA, Hellweg S, Frischknecht R, Hendriks AJ, van de Meent D, Ragas AMJ, Reijnders L, Struijs J (2006) Is cumulative fossil energy demand a useful indicator for the environmental performance of products? Environ Sci Technol 40:641–648
Ingwersen WW (2011) Emergy as a life cycle impact assessment indicator. J Ind Ecol 15:550–567
ISO (2006) ISO international standard 14040: environmental management — life cycle assessment — principles and framework. International Organization for Standardisation, Geneva
Jolliet O, Margni M, Charles R, Humbert S, Payet J, Rebitzer G, Rosenbaum R (2003a) IMPACT 2002+: a new life cycle impact assessment methodology. Int J Life Cycle Assess 8:324–330
Jolliet O, Müller-Wenk R, Brent A, Goedkoop M, Itsubo N, Pena C, WB P, Schenk R, Stewart M (2003b) Final report of the LCIA definition study. UNEP/SETAC Life Cycle Initiative, Paris
Liao W, Heijungs R, Huppes G (2012a) Natural resource demand of global biofuels in the anthropocene: a review. Renew Sust Energ Rev 16:996–1003
Liao W, Heijungs R, Huppes G (2012b) Thermodynamic resource indicators in LCA: a case study on the titania produced in Panzhihua city, southwest China. Int J Life Cycle Assess 17:951–961
Lindeijer E, Müller-Wenk R, Steen B (2002) Impact assessment of resources and land use. In: Udo de Haes HA (ed) Life cycle impact assessment: striving towards the best practice. SETAC, Pensacola, pp 11–64
Milà i Canals L, Chenoweth J, Chapagain A, Orr S, Antón A, Clift R (2009) Assessing freshwater use impacts in LCA: part I – inventory modelling and characterisation factors for the main impact pathways. Int J Life Cycle Assess 14:28–42
Miyazaki N (2006) The JEPIX Initiative in Japan. A new ecological accounting system of a better measurement of eco-efficiency. In: Schaltegger S, Bennett M, Burritt R (eds) Sustainability accounting and reporting. Springer, Dordrecht, pp 339–354
Müller-Wenk R (1998) Depletion of abiotic resources weighted on the base of 'virtual' impacts of lower grade deposits in future. IWO Diskussionsbeitrag Nr. 57
Odum HT (1996) Environmental accounting: emergy and environmental decision making, 1st edn. Wiley, New York
OECD (2008a) Measuring material flows and resource productivity, vol III, Inventory of country activities. OECD Publishing, Paris
OECD (2008b) Measuring material flows and resource productivity, vol II, The accounting framework. OECD Publishing, Paris
OECD (2008c) Measuring material flows and resource productivity. The OECD guide, vol I. OECD Publishing, Paris
OECD/IEA (2013) Resources to reserves 2013 – oil, gas and coal technologies for the energy markets of the future. Organisation for Economic Co-operation and Development, International Energy Agency, Paris
PE International (2012) http://www.gabi-software.com. Accessed 20 Apr 2012
Pfister S, Koehler A, Hellweg S (2009) Assessing the environmental impacts of freshwater consumption in LCA. Environ Sci Technol 43:4098–4104
Pimentel D, Hurd LE, Bellotti AC, Forster MJ, Oka IN, Sholes OD, Whitman RJ (1973) Food production and the energy crisis. Science 182(4111):443–449
Ritthoff M, Rohn H, Liedtke C (2002) MIPS Berechnen: Ressourcen produktivität von Produkten und Dienstleistungen. Wuppertal Spezial. Wuppertal Institut für Klima, Umwelt, Energie, Wuppertal
Rugani B, Benetto E (2012) Improvements to emergy evaluations by using life cycle assessment. Environ Sci Technol 46:4701–4712

Rugani B, Huijbregts MAJ, Mutel C, Bastianoni S, Hellweg S (2011) Solar energy demand (sed) of commodity life cycles. Environ Sci Technol 45:5426–5433

Schmidt-Bleek F, Bringezu S, Hinterberger F, Liedtke C, Spannenberg J, Stiller H, Welfens MJ (1998) MAIA Einführung in die Material-Intensitäts-Analyse nach dem MIPS-Konzept, Berlin

Schneider L, Berger M, Finkbeiner M (2011) The anthropogenic stock extended abiotic depletion potential (AADP) as a new parameterisation to model the depletion of abiotic resources. Int J Life Cycle Assess 16:929–936

Spangenberg JH, Hinterberger F, Moll S, Schütz H (1999) Material flow analysis, TMR and the mips-concept: a contribution to the development of indicators for measuring changes in consumption and production patterns. Wuppertal Institute for Environment, Climate and Energy; Department for Material Flows and Structural Change, Wuppertal

Steen B (1999a) A systematic approach to environmental priority strategies in product development (EPS). Version 2000 – general system characteristics. Chalmers University of Technology, Gothenburg

Steen B (1999b) A systematic approach to environmental priority strategies in product development (EPS). Version 2000 – models and data of the default method. Chalmers University of Technology, Gothenburg

Steen B (2006) Abiotic resource depletion different perceptions of the problem with mineral deposits. Int J Life Cycle Assess 11:49–54

Steen B, Borg G (2002) An estimation of the cost of sustainable production of metal concentrates from the earth's crust. Ecol Econ 42:401–413

Stewart M, Weidema BP (2005) A consistent framework for assessing the impacts from resource use – a focus on resource functionality. Int J Life Cycle Assess 10:240–247

Szargut J, Morris DR, Steward FR (1988) Exergy analysis of thermal, chemical, and metallurgical processes. Springer, Berlin

Udo de Haes HA, Jolliet O, Finnveden G, Hauschild M, Krewitt W, Miiller-Wenk R (1999) Best available practice regarding impact categories and category indicators in life cycle impact assessment. Int J Life Cycle Assess 4:167–174

United States Geological Survey (2010) Mineral commodity summaries 2010. Geological Survey, Washington, DC

Urban RA, Bakshi BR (2009) 1,3-Propanediolfrom fossils versus biomass: a life cycle evaluation of emissions and ecological resources. Ind Eng Chem Res 48:8068–8082

Van Oers L (2012) CML spreadsheets. CML – Institute of Environmental Sciences, Leiden University, Leiden. http://www.cml.leiden.edu/software/data-cmlia.html

Van Oers L, de Koning A, Guinee J, Huppes G (2002) Abiotic resource depletion in LCA – improving characterisation factors for abiotic resource depletion as recommended in the new Dutch LCA Handbook. Road and Hydraulic Engineering Institute, Leiden University. http://www.leidenuniv.nl/cml/ssp/projects/lca2/report_abiotic_depletion_web.pdf

VDI (1997) Cumulative energy demand – terms, definitions, methods of calculation.VDI guideline 4600. Verein Deutscher Ingenieure, Düsseldorf

Vieira MDM, Goedkoop MJ, Storm P, Huijbregts MAJ (2012) Ore grade decrease as life cycle impact indicator for metal scarcity: the case of copper. Environ Sci Technol 46:12772–12778

Wackernagel M, Rees W (1996) Our ecological footprint: reducing human impact on the earth. New Society Publishers, Gabriola Island

Weidema BP, Finnveden G, Stewart M (2005) Impacts from resource use: a common position paper. Int J Life Cycle Assess 10(6):382

Yellishetty M, Mudd GM, Ranjith PG (2011) The steel industry, abiotic resource depletion and life cycle assessment: a real or perceived issue? J Clean Prod 19:78–90

Zhang Y, Baral A, Bakshi BR (2010) Accounting for ecosystem services in life cycle assessment, part ii: toward an ecologically based LCA. Environ Sci Technol 44:2624–2631

Chapter 14
Normalisation

Alexis Laurent and Michael Z. Hauschild

Abstract Defined as an optional step in the ISO14044 requirements, normalisation in LCA relates the characterised impact indicator scores of an analysed system to those of a reference system. By putting the LCA results in a broader perspective, it can facilitate their interpretation and communication, and allow checking whether their magnitude looks reasonable. This chapter provides a comprehensive overview of the two major normalisation approaches, internal and external normalisation, encompassing for the latter both production-based and consumption-based inventory methods. Pros and cons are addressed for each approach. Because of its wide use and usefulness, emphasis is put on external normalisation. The chapter details the calculation of external normalisation references, including their scoping, the collection of data for building their associated production-based or consumption-based normalisation inventories, their computation by use of adequate sets of characterisation factors, and their resulting uncertainties. The chapter provides insights in the application of normalisation in practice. After listing the past efforts of establishing external normalisation references for different regions in the world, the use of different normalisation approaches and their possible interpretation in relation to the goals and scope of an LCA study are discussed, along with the potential uncertainties and biases in the normalised scores.

Keywords Consumption-based inventory • External normalisation • Internal normalisation • LCA • LCIA • Life cycle assessment • Life cycle impact assessment • Production-based inventory • Reference system • Result communication • Substance coverage

A. Laurent (✉) • M.Z. Hauschild
Division for Quantitative Sustainability Assessment, Department of Management Engineering, Technical University of Denmark (DTU), Produktionstorvet, Building 424, Lyngby 2800, Denmark
e-mail: alau@dtu.dk; mzha@dtu.dk

M.Z. Hauschild, M.A.J. Huijbregts (eds.), *Life Cycle Impact Assessment*,
LCA Compendium – The Complete World of Life Cycle Assessment,
DOI 10.1007/978-94-017-9744-3_14

1 Definition and Purpose of Normalisation

After the characterisation step, the results of an LCA may include a large number of impact indicator scores, all expressed in individual metrics specific to the impact category. For example, climate change at midpoint level would typically be expressed in a unit of mass CO_2-equivalent; photochemical ozone formation could be expressed in a unit of mass NMVOC[1]-equivalent or mass C_2H_4-equivalent. Also at endpoint level, the different areas of protection are commonly expressed in different metrics and a comparison is not straightforward. Therefore, the interpretation of the results can be difficult if the LCA practitioner seeks to analyse the obtained characterised profile as a whole and identify the most relevant impacts to address in the decision-making process. To facilitate the interpretation of the LCA results and their communication to decision- and policy-makers, two additional steps can be performed, namely normalisation and weighting, which are both optional in the conduct of LCA (ISO 14044 2006). Normalisation is addressed in this chapter; weighting is discussed in Chap. 15 of this volume.[2]

1.1 Definition and Purpose

According to the ISO standard, normalisation is the calculation of the magnitude of an impact indicator score relative to reference information with the aim to better understand the relative magnitude for each indicator result of the product system under study (ISO 14044 2006). In practice, each characterised impact indicator score is divided by a corresponding impact indicator score reflecting the impact of the reference system, i.e. the normalisation reference (Eq. 14.1). This reference system can take many forms: it can be a product or a service, or it can be the annual activities of a company, an industrial or societal sector, a nation, a larger region or the whole world. The choice of the reference system should be consistent with the goal of the study, and it is thus dependent on the context, in which the LCA study is performed. For example, global normalisation references could be used as default as, in the current globalised economy, most product systems stretch out worldwide, but the application of national or regional weighting factors in the LCA study may call for the use of regional normalisation references. It is subject to a categorisation of normalisation approaches, which is structuring this chapter. In accordance with the chosen form of characterisation, normalisation can be performed at both midpoint and endpoint level.

[1] NMVOC = non-methane volatile organic compounds.

[2] Chapter 15 "Weighting" by Norihiro Itsubo.

$$NS_i^{sys} = \frac{CS_i^{sys}}{CS_i^{ref}} \tag{14.1}$$

Where:

NS_i^{sys}: Normalised impact indicator score for impact category i of the system (sys) under study

CS_i^{sys}: Characterised impact indicator score for impact category i of the system (sys) under study

CS_i^{ref}: Characterised impact indicator score for impact category i of the reference system (ref), also called the *normalisation reference* for the impact category i.

Performance of normalisation can fulfil two major purposes:

1. Purpose 1: Facilitate the interpretation and communication of the indicator results by (i) appraising the magnitude of the system's impact indicator scores relative to the impacts of the reference system, (ii) expressing the different impact indicator scores on a common scale, which supports comparisons of results across impact categories (keeping in mind that any differences in the severity or seriousness of the impact categories need to be dealt with separately in a weighting step), and (iii) preparing for weighting or valuation by expressing the scores in a form that is compatible with the intended weighting factors.
2. Purpose 2: Check the sanity of the impact indicator scores, i.e. whether or not the results look reasonable, both in terms of the pattern across the categories ("is it reasonable that the normalised eutrophication impact is higher than the climate change impact?") and in the absolute value of the normalised results (e.g. the assessment of a pen leading to a climate footprint equivalent to the annual contribution of ten global average persons would reflect errors somewhere in the conduct of the LCA).

These purposes can be fulfilled to different extent depending on the selected normalisation approaches. In particular, the sanity check performed after normalisation can add reliability to an LCA outcome as it can reveal errors or inconsistencies both in the inventory and impact assessment phases of the LCA (purpose 2).

1.2 Internal and External Normalisation

Normalisation approaches can be classified into two distinct families: internal and external normalisation (Norris 2001). The terms 'internal' and 'external' refer to the selected reference system, which can either be contained within the study or be independent from it.

In an internal normalisation, the primary objective is to eliminate the specificity of the impact indicator units so that the obtained normalised scores can undergo

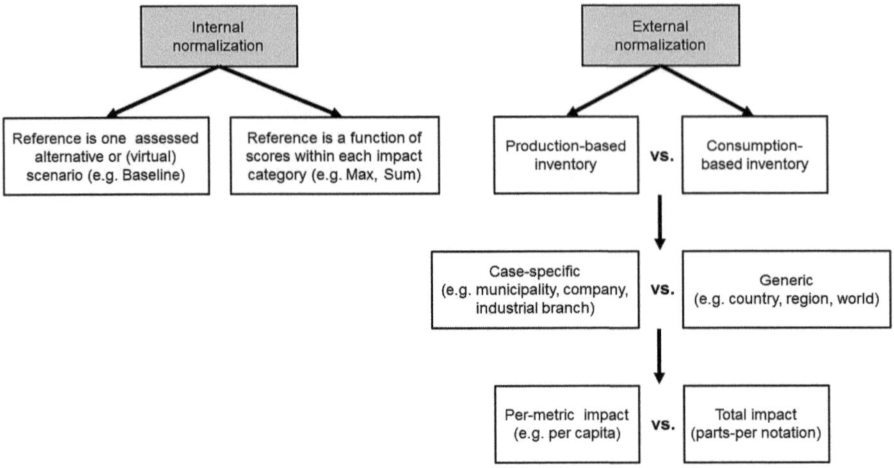

Fig. 14.1 Overview of the different normalisation approaches

further valuation (Norris 2001). The reference system is often a function of the impact indicator scores (e.g. maximum or sum within each impact category) or one of the assessed alternatives. In an external normalisation, the reference system corresponds to a given entity independent of the object of the LCA, e.g. a company or a region, in a given time period.

Figure 14.1 illustrates the two families of approaches as well as the subsequent divisions within each. They are all detailed in Sects. 2 and 3.

1.3 Structure of the Chapter

Following the above introduction of normalisation in LCA, the chapter is structured into three major parts. The first (Sect. 2) addresses internal normalisation, i.e. the different forms of internal normalisation, their uses in practice and their limitations. Sect. 3 focuses on the external normalisation references taken as stand-alone systems (without considering their use in practice). It thus addresses the calculation of external normalisation references and includes the selection and delimitation of the reference system, the building of the emission and resource consumption inventories, and the uncertainties associated with the values of the normalisation references. The final part, Sect. 4, positions the external normalisation references in their context of use and addresses the normalisation step as a whole, i.e. how have the normalisation references been applied in practice? What benefits can be brought by the use of normalisation in LCA case studies? And what overall uncertainties characterise the normalisation step?

2 Internal Normalisation

The origins of internal normalisation can be tracked to the field of multi-attribute decision analysis, which includes different approaches to evaluate project alternatives from a variety of economic and non-economic (e.g. environmental) attributes (Norris and Marshall 1995). In this setting, internal normalisation is used as a first step to bring non-commensurate impact indicator scores into a common metric and ultimately allow their valuation.

2.1 *Definition and Application*

Internal normalisation can be performed in many ways. Among the ones used in LCA, one approach consists of selecting a reference alternative or a baseline scenario among the ones under study, termed 'division by baseline' (DBB) in the following. For each impact category, the characterised impact indicator scores of all alternatives are divided by the score obtained for the selected reference alternative. Comparisons of the resulting ranking, relative to the reference alternative, can thus be performed across impact categories, e.g. expressing the obtained ratios as percentages.

Rather than selecting a baseline alternative, other approaches define the normalisation reference as a function of the characterised impact indicator scores of the different alternatives. For example, in the 'division-by-maximum' (DBM), the normalisation reference for each impact category is the maximum impact indicator score obtained across all alternatives in the study. All resulting scores thus range from 0 to 1, where the option scoring closest to 0 is the best and 1 is the worst. Another example can be the 'division by sum' (DBS), in which the normalisation reference for each impact category is the sum of the impact indicator scores for all the studied alternatives within that impact category. The obtained normalised scores also range from 0 to 1, but the sum of normalised impact scores across the alternatives' scores equals 1 within each impact category.

2.2 *Illustrative Applications and Limitations*

To illustrate the principles and limitations of the different internal normalisation approaches, a simple example has been constructed, in which characterised scores were determined for three impact categories at midpoint level, i.e. climate change (CC), photochemical ozone formation (PO) and metal depletion (MD). Table 14.1 shows the characterised impact indicator scores for the three studied alternatives A, B and C.

Table 14.1 Illustrative case study with 3 alternatives A, B and C (characterised impact indicator scores)

Alternatives	Climate change (CC) (kg-CO_2-eq)	Photochemical ozone formation (PO) (kg-NMVOC-eq)	Metal depletion (MD) (kg-Fe-eq)
A	4,000	0.2	30
B	500	12	15
C	1,000	2	55
Weights[a]	0.45	0.25	0.30

[a]Arbitrarily-set weights (which are used to illustrate the limitations of internal normalisation – see text)

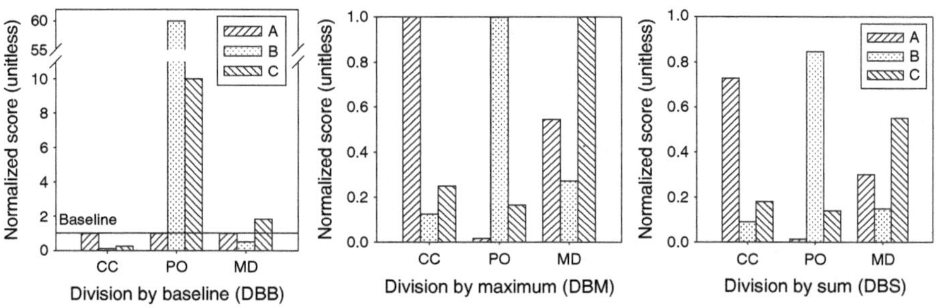

Fig. 14.2 Use of different internal normalisation approaches in assessment of 3 alternatives *A*, *B* and *C*. *CC* climate change, *PO* photochemical ozone formation, *MD* metal depletion

The three approaches, DBB, DBM and DBS, are the most frequent internal normalisation approaches in LCA (Norris 2001). When applied to the characterised scores in Table 14.1, they give the normalised scores shown in Fig. 14.2. The graphic expression of the scores enables to grasp more easily the environmental problems of one alternative over another across impact categories. It may thus contribute to facilitate the communication of the LCA results to relevant stakeholders, but it is clear from the figure that the somewhat arbitrary choice of internal normalisation reference has a strong influence on the relative magnitudes of the scores for the three impact categories and hence on their interpretation.

Several limitations and drawbacks exist in the use of internal normalisation approaches. Most of them have been well described in Norris (2001) and Norris and Marshall (1995). The main ones are:

- The DBB approach runs the risk of including a division-by-zero in the event that the baseline alternative has no impact contribution for some of the impact categories (Norris 2001). In common LCA practice, this risk is limited because it is rare to see an alternative with no contribution to an impact category.
- Using the same set of weighting factors, DBM and DBS may lead to different rankings (Norris and Marshall 1995). For example, taking the normalised results presented in Fig. 14.2 and assuming generic weighting factors of 0.45, 0.25 and 0.35 for CC, PO and MD, respectively, leads to the results shown in

Table 14.2 Weighted scores using different internal normalisation approaches and generic weights[a]

Alternatives	Assessing A, B and C		Assessing A and C only	
	DBM	DBS	DBM	DBS
A	0.62	0.42	0.64	0.49
B	0.39	0.30	-	-
C	0.45	0.28	0.66	0.51

DBM division by maximum, *DBS* division by sum
[a]Ranking of alternatives A, B and C is indicated by color-coding for each internal normalisation approach (dark grey > light grey > white: highest to lowest impact)

columns 2 and 3 of Table 14.2. Alternative A ranks worst in both approaches, but whether B or C ranks best depends on the selected normalisation approach.

- In any internal normalisation, where generic weights are applied (although this approach is not recommended, it was commonly applied in North America in the 90s; see Norris 2001), the results are insensitive to the magnitude of the impact indicator scores. For example, taking the impact assessment results in Table 14.1 and assuming that all scores for climate change are in ng CO_2-equivalent instead of kg CO_2-equivalent would not change the results in Table 14.2 (column 2 and 3). For the same reason, the additions or removal of one alternative may reverse the ranking (Norris 2001). Taking again the example in Table 14.1 and leaving alternative B out of the comparison results in the new scores shown in columns 4 and 5 of Table 14.2. Although nothing has changed in either alternative, alternative A now ranks better than alternative C.

In the context of decision-making, some of these limitations are critical and the LCA practitioner should therefore be aware of them. To ensure more robust results, it is advocated to run several approaches, e.g. DBS and DBM (Norris and Marshall 1995); any potential discrepancies in the obtained results should then be investigated before conclusions are drawn.

2.3 Use in LCA Practice

In practice, internal normalisation can only address some of the normalisation purposes described in the introduction, and can do so only to a limited extent, as all reference information is contained within the assessed system. Results cannot be put in a broader perspective, thus preventing an absolute appraisal of the magnitude of the impacts (purpose 1). Comparisons across impact categories cannot be performed either. A consistency check can be performed (purpose 2), but errors may be difficult to unveil, e.g. if the error is nested in the baseline scenario used as normalisation reference or if it is common to all assessed scenarios. Internal normalisation can, however, serve as a preparation for a valuation or weighting step (purpose 1).

The latter purpose has been the main reason for using internal normalisation. It was widely used in LCAs in the 90s in North America (see, e.g., Lippiatt 2000). At that time, external normalisation was not yet developed in North America and internal normalisation was the best approach for LCA practitioners to perform normalisation. Since then, the interest for internal normalisation appears to have decreased primarily because of (1) the development of external normalisation, which can address the two purposes in a more complete way, and (2) its important drawbacks, which may compromise its use in support for decision-making (see Sect. 2.2). Although some LCA studies using internal normalisation can still be found (e.g. DBM in Chevalier et al. 2003; DBB in Boughton and Horvath 2006), the users rarely refer to it as normalisation and, even less, as internal normalisation.

3 External Normalisation

In contrast to the use of internal references for normalisation, the application of external normalisation implies the use of normalisation references which represent the environmental profile of an external reference system that is independent from the system under study, and often at a much larger scale.

3.1 *Definition and Fundamental Concepts*

The determination of external normalisation references can be seen as a life cycle impact assessment of the inventory for a large-scale system, for which the input and output of resources and emissions over a defined period of time are inventoried and characterised. The scoping of the normalisation references is dependent on the choice of the LCA practitioner, who needs to evaluate what reference system would be the most relevant to compare his/her assessed system with in relation to the goals of the LCA. This scoping must address two major points, namely the boundaries of the reference system and the period of time or reference duration of the included activities (see Sect. 3.2).

The mapping of environmental flows related to the activities included within the system boundaries can apply two different perspectives, following either a production-based or a consumption-based inventory approach. The differences between the two are described in Fig. 14.3. The production-based approach aims to reflect the level of environmental impact associated with the total production activities of the reference system within each impact category and therefore inventories the flows from all activities occurring within the physical or geographical boundaries of the reference system over the reference duration. In contrast, the consumption-based approach aims to reflect the level of environmental impact associated with the total consumption of the reference system and hence quantifies the flows from all processes needed to support the consumption activities of the

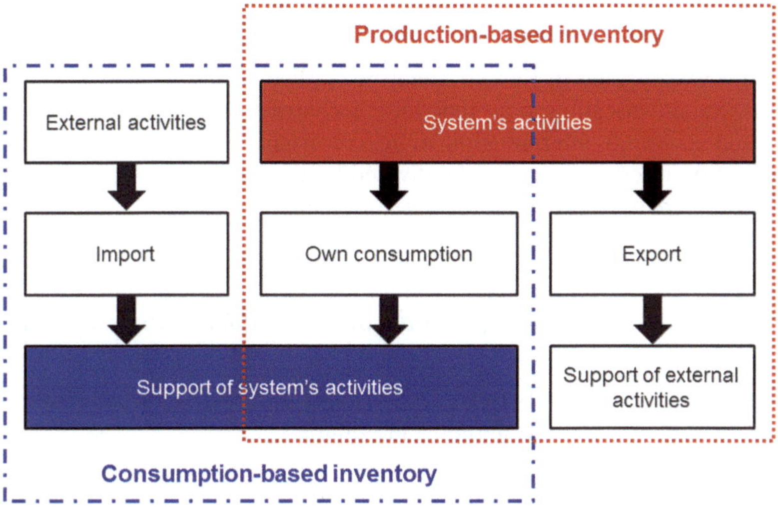

Fig. 14.3 Distinction between production-based and consumption-based inventories

reference system over the reference duration, including those that occur outside its physical or geographical boundaries. For example, production-based normalisation references for Europe reflect the impacts from all activities taking place within the European territory while consumption-based normalisation references for Europe reflect the impacts from processes that are needed to support the consumption taking place in Europe, and thus include several impacts that occur in other regions than Europe, e.g. stemming from manufacture of products in China or the USA. In both approaches, the collection of elementary flow data is the most challenging task to accomplish because the data availability, more limited than in most product LCAs due to the macro-scale perspective, can preclude the achievement of an inventory sufficiently complete to be used in practice (see Sects. 3.3, 3.4, and 3.5).

The inventory associated with the reference system and aggregated over the selected reference duration (i.e. normalisation inventory) is translated into scores for each impact category (i.e. normalisation references) using the set of characterisation factors that stems from the same LCIA method or characterisation model (at the same midpoint or endpoint level) as that used in the characterisation of the inventory for the system under study. The resulting set of normalisation references are thus calculated in the same way and expressed in the units of the characterised impact indicator scores of the assessed system over the reference duration, hence expressing the normalised profile in a time metric. The normalised profile can be expressed in parts-per notation and be interpreted as the share of the total reference system's impact. To facilitate the interpretation and communication of the results, other final metrics can be used, such as the Person Equivalent (PE) or person-years, which represents the time-limited contribution of an average person to the impacts occurring within the chosen reference system and which is used by a number of existing LCIA methodologies (e.g. EDIP, Wenzel et al. 1997, IMPACT 2002+, v.

Q2.2, Humbert et al. 2012, ReCiPe 2008, v.1.08, Goedkoop et al. 2009, 2013). An additional reason for using the PE metric is that marginal characterisation factors, which are widely used in LCIA, cannot be applied to assessments of large-scale systems; the downscaling to an individual's contribution thus makes the methodology and its result interpretation more consistent. For expression in PEs, all normalisation references need to be divided by the population consistent with the chosen geographic scope of the reference system.

3.2 Scoping of the Background Load of the Reference System

The reference system needs to be scoped both in terms of its boundaries (geographic or economic) and its duration and timing.

3.2.1 System Boundaries

The system boundaries can be classified in two types: generic or case-specific. Generic system boundaries are commonly defined by geographical boundaries at a large scale, where the system is a region, a country, a continent or the entire world. Case-specific boundaries are defined by case-study-specific parameters, which for example can result in entities being a municipality to which the assessed system belongs (e.g. waste management), a company manufacturing the product under study or an entire branch of manufacture and trade related to the product (e.g. the textile industry). Another distinction of the system boundaries can also be made between the physical or geographical boundaries and the actual system boundaries of the reference system. In a production-based inventory, the two overlap. However, in a consumption-based approach, they generally differ, as illustrated in Fig. 14.3 (see also Sect. 3.1).

3.2.2 Reference Duration and Reference Time

The selected duration, for example a month or a business year, is dependent on the context of the LCA under study (goal and scope) and on the type of selected reference system. Very often a given year is taken as reference duration. This year is typically defined as the latest year for which reliable data are available (EC 2010) for all impact categories. The same reference time or reference year should be chosen for all impact categories considered in the set of normalisation references to avoid the bias that may otherwise be introduced into the set of normalisation references, if emissions show a consistent trend over time (typically they increase, but, e.g., for stratospheric ozone depletion we have witnessed a strong decline due to the efficient regulation of ozone depleting substances). Variations in reference time may have to be accepted for some elements of a normalisation inventory, e.g. if some substance emissions are only known for another year than the selected reference year. However, these data should be adjusted (e.g. via extrapolations) or justified to not vary significantly between the reported year and the chosen reference year.

3.2.3 Choice of the Reference System

The choice of the set of normalisation references is typically made during the initial scope definition. It is an important step because this choice can alter the normalised profile and its subsequent interpretation (see Sect. 4.1.3). For this reason, the ISO standards also suggest the use of several sets of normalisation references, reflecting different reference systems, to assess the sensitivity of the final results to the normalisation step. All the same, the choice needs to be justified in relation to (i) the compatibility of the available normalisation references with the goal and scope of the LCA study, the applied characterisation models and the selected weighting approach (if any), (ii) the scope of the decisions that may be taken following the results of the LCA study, e.g. where do they apply?, and (iii) the relevance for the intended application and target audience of the LCA study (ISO 14044 2006; EC 2010).

The first criterion (i) can be judged by the LCA practitioner alone as he or she has the best knowledge of the LCA study itself. The scope of the selected normalisation references can thus be checked for compatibility with that of the product system under study. If emissions and resource consumptions occur in a specific region or country, or if the weighting step requires the use of regional or national normalisation references, the geographical scope may be chosen to match that region or country. In most situations, products or services would, however, not be regionally defined and their system boundaries would encompass a global scope. Therefore, the use of global normalisation references is generally advisable (Huijbregts et al. 2003). With respect to production-based normalisation references, the practitioner should also be aware of the risk of division-by-zero in case of improper choice. For example, in the assessment of a product heavily relying on fossil-fuelled transportation, the use of a set of resource depletion normalisation references for a region, where no oil is extracted, leads to a division-by-zero for the fossil depletion impact category. Unlike the first criterion, the last two criteria ((ii) and (iii)) require a good understanding of post-LCA study aspects. To secure this knowledge and ultimately identify the most adequate set of normalisation references, the involvement of the different stakeholders, including decision-makers, can be a good help (see, e.g., Dahlbo et al. 2013).

3.3 Production-Based Inventory

The production-based inventory requires gathering emission and resource consumption data related to the reference system activities within its delimitations (system boundaries, duration). Depending on the selected scope of the normalisation reference, i.e. case-specific or generic, and the context of the LCA, the data sources to be used may differ considerably. Because of the high level of aggregation in the data needed, i.e. typically referring to a reference region, country or continent, generic normalisation references can be calculated using publicly available data sources reporting on the annual emissions or resource consumptions of the chosen reference region. Several pre-calculated generic normalisation references are available to LCA practitioners (see Sect. 4.1). In contrast, the determination of case-specific normalisation references for a company's activities or for an entire industrial sector may require the collection of specific data, which often are not publicly reported for the chosen system. In such situations, the LCA practitioner, who performs the LCA study, has to calculate the case-specific normalisation references.

In the determination of generic normalisation references, public databases from different organisations or monitoring bodies are often used. Their coverage of flows causing the impacts (emissions or resource consumptions) can differ considerably across impact categories depending on (i) the recognition of the environmental problem, i.e. the more recognised, the more studied and monitored, and (ii) the number of substances contributing to the environmental problem (see Sect. 3.5.4 for in-depth discussion). Extreme examples are on the one side climate change, a globally recognised and well-studied environmental problem, mainly caused by a limited number (<100) of greenhouse gas emissions, and on the other hand ecotoxicity, with large local variations in the impact pathway, potentially caused by contributions from tens of thousands of substances and less widely recognised due to its local character. Table 14.3 provides a non-exhaustive overview of typical data availability for some of the impact categories typically addressed in a midpoint characterisation.

3.4 Consumption-Based Inventory

The major difference between the production-based inventory and the consumption-based inventory is the consideration of trade in the system boundaries, i.e. the exports and imports of goods or services (see Fig. 14.3). Apart from closed systems that are totally self-sufficient, the global scale is nearly the only situation, for which production equals consumption (including international transportation activities, which are typically not assigned to any country). On the global scale, data collection is relatively straightforward (everything needs to be included). In all other situations, there is a need for identifying and characterising the exports and

Table 14.3 Examples of substance emission/consumption coverage in public sources (non-exhaustive list)[a]

Substances or substance groups	Related environmental impact category	Type of coverage	Example of data sources
CO_2, CH_4, N_2O, SF_6, HFCs, PFCs	Climate change	Global coverage	UNFCCC (Global)
ODS (CFCs, HCFCs, Halons, . . .)	Stratospheric ozone depletion, climate change	Global coverage	UNEP Ozone secretariat (Global), WMO (2011)
NO_X, NH_3, SO_X, NMVOC, CO, PM	Acidification, eutrophication, photochemical ozone formation, respiratory inorganics	Regional coverage	UNFCCC (Global), OECD (global), EMEP/CEIP (Europe), EEA (EU-27), NITE-PRTR (Japan), US-EPA TRI (United States of America), EC-NPRI (Canada), AG-NPI (Australia)
Heavy metals to air	Ecotoxicity, human toxicity	Ca. 10 heavy metals, limited country coverage	
Persistent organic pollutants to air	Ecotoxicity, human toxicity	Limited to few substances, limited country coverage	
N and P-compounds to water and soil	Eutrophication	Limited country coverage	FAOSTAT, EEA (E-PRTR; EU-27)
Toxic emissions to water or soil	Ecotoxicity, human toxicity	Limited coverage of heavy metals and organics including pesticides, limited country coverage	EEA (E-PRTR; EU-27), NITE-PRTR (Japan), US-EPA TRI (United States of America), EC-NPRI (Canada), AG-NPI (Australia)
Resources	Metal/fossil depletion	Global coverage	USGS, IEA, OECD (Global)
Land use	Land use	Global coverage	FAOSTAT, OECD (global)
Water use	Water use	Global coverage	Outcomes of publicly-funded projects, e.g. WATCH project (http://www.eu-watch.org/)

[a]A list of abbreviations with their expanded names is provided below

Abbreviations	Expanded names	Links
UNFCCC	United Nations Framework Convention on Climate Change	http://www.unfccc.int/
UNEP	United Nations Environment Programme	http://www.unep.org/
OECD	Organisation for Economic Co-operation and Development	http://www.oecd.org/

(continued)

Table 14.3 (continued)

Abbreviations	Expanded names	Links
EMEP/CEIP	European Monitoring and Evaluation Programme (EMEP)– Centre on Emission Inventories and Projections (CEIP)	http://www.ceip.at/
EEA	European Environment Agency	http://www.eea.europa.eu/
NITE-PRTR	National Institute of Technology and Evaluation – Pollutant Release and Transfer Register (Japan)	http://www.nite.go.jp/
US-EPA TRI	United States Environmental Protection Agency (US-EPA) – Toxics Release Inventory	http://www2.epa.gov/toxics-release-inventory-tri-program
AG-NPI	Australian Government – National Pollutant Inventory	http://www.npi.gov.au/
E-PRTR	European Pollutant Release and Transfer Register	http://prtr.ec.europa.eu/
EC-NPRI	Environmental Canada – National Pollutant Release Inventory	http://www.ec.gc.ca/inrp-npri/
USGS	United States Geological Survey	http://www.usgs.gov/
IEA	International Energy Agency	http://www.iea.org/
FAOSTAT	Food and Agriculture Organization of the United Nations	http://faostat.fao.org/

imports, which may be difficult to address consistently. The exports relate to activities that take place within the physical boundaries of the reference system and result in products or services which are exported out of it. The corresponding inventory data are therefore already covered in a production-based inventory, but they need to be isolated for exclusion (see Fig. 14.3). In contrast, the imports stem from activities that take place outside the physical boundaries of the reference system. Besides their potentially large number and diversity, which may render the data collection task very resource-demanding, they may not be well-characterised with respect to types of technologies and associated efficiencies and with respect to consumption of resources and pollutant emissions. For example, companies typically have less insight into activities that are located outside their direct financial or operational control (e.g. sub-suppliers). Likewise, in the determination of generic normalisation references for all impact categories but those that relate to globally recognised and thus well monitored environmental problems (climate change, ozone depletion, non-renewable resource depletion), activities located outside the system's physical boundaries, e.g. in newly-industrialised regions can be expected to entail the problem that the needed data is simply not available.

To overcome these hurdles, Input/Output (IO) analysis is often applied. The IO analysis relies on input/output statistics reporting how each sector of the society exchanges inputs and outputs with the other sectors of the economy (Suh and Huppes 2009). I/O statistics can thus provide matrices representing the sector-disaggregated monetary transaction flows among the industrial sectors and from the industrial sectors to final demand (e.g. private households, government, exports). Imports and exports to/from outside the economy system can therefore be easily identified. Because of their relevance in policy information, these matrices are often constructed at country level, but also regional I/O statistics exist coupling national I/O tables reflecting exchanges between the regions. In order to be useful for the development of normalisation references, the economic flows need to be translated into environmental flows to result in environmental-extended Input/ Output (EIO) tables (e.g. national accounting matrices with environmental accounts (NAMEA)). The EIO analysis assumes that the pollutant emissions and resource consumptions of a sector are homogeneous for all activities inside the sector and proportional to the amount of output from that sector (Suh and Huppes 2009). Table 14.4 shows examples of studies that aimed at building generic consumption-based inventories for impact assessment; note that only few of them were built with the intent to support the calculation of normalisation references (see Sect. 4.1.1).

With an impact coverage limited to a non-toxic impact category, like climate change, the use of EIO can be straightforward as the most important environmental flows causing those impacts are available in databases at a sectoral level for a large number of countries. Because most of these countries are also covered with consistent IO tables (e.g. see Lenzen et al. 2012), the determination of environmental flows from imports in a given country or region can easily be performed. The review made by Wiedmann (2009) illustrates this point by showing a large amount of studies investigating the environmental pressures embodied in trade or associated with consumption in regions such as Japan, China, the USA or Europe. Most of these studies limit their scope to the assessment of CO_2 or GHGs, a very few also including acidifying substance emissions (e.g. Lenzen et al. 2012).

However, in the event that other impact categories, e.g. ecotoxicity or human toxicity, need to be assessed at a country or region level, the use of EIO alone is inadequate due to the lack of sectorial emission data for most of the substances that contribute to these impacts; but also to the heterogeneity of many of the industrial sectors applied in the IO statistics which makes the use of sectorial average toxic emissions meaningless. The few studies which conducted such assessments, often combine the EIO analysis with the use of process-based LCI data sets (similar to the hybrid analysis applied to products, Suh et al. 2004) and/or make strong assumptions for characterising the environmental flows associated with the imported goods and services, e.g. considering the same production technology as in domestic production (see examples in Table 14.4; further readings can be found in Wiedmann 2009; Tukker et al. 2006 and Tukker et al. 2013).

3.5 Uncertainties of the Normalisation References

In this section, the uncertainties accompanying the normalisation references are addressed, i.e. the degree to which the determined normalisation references represent the complete environmental impacts of the chosen reference system. Uncertainties arising in the use of the references for normalisation of characterised impact scores are discussed in Sect. 4.3.

Table 14.4 Different approaches to build consumption-based inventories and assess the related environmental impacts (non-exhaustive list)

Covered impacts[a]	System scope	Approach description	References
Climate change	The Netherlands, EU-15 (extended)	Use of EIO tables (incl. for production-based inventory) with differentiation of imports according to the type of technologies (using categorization of three import regions).	Wilting and Ros (2009)
Climate change	World (73 nations, 14 regions)	Multi-region EIO tables using the economic database from the Global Trade Analysis Project (GTAP) database and the GHG data from environmental databases and literature sources.	Hertwich and Peters (2009)
Non-toxic impacts, resources (partial)	World (multi-regions; 187 countries)	Multi-region EIO tables (Eora), using available IO tables (for countries without an IO table, a proxy is built combining the structure of US, JP and AU IO tables with country-specific macro-economic data) and environmental flows from national and international databases.	Lenzen et al. (2012)
Non-toxic and toxic impacts, resource depletion	The Netherlands	Start from production-based inventory with data from available emission databases + use of IO tables for characterising exports and imports using the "domestic technology assumption".[b]	Breedveld et al. (1999)
Non-toxic and toxic impacts, resource depletion	Switzerland	Start from production-based inventory with data compiled in earlier work based on emission databases and activity data + use of IO tables for characterising exports and imports assigning process-based LCI data (e.g. ecoinvent data) to import flows.	Jungbluth et al. (2011)
Non-toxic and toxic impacts, resource depletion	Finland	Use of national emission inventories + use of IO tables for characterising exports and imports combining process-based LCI data (e.g. ecoinvent data) for largest volumes of imported flows and the "domestic technology assumption" for the remaining part of the imports.[b]	Seppala et al. (2011), Koskela et al. (2011), Dahlbo et al. (2013)

[a]Impact coverage is categorised into three major groups: (1) non-toxic impacts (including climate change, stratospheric ozone depletion, acidification, photochemical ozone formation and eutrophication); (2) toxic impacts (including ecotoxicity and human toxicity); (3) resource depletion (including fossils and minerals depletion)

[b]The "domestic technology assumption" replaces missing information about the flows associated with imported goods by the flows for similar goods made with domestic production, hence assuming the same emission resource intensities, and ultimately the same environmental impact, per monetary unit (Tukker et al. 2013)

3.5.1 Overview

Uncertainties of normalisation references can be divided into three major types, which are detailed in the following sub-sections: (1) the uncertainties related to the raw data, assumptions and extrapolations used to build the inventory, (2) the uncertainties related to the characterisation model and the characterisation factors used for the impact assessment, and (3) the uncertainties related to the incomplete coverage of environmental flows in both the inventory and the sets of characterisation factors. Only the environmental flows present in both the inventory and in the characterisation factor base can be represented in the normalisation references. Because of the lack of uncertainty information on the raw data and the inter-twined assumptions, models and extrapolations used in the development of the normalisation inventories, the quantification of the overall uncertainties related to the normalisation references can become a very challenging task, which, until now, no work has undertaken.

3.5.2 Uncertainties Related to the Inventory Building

The uncertainties related to the building of the inventory come from two sources: the uncertainties in the raw data used, and the uncertainties related to the modelling, assumptions and extrapolations performed to improve the consistency and coverage of the inventory. While the first source is the same for production-based and consumption-based inventories, the second is more specific to each approach. In the calculation of generic normalisation references (e.g. national, regional or global levels), the raw data may be found in databases compiled for other purposes by international and national, governmental and non-governmental monitoring bodies. Quantified estimates of the data uncertainties are often not available. The completeness and accuracy in a data set are strongly dependent on:

- *Substance considered.* Due to the focus on the climate change impact, fuel combustion leading to CO_2 emissions is well monitored. For acidification, emissions of NO_x and SO_x have also been monitored and regulated for years. For many other substance emissions, e.g. toxic substances, there is little consistent monitoring hence resulting in data sets of a lower quality.
- *Country or region considered.* For most substances or flows, the principles according to which the data are gathered at a national or regional level (e.g. conditions and thresholds above which industries are obliged to report their emissions) often differ across countries, thus resulting in different completeness in the data sets for a same substance.
- *Reference year selected.* With the generally increasing awareness of environmental problems, regulations and monitoring bodies have made data more refined, more frequently updated and more available. These developments are expected to have contributed to a higher quality in the more recent raw data.

To remedy the lack of directly usable data and bring as much consistency and completeness as possible, the use of estimation techniques and extrapolations are commonly required in the calculation of the normalisation reference inventories for some impact categories, particularly the toxicity-related impact categories. Estimation techniques are developed to address two types of emission data issues: (1) releases of single substances from diffuse sources, which are not easily traceable because of their diffuse nature and therefore not well inventoried, e.g. emissions related to agricultural practice like ammonia or phosphorus; and (2) releases of substances which are reported in emission inventory databases as substance groups, e.g. non-methane volatile organic compounds (NMVOCs) and pesticides. In the latter, it is often necessary to disaggregate these groups into single substances as the characterisation step is typically performed at this level and as characterisation factors may differ strongly between individual substances within the group. For both types of data issues, estimation techniques have been developed and applied in the calculation of normalisation references. These, however, are associated with uncertainties and vary depending on the type of emission sources and the substance (or substance group) considered (e.g. Huijbregts et al. 2003 or Wegener Sleeswijk et al. 2008 for ammonia and phosphorus; Laurent and Hauschild 2014 for NMVOCs; Huijbregts et al. 2003 or Lautier et al. 2010 for pesticides).

To extend the substance coverage in the normalisation reference inventory, extrapolations (and sometimes interpolations) in time and geography can also be useful when complementary data are available for other regions than the one framed by the system boundaries or when complementary data are available in other years than the one selected as reference year. Different parameters can be used as basis of the extrapolations depending on the type of substance and the type of activities driving the substance emission or consumption. A frequently used parameter (often used per default) is the gross domestic product (GDP). A direct extrapolation based on GDP inherently assumes that the types of activities involved in the GDP are the same in the concerned regions and that this also goes for the level of prevention of environmental impact. However, GDP has not been shown to be always a robust proxy (e.g. Wegener Sleeswijk et al. 2008). Any performed extrapolations and assumptions should thus systematically be checked for validity. Extrapolations in time should be performed with care to avoid the risk that out-of-date or strongly deviant data are being introduced in the inventory, ultimately leading to unwanted bias in the final normalisation references. Likewise, geographic extrapolations should be realised without introducing biases, e.g. if releases linked to an industrial activity in a region is used to estimate releases in another region, where this industrial activity is absent or where such releases are differently regulated.

With respect to consumption-based inventories, uncertainties mainly consist of the uncertainties of the raw data and those inherent to the use of EIO tables. Unless the inventory is intended to include environmental flows not well-covered in international or national databases (e.g. toxic substances), neither extrapolation nor modelling for specific substances (e.g. pesticide emissions) is required. Several limitations of the EIO models have been highlighted in the literature (e.g. Suh et al. 2004; Suh and Huppes 2009) and nearly all of them are relevant to normalisation purposes.

3.5.3 Uncertainties Associated With Available CFs

As discussed in previous chapters, the characterisation models used to translate the normalisation inventory flows into their impacts on the environment are also associated with uncertainties. These uncertainties vary considerably between the impact categories, and they are particularly high for the categories addressing human toxicity and ecotoxicity. They are discussed in more detail in the chapters addressing each of the impact categories. The characterisation models and factors applied in calculation of the normalisation references must be the same as applied in the characterisation of the impacts from the studied product system. Hereby the model uncertainty that is reflected in the choice among existing characterisation models is eliminated and the remaining source of uncertainty is associated with the parameter values used in the calculation of the characterisation factors.

3.5.4 Uncertainties Related to Limited Coverage of CFs/Emissions

In particular for the toxicity impacts there can be problems with a limited coverage of the relevant flows, both in the inventory and in the characterisation factor database. Estimation of the incompleteness of the coverage in both the inventory and the sets of characterisation factors is a challenging task because of numerous unknowns. In short, the problem comes down to estimating the omitted contribution resulting from (i) the unknown intensity of the pollutant emissions or resource consumptions, and/or (ii) the unknown magnitude of their contribution to impact potentials. Both may vary independently across an unknown number of environmental flows. Taking a single environmental flow, three situations with regard to its coverage may thus occur:

- Coverage in the inventory but not in the set of characterisation factors.
- Coverage in the set of characterisation factors but not in the inventory.
- No coverage by either inventory or characterisation factors.

While the number of environmental flows related to the first two situations can easily be estimated, the number of flows in the third situation can constitute an important unknown. The toxicity-related impacts may stem from tens of thousands of substance that are potentially released to the environment. In contrast to these staggering numbers, currently existing normalisation inventories only cover hundreds of substances (e.g. Wegener Sleeswijk et al. 2008; Lautier et al. 2010), and the most comprehensive existing sets of characterisation factors cover a few thousand substances (e.g. USEtox model, Hauschild et al. 2008; Rosenbaum et al. 2008).

To estimate how the normalisation references may deviate from the true situation due to uncovered substances, the emission intensities and the magnitude of the impact potentials (i.e. CFs) should be treated together. For many impact categories

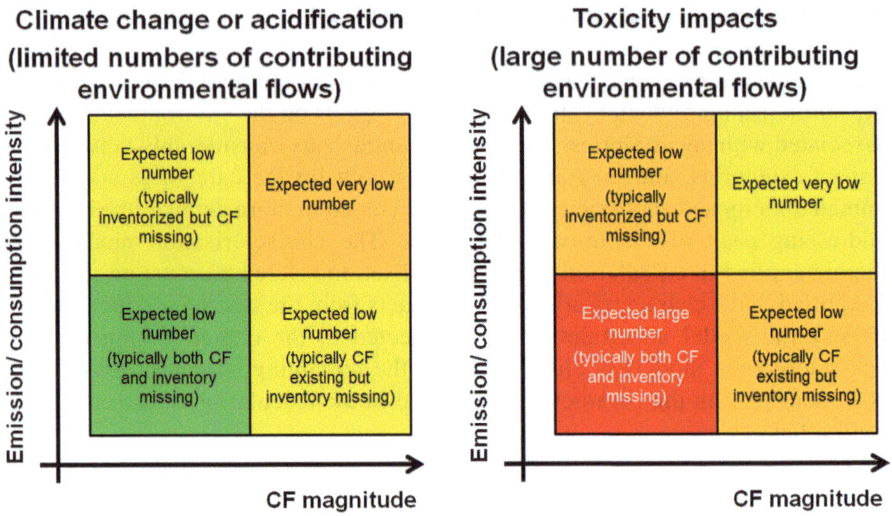

Fig. 14.4 Examples of uncertainty characterisation as a function of the number of uncovered substances (decreasing magnitude of impact contribution: red > orange > yellow > green)

with a limited number of contributing environmental flows (e.g. acidification), the largest uncertainties can be expected to stem from the unmapped compounds with relatively high CFs and/or high emission intensities –see dark grey cell in the left inset of Fig. 14.4. However, for impact categories with a high number of potentially-contributing flows and a very incomplete coverage in the inventory and/or the sets of characterisation factors (e.g. toxicity-related impacts), the uncertainties may also stem from low-CF, low-emission environmental flows, for which their large number may compensate their small individual contributions once aggregated – see black cell in the right inset of Fig. 14.4.

4 Application of Normalisation in Practice

Although an optional step according to the ISO 14044 standard, normalisation has been widely used in practice in a large variety of fields, with LCA case studies ranging from single commodities to large-scale services like waste management systems. As stated in Sect. 2.3, the increasing availability of external normalisation references and the limitations related to internal normalisation has contributed to internal normalisation being marginalised. Because of its limited application compared to external normalisation, the following sections only address external normalisation.

4.1 *Common Practice and use in Decision-Making*

4.1.1 Current Status

With respect to generic normalisation references, the practitioner can rely on those that have already been calculated and made publicly available in literature (often by method developers). Table 14.5 provides an overview of these as of 2013.

As visible from Table 14.5, nearly all existing normalisation references have been calculated using a production-based approach. EIO tables have been used to obtain consumption-based inventories because of their relevance to national/ regional policies (e.g. Tukker et al. 2006), but due to their important uncertainties and limitations, particularly when toxic impact categories must be included (see Sects. 3.4 and 3.5), they have rarely been used for LCA normalisation purposes. The geographical coverage of normalisation references reflects the development of LCA methods worldwide, as Europe, North America, Australia and Japan are the continents or countries, where most method developments have taken place, often delimitating the geographical applicability of the characterisation models. Global normalisation references have been calculated in three studies, all applying extensive extrapolations from a limited number of regions, i.e. developed countries (EU-15 in Stranddorf et al. 2005 and Huijbregts et al. 2003; EU, JP, US and CA to different extents in Wegener Sleeswijk et al. 2008). However, as most products currently have their life cycle stretched all over the globe, the relevance of the global references for performing normalisation in LCA case studies is high. Table 14.5 also illustrates the important delays that exist between the reference year and the year of publication of the normalisation references. Potential consequences for the consistency of the normalised results are discussed in Sect. 4.2.

4.1.2 Fulfilment of the Normalisation Purposes

External normalisation can fulfil all purposes of normalisation outlined in Sect. 1.1 of this chapter. The interpretation of normalisation results in context with these is addressed in the following; the use of normalisation as support for weighting is separately discussed in Sect. 4.4.

To demonstrate the fulfilment of the purposes, consider the characterised impact scores provided in Table 14.1 (Sect. 2) and assume that the product is consumed in Finland. Its entire life cycle is assumed to span many regions worldwide. A normalised profile can then be calculated according to different normalisation references. Fig. 14.5 shows the normalised results (in Person Equivalents; see Sect. 3) using ReCiPe normalisation references for Finland following a consumption-based approach (taken from Dahlbo et al. 2013 with correction for CC based on Hertwich and Peters 2009), Europe and the world (taken from Goedkoop et al. 2013; based on Sleeswijk et al. 2008). Both approaches are

Table 14.5 Publicly available generic normalisation references[a]

Regions	Reference year[a]	LCIA method/characterisation model/impact categories[b]	Reference
Consumption-based normalisation references			
The Netherlands	1993/1994	CML	Breedveld et al. (1999)
Finland	2005	ReCiPe	Dahlbo et al. (2013)
Production-based normalisation references			
Denmark	1990	EDIP	Wenzel et al. (1997)
Denmark	1994	EDIP	Stranddorf et al. (2005)
The Netherlands	1993/1994	CML	Breedveld et al. (1999)
The Netherlands	1997/1998	Combination of characterisation models, USES-LCA (toxic impacts)	Huijbregts et al. (2003)
Finland	2005	ReCiPe	Dahlbo et al. (2013)
South Africa	2001	Abiotic resource extraction	Strauss et al. (2006)
Australia	2002/2003	USES-LCA (toxic impacts)	Lundie et al. (2007)
Australia	2005/2006	CML2001, IMPACT2002+	Foley and Lant (2009)
Japan	NS	LIME 1	Itsubo et al. (2003, 2004)
Japan	NS	LIME 2	Itsubo et al. (2012)
United States	1999	TRACI	Bare et al. (2006)
United States	2002–2008	IMPACT2002+	Lautier et al. (2010)
United States	2006	TRACI	Kim et al. (2013)
United States	2008	TRACI	Ryberg et al. (2014)
Canada	2005	IMPACT2002+	Lautier et al. (2010)
North America (USA, Canada)	2002–2008	IMPACT2002+, USEtox	Lautier et al. (2010), Laurent et al. (2011b)
North America (USA, Canada)	2005–2008	TRACI	Ryberg et al. (2014)
EU-15 + 3	1990/1994	CML	Breedveld et al. (1999)
EU-15	1995	Combination of characterisation models, USES-LCA (toxic impacts)	Huijbregts et al. (2003)
EU-15	1994	EDIP	Stranddorf et al. (2005)
EU-25 + 3	2000	ReCiPe, IMPACT 2002+[c]	Wegener Sleeswijk et al. (2008), Lautier et al. (2010)[c]
Europe	2004	EDIP, USEtox	Laurent et al. (2011a, b)
World	1990, 1995	Combination of characterisation models, USES-LCA (toxic impacts)	Huijbregts et al. (2003)
World	1994	EDIP	Stranddorf et al. (2005)
World	2000	ReCiPe	Wegener Sleeswijk et al. (2008)
World	NS	LIME 3	Murakami et al. (2013)

[a]*NS* not specified

[b]When the name of the LCIA method is given, the reader is referred to check the reference to see the impact coverage of the method

[c]In Lautier et al. (2010), European normalisation references for IMPACT2002+ were also calculated using the inventory for Europe by Wegener Sleeswijk et al. (2008)

consistent with the presumed geographical scope of the product system (see Sect. 3.2.3).

External normalisation does not change the ranking of the alternatives because the same set of normalisation references is applied for all three alternatives. For the same reason, the relative process and substance contributions remain the same. Taking the set of global normalisation references as an example, it can be seen that system A is contributing 0.58 PE_{Glo} for climate change, meaning that its climate change impact is equivalent to 58 % of the annual contribution of an average global citizen to climate change. When interpreting, this figure could be brought into perspective with other known contributions to climate change, such as food, energy supply, transportation or simply another commodity. Purpose 1, appraising the environmental burden in relation to that of the reference system, can thus be met. At the same time, checking the sanity of the result (purpose 2) can easily be done. For example, a normalised score of ca. 0.6 PE for climate change could qualify for a service like the annual supply of electricity and heat to an average person living in a developed country, but it would be too big for the serving of 1 l of coffee in a restaurant or too small for the treatment of the municipal waste generated in the city of Copenhagen in 1 year.

4.1.3 Use of Different Sets of Normalisation References

As indicated in Sect. 3.2.3, the LCA practitioner should be aware that the use of global normalisation references or consumption-based regional normalisation references, if available, will often be the most consistent approach when the analysed system stretches over the world. However, there may also be situations, where a regional, production-based normalisation reference is required, e.g. if the choice of weighting factors of the study are based on the EU-27 policy context. This reflects the dependence of the selection of the normalisation references on the goal and scopes of the study, which, in turn, have influence on the magnitude of the obtained normalised results. For example, taking alternative C, the globally-normalised score for climate change is 3.6 times higher than that for metal depletion, while it is only 2.4 times higher when using the production-based normalisation references for Europe (Fig. 14.5). In this example, it does not change the ranking of the impact categories (assuming equal weights), but this may happen depending on the compared sets of normalisation references (see e.g. Dahlbo et al. 2013) and on the choice of weighting factors, and it may thus have consequences for the interpretation of the results with respect to purpose 1 (appraisal of the environmental burden in relation to that of the reference system and comparisons of impact categories between each other).

However, from a decision-making perspective, it does not constitute a problem because the intent of the external normalisation step is to bring the results into a broader context, which is selected through the choice of the normalisation references by the LCA practitioner and/or the different stakeholders (see Sect. 3.2.3). The LCA practitioner thus needs to analyse the results in relation to that specific

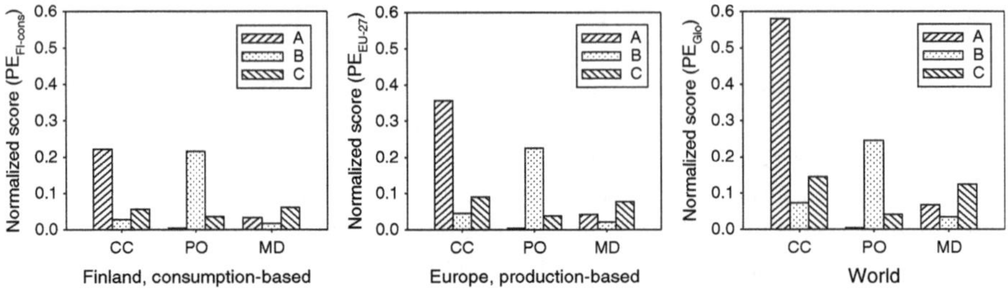

Fig. 14.5 Use of different external normalisation approaches in assessment of 3 alternatives *A*, *B* and *C*. *CC* climate change, *PO* photochemical ozone formation, *MD* metal depletion. *PE* person-equivalent (with indication of the geographical scope of the system: Finland, EU-27 or world). ReCiPe normalisation references for Finland (Dahlbo et al. 2013; CC adapted from Hertwich and Peters 2009; reference year: 2005), EU-27 and World (Goedkoop et al. 2013; based on Sleeswijk et al. 2008; reference year: 2000) were applied

context. For example, the normalised score for climate change is lower when using the Finnish consumption-based normalisation reference (i.e. 18 t CO_2-eq/person; Hertwich and Peters 2009) because the Finnish consumption per capita is higher than the global average per person. If the products were consumed in a country with low average GHG emissions per person and limited imports, like China (i.e. 3.1 t CO_2-eq/person; Hertwich and Peters 2009), the normalised score for climate change following the consumption-based approach, would have been much higher than that using global normalisation references (due to lower normalisation references). Likewise, the use of production-based rather than consumption-based normalisation references can give very different results. For example, consumption-based normalisation references for non-renewable resource depletion will always be significant because most national activities heavily draw on fossils, minerals and metals (e.g. for energy, manufacturing, etc.). However, there exist countries or regions, where those resources are not or very little present in the ground, hence resulting in very small production-based normalisation references for resource depletion. Applying production-based in lieu of the consumption-based normalisation references will thus yield very high normalised scores (due to very low normalisation references). In such situation, the LCA practitioner needs to carefully relate this score to the analysed system and its context.

4.1.4 Communication to Stakeholders

The complex and heterogeneous metrics applied in the characterised impact indicator scores are not immediately understandable and can impede communication to different stakeholders. Through the conduct of workshops and stakeholder interviews, primarily targeting industry partners, Dahlbo et al. (2013) have shown that stakeholders, who are not LCA experts, wish to be supplied with a support that is concise, attractive and sufficiently documented, and that also brings the results into a wider perspective, i.e. reflecting the magnitude of the LCA results in a context.

The latter can be achieved through normalisation, and the results can thus be presented in a format that is agreed among the different stakeholders (e.g. use of fact sheets in Dahlbo et al. 2013). The expression of normalised impact scores in PEs has the communicative advantage of presenting the impacts of the studied product or service in a unit that is easy to relate to for many lay people – as a share of an average person's annual impact. Although Dahlbo et al. (2013) recognise the beneficial use of normalisation, they also mention the difficulty for stakeholders to understand the caution that needs to be taken when interpreting the normalised profiles, which do not include information on the importance or the seriousness of the impact categories, and thus prevent their comparative ranking (see Sect. 4.4).

4.2 Data Availability and Importance of Updating NR

Although the aim is to calculate as recent normalisation references as possible, it is typical to have a gap of 5–10 years between the reference year and the year of release of normalisation references (see Table 14.5). Because updates are infrequent, normalisation in LCA case studies is often performed with older sets of normalisation references. This often does not pose problems for the interpretation of the normalised profiles because important changes in emission patterns are rarely seen[3] over the period of time separating two updates. Three situations that can trigger the need for updating normalisation references are: (1) an important change in the LCIA method; (2) a change in the methodology to build the inventory that results in significant changes in the emission or resource consumption of most contributing compounds, and (3) a large change in the emission patterns (e.g. due to new regulations) and/or in the data availability and quality (e.g. increased substance coverage) for the reference system. For example, Laurent et al. (2011a) have shown that, by moving from a pesticide inventory modelling based on direct emissions to agricultural soil to an inventory modelling only considering the fractions of pesticides that reach the biosphere, the ranking of the normalised results of terrestrial ecotoxicity and freshwater ecotoxicity could be reversed. Kim et al. (2013) also demonstrated how a refinement of an existing normalisation inventory by increasing the substance coverage and filling in important gaps can lead to tremendous changes (several orders of magnitude) for the ecotoxicity and human toxicity impact categories.

[3] Stratospheric ozone depletion represents an example of an unusually rapid decline in the level of man-made environmental impact during the 1990s. The reason was the global agreement on phasing out most of the contributing gases between 1986 and 2006 under the auspices of the Montreal Protocol.

4.3 *Uncertainties and Biases in Practice*

A number of uncertainties accompanying the normalisation references have been discussed. However, these do not represent the overall uncertainty of normalised impact scores because the latter arises from the entire normalisation process and are also affected by the uncertainties of the characterised scores and the combination of the two when the characterised scores are divided by the normalisation references.

Heijungs et al. (2007) described different types of biases that may occur when applying normalisation. Essentially, if the characterisation factor database used in the determination of both the characterised results and the normalisation references is assumed to be the same, the main biases occur because of (i) substance emissions or resource consumptions covered in the inventory of the analysed system but not in the normalisation inventory, (ii) substance emissions or resource consumptions covered in the normalisation inventory but not in the inventory of the analysed system (despite knowing that some are caused by the system), (iii) missing characterisation factors for substance emissions or resource consumptions. In the first two situations, the magnitude of the biases will depend on the potential contribution that the missing substances would have in either the normalisation reference or the characterised score of the analysed system. For example, if a substance emission largely contributing to the normalisation reference of a given impact could not be estimated in the LCI for the product (despite knowing its occurrence), the normalised score would likely be underestimated. Conversely, if a substance emission takes place in the analysed system, but was not inventoried in the calculation of the normalisation reference, the normalised score would likely be overestimated; the extent of this overestimation depending on the potential contribution of the emission to the total normalisation references. The same reasoning applies when characterisation factors are missing, except that the bias would depend on the potential contribution of the missing substances in both the normalisation reference and the characterised score of the analysed system. The resulting normalised score for an impact category may thus be either too high or too low. Because each impact category can be differently affected by these types of biases, the relative magnitude of the normalised impact scores between each other may also be biased.

In general, the combination of the uncertainties of the characterised scores (in the numerator), those inherent to the normalisation references (in the denominator) and the potential biases between the two make it very difficult to predict the extent of the overall uncertainty in the normalised score because these uncertainties/biases are not additive and some of them may compensate each other or cancel out. For example, potential errors associated with a characterisation model may be neutralised when normalising since the errors will appear in both the numerator (characterised score) and the denominator (normalisation reference). Regardless of other uncertainties and biases, in such situations, the use of normalisation may thus reduce the overall uncertainty of the impact scores, and the normalised impact score may become less uncertain than the characterised impact score.

4.4 *Normalisation as Support for Weighting*

According to the ISO 14044 standard, one of the purposes of normalisation is to prepare for another optional element of the impact assessment, namely the ranking, grouping or weighting, which further prepare for the comparison of results across impact categories that may occur as part of the interpretation. Comparisons of results across impact categories after normalisation inherently assume that an equal weight is put on one normalised unit (be it a PE or a ppt of the impact of the reference system) for each impact indicator, and this will often not be a relevant assumption. The damage modelling which is part of the foundation for endpoint characterisation factors quantifies the ability of the midpoint indicator to contribute to damage to the area of protection. The resulting midpoint-to-endpoint characterisation factors can vary strongly between the impact categories illustrating that the midpoint indicators can have very different ability to cause damage to human health or functioning of ecosystems. The ranking, grouping or quantitative weighting requires a deliberation of the relative importance of the different impact indicators in accordance with the goal of the LCA study.

Some weighting approaches pose specific requirements to the choice of normalisation references. For example, weighting factors based on political reduction targets may require that the normalisation reference represents the geographical scale which is relevant for the impact category (global for global categories and regional for regional categories). There are also examples of impact assessment methods that perform weighting without prior normalisation. Chap. 15 discusses different approaches to weighting and also their requirements to the preceding normalisation step, but in general terms it can be said that an important purpose of normalisation in preparation of a weighting scheme is to eliminate the potential bias inherent in the arbitrarily chosen different units of the characterised indicator results by expressing them on a common scale, in a common metric.

References

AG – Australia's National Pollutant Inventory (NPI). Australian Government; Department of Sustainability, Environment, Water, Population and Communities, Canberra. http://www.npi.gov.au

Bare J, Gloria T, Norris G (2006) Development of the method and U.S. normalisation database for life cycle impact assessment and sustainability metrics. Environ Sci Technol 40 (16):5108–5115

Boughton B, Horvath A (2006) Environmental assessment of shredder residue management. Resour Conserv Recycl 47:1–25. doi:10.1016/j.resconrec.2005.09.002

Breedveld L, Lafleur M, Blonk H (1999) A framework for actualizing normalisation data in LCA: experiences in the Netherlands. Int J Life Cycle Assess 4:213–220

Chevalier J, Rousseaux P, Benoit V, Benadda B (2003) Environmental assessment of flue gas cleaning processes of municipal solid waste incinerators by means of the life cycle assessment approach. Chem Eng Sci 58:2053–2064. doi:10.1016/S0009-2509(03)00056-3

Dahlbo H, Koskela S, Pihkola H, Nors M, Federley M, Seppälä J (2013) Comparison of different normalised LCIA results and their feasibility in communication. Int J Life Cycle Assess 18:850–860

EC (2010) International Reference Life Cycle Data System (ILCD) Handbook: general guide for life cycle assessment – detailed guidance. European Commission, Joint Research Centre, Institute for Environment and Sustainability, 1st edn, Publications Office of the European Union, Luxembourg

EC (2014) National Pollutant Release Inventory (NPRI). Environment Canada (EC), Gatineau. http://www.ec.gc.ca/inrp-npri/. Accessed Feb 2014

EEA (2014) European Pollutant Release and Transfer Register (E-PRTR). European Environmental Agency (EEA), Copenhagen. http://prtr.ec.europa.eu/. Accessed Feb 2014

EMEP/CEIP (2014) Centre on Emission Inventories and Projections (CEIP) providing country- and sector-specific pollutant emission data. Umweltbundesamt, Vienna. http://www.ceip.at/. Accessed Feb 2014

FAOSTAT (2014) Expected to provide statistical data related to food and agriculture. Food and Agriculture Organization of the United Nations (FAO), Rome. http://faostat.fao.org/. Accessed Feb 2014

Foley J, Lant P (2009) Regional normalisation figures for Australia 2005/2006—inventory and characterisation data from a production perspective. Int J Life Cycle Assess 14:215–224

Goedkoop MJ, Heijungs R, Huijbregts MAJ, De Schryver A, Struijs J, Van Zelm R (2009) ReCiPe 2008, a life cycle impact assessment method which comprises harmonised category indicators at the midpoint and the endpoint level, 1st edn. Report I: characterisation. First edition (revised). Ministry of Housing, Spatial Planning and the Environment, Netherlands, May 2013. http://www.lcia-recipe.net. Accessed June 2013

Goedkoop MJ, Heijungs R, Huijbregts MAJ, De Schryver A, Struijs J, Van Zelm R (2013) ReCiPe Mid/Endpoint method, version 1.08, February 2013. Available at: http://www.lcia-recipe.net/. Accessed Mar 2013

Hauschild MZ, Huijbregts MAJ, Jolliet O, Macleod M, Margni M, Rosenbaum RK, van de Meent D, McKone TE (2008) Building a model based on scientific consensus for life cycle impact assessment of chemicals: the search for harmony and parsimony. Environ Sci Technol 42:7032–7037

Heijungs R, Guinée J, Kleijn R, Rovers V (2007) Bias in normalisation: causes, consequences, detection and remedies. Int J Life Cycle Assess 12:211–216

Hertwich EG, Peters G (2009) Carbon footprint of nations: a global, trade-linked analysis. Environ Sci Technol 43:6414–6420

Huijbregts MAJ, Breedveld L, Huppes G, De Koning A, Van Oers L, Suh S (2003) Normalisation figures for environmental life-cycle assessment: The Netherlands (1997/1998), western Europe (1995) and the world (1990 and 1995). J Clean Prod 11(7):737–748

Humbert S, De Schryver A, Margni M, Jolliet O (2012) IMPACT 2002+: user guide. Draft for version Q.2.2. Quantis, Lausanne

IEA (2014) Expected to provide data on energy worldwide. International Energy Agency, Paris. http://www.iea.org/. Accessed Feb 2014

ISO 14044 (2006) Environmental management – life cycle assessment – requirements and guidelines, Geneva

Itsubo N, Sakagami M, Kuriyama K, Washida T, Kokubu K, Inaba A (2003) Development of weighting factor for LCIA based on conjoint analysis. J Environ Sci 15:357–368 (in Japanese)

Itsubo N, Sakagami M, Washida T, Kokubu K, Inaba A (2004) Weighting across safeguard subjects for LCIA through the application of conjoint analysis. Int J Life Cycle Assess 9:196–205

Itsubo N, Sakagami M, Kuriyama K, Inaba A (2012) Statistical analysis for the development of national average weighting factors—visualization of the variability between each individual's environmental thoughts. Int J Life Cycle Assess 17:488–498

Jungbluth N, Nathani C, Stucki M, Leuenberger M (2011) Environmental impacts of Swiss consumption and production. A combination of input-output analysis with life cycle assessment. Environmental studies no. 1111. Federal Office for the Environment, Bern, p 171

Kim J, Yang Y, Bae J, Suh S (2013) The importance of normalisation references in interpreting life cycle assessment results. J Ind Ecol 17:385–395

Koskela S, Mäenpää I, Seppälä J, Mattila T, Korhonen M-R (2011) EEIO modeling of the environmental impacts of finnish imports using different data sources. Ecol Econ 70:2341–2349

Laurent A, Olsen SI, Hauschild MZ (2011a) Normalisation in EDIP97 and EDIP2003: updated European inventory for 2004 and guidance towards a consistent use in practice. Int J Life Cycle Assess 16:401–409

Laurent A, Lautier A, Rosenbaum RK, Olsen SI, Hauschild MZ (2011b) Normalisation references for Europe and north America for application with USEtox characterisation factors. Int J Life Cycle Assess 16:728–738

Laurent A, Hauschild MZ (2014) Impacts of NMVOC emissions on human health in European countries for 2000–2010: use of sector-specific substance profiles. Atmos Environ 85:247–255

Lautier A, Rosenbaum RK, Margni M, Bare J, Roy P, Deschênes L (2010) Development of normalisation factors for Canada and the United States and comparison with European factors. Sci Total Environ 409(1):33–42

Lenzen M, Kanemoto K, Moran D, Geschke A (2012) Mapping the structure of the world economy. Environ Sci Technol 46:8374–8381

Lippiatt B (2000) BEES 2.0: building for environmental and economic sustainability. Technical manual and user guide. NISTIR 6520. National Institute of Standards and Technology, Gaithersburg

Lundie S, Huijbregts MAJ, Rowley HV, Mohr NJ, Feitz AJ (2007) Australian characterisation factors and normalisation figures for human toxicity and ecotoxicity. J Clean Prod 15:819–832

Murakami K, Itsubo N, Kuriyama K, Yoshida K, Tokimatsu K (2013) Development of global scale weighting factors in LIME3. Conference proceedings of the international conference on Eco Balance 2012. Yokohama, pp 20–23. Nov 2012

NITE – Expected to provide data on pollutant releases (PRTR). National Institute of Technology and Evaluation, Tokyo. http://www.safe.nite.go.jp/

Norris G, Marshall H (1995) Multi-attribute decision analysis method for evaluating buildings and building systems. NISTIR 5663. National Institute of Standards and Technology, Gaithersburg

Norris GA (2001) The requirement for congruence in normalisation. Int J Life Cycle Assess 6:85–88

OECD (2014) Expected to provide statistical data and metadata for OECD countries and selected non-member economies. The Organisation for Economic Co-operation and Development, Paris. http://stats.oecd.org/. Accessed Feb 2014

Rosenbaum RK, Bachmann TM, Gold LS, Huijbregts MAJ, Jolliet O, Juraske R, Koehler A, Larsen HF, Macleod M, Margni M, McKone TE, Payet J, Schuhmacher M, van de Meent D, Hauschild MZ (2008) USEtox – The UNEP-SETAC toxicity model: recommended characterisation factors for human toxicity and freshwater ecotoxicity in life cycle impact assessment. Int J Life Cycle Assess 13:532–546

Ryberg M, Vieira MDM, Zgola M, Bare J, Rosenbaum RK (2014) Updated US and Canadian normalization factors for TRACI 2.1. Clean Techn Environ Policy 16:329–339

Seppälä J, Mäenpää I, Koskela S, Mattila T, Nissinen A, Katajajuuri JM, Härmä T, Korhonen M-R, Saarinen M, Virtanen Y (2011) An assessment of greenhouse gas emissions and material flows caused by the Finnish economy using the ENVIMAT model. J Clean Prod 19:1833–1841

Stranddorf HK, Hoffmann L, Schmidt A (2005) Update on impact categories, normalisation and weighting in LCA. Selected EDIP97-data. Environmental project nr.995 2005. Miljøprojekt. Danish Ministry of the Environment. Danish Environmental Protection Agency, Copenhagen, p 290

Strauss K, Brent A, Hietkamp S (2006) Characterisation and normalisation factors for life cycle impact assessment mined abiotic resources categories in South Africa: the manufacturing of catalytic converter exhaust systems as a case study. Int J Life Cycle Assess 11(3):162–171

Suh S, Lenzen M, Treloar GJ, Hondo H, Horvath A, Huppes G, Jolliet O, Klann U, Krewitt W, Moriguchi Y, Munksgaard J, Norris G (2004) System boundary selection in life-cycle inventories using hybrid approaches. Environ Sci Technol 38:657–664

Suh S, Huppes G (2009) Methods in the life cycle Inventory of a product. In: Suh S (ed) Handbook of input–output economics in industrial ecology. Springer, New York, pp 263–282. ISBN 978-1-4020-4083-2

Tukker A, Huppes G, Guinée J, Heijungs R, de Koning A, van Oers L, Suh S, Geerken T, Holderbeke MV, Jansen B, Nielsen P (2006) Environmental impact of products (EIPRO): analysis of the life cycle environmental impacts related to the final consumption of the EU-25. European Commission. Technical report EUR 22284 EN. European Commission Joint Research Centre (DG JRC), Institute for Prospective Technological Studies, Seville

Tukker A, de Koning A, Wood R, Moll S, Bouwmeester MC (2013) Price corrected domestic technology assumption: a method to assess pollution embodied in trade using primary official statistics only. With a case on CO_2 emissions embodied in imports to Europe. Environ Sci Technol 47:1775–1783

UNEP Ozone secretariat (Data Access Centre) (2014) Expected to provide yearly updated data for ODS (world scale). United Nations Environment Programme, Nairobi. http://ozone.unep.org/. Accessed Feb 2014

USFCCC (2014) Expected to provide most recent GHG reports from world countries (Annex I & non-Annex I). United Nations Framework Convention on Climate Change, Bonn. http://unfccc.int/. Accessed Feb 2014

USGS (2014) Expected to provide data on natural resources (among others). U.S. Geological Survey, Reston. http://www.usgs.gov/. Accessed Feb 2014

US-EPA (2014) Toxics release inventory (TRI) Program. TRI explorer. United States Environmental Protection Agency (US-EPA), Washington, DC. http://www.epa.gov/triexplorer/. Accessed Feb 2014

Wegener Sleeswijk A, Van Oers LFCM, Guinée JB, Struijs J, Huijbregts MAJ (2008) Normalisation in product life cycle assessment: an LCA of the global and European economic systems in the year 2000. Sci Total Environ 390(1):227–240

Wenzel H, Hauschild M, Alting L (1997) Environmental assessment of products, vol 1; methodology, tools and case studies in product development. Chapman and Hall/Thomson Science, London. ISBN 0-792-37859-8

Wiedmann T (2009) A review of recent multi-region input-output models used for consumption-based emission and resource accounting. Ecol Econ 69:211–222

Wilting HC, Ros JPM (2009) Comparing the environmental effects of production and consumption in a region—a tool for policy. In: Suh S (ed) Handbook of input-output economics in industrial ecology. Springer, New York, pp 379–395. ISBN 978-1-4020-4083-2

WMO (2011) Scientific assessment of ozone depletion: 2010, Global ozone research and monitoring project–Report no 52. World Meteorological Organization, Geneva, p 516

Chapter 15
Weighting

Norihiro Itsubo

Abstract In the ISO 14044 standard 2006, weighting is an optional step in life cycle impact assessment (LCIA). It enables the user to integrate various environmental impacts in order to facilitate the interpretation of the life cycle assessment (LCA) results. Many different weighting methodologies have been proposed and several are currently being used regularly. Most existing studies apply the average of the responses obtained from the people (i.e. the decision makers) that were sampled. Others believe that weighting factors should be based on the preferences of society as a whole so that LCA practitioners can successfully apply them to products and services everywhere. This chapter classifies methods of weighting into three categories: proxy, midpoint, and endpoint methods. Results using proxy methods, such as MIPS (Material Input Per Service), CED (Cumulative Energy Demand), TMR (Total Material Requirement), Ecological Footprint, and CExD (Cumulative Exergy Demand), are fairly easy to understand because physical quantities such as weight and energy are used. The advantages of midpoint methods include compliance with the ISO framework and how it permits weighting that uses characterisation results. Endpoint methods allocate weights to Areas of Protection (AoP) rather than at midpoints, reducing the number of subject items and simplifying interpretation. Recently, weighting with endpoint methods has attracted attention due to the advancement of characterisation methodologies of this type. This chapter presents the different features of weighting and integration approaches applied in LCIA. The important differences and future problems concerning five key endpoint weighting methods are described. It concludes with a brief summary of the key features of the weighting methods introduced herein.

Keywords Economic assessment • Endpoint approach • LCA • LCIA • Life cycle assessment • Life cycle impact assessment • Midpoint approach • Weighting

N. Itsubo (✉)
Department of Environmental Studies, Tokyo City University, Yokohama,
3-3-1, Ushikubo-Nishi, Tsuzuki-ku, Yokohama, Japan
e-mail: itsubo-n@tcu.ac.jp

M.Z. Hauschild, M.A.J. Huijbregts (eds.), *Life Cycle Impact Assessment*,
LCA Compendium – The Complete World of Life Cycle Assessment,
DOI 10.1007/978-94-017-9744-3_15

1 Introduction

The types of potential environmental impact associated with a product life cycle vary widely from highly local, such as indoor air pollution and noise, to global, such as global warming and resource depletion, as discussed in the previous chapters in this book on characterisation (Chaps. 3, 4, 5, 6, 7, 8, 9, 10, 11, 12, and 13). However, products or systems developed to improve the environment often focus on one or only a few environmental problems in order to alleviate or reduce their impact. This is the case in biofuels which are expected to help reduce the risk of global warming because their combustion is considered carbon-neutral. However, while potentially reducing the risk of global warming, biofuels create a relatively greater impact to water resources by using crops as their raw materials and the occupation of land may impact biodiversity. Realising that products and services are associated with diverse environmental impacts, we need to assess them by considering the balance among their environmental impacts in an explicit manner and in accordance with the defined goal of the study, to ensure that the conclusions take the whole relevant spectrum of impacts into account.

Life cycle impact assessment (LCIA) covers multiple impact categories in the characterisation phase. The ISO standard requires that "The selection of impact categories shall reflect a comprehensive set of environmental issues related to the product system being studied, taking the goal and scope into consideration" (ISO 2006). In Europe, a default list of 14 impact categories is recommended for studying environmental footprints, and where any one of these is excluded, validity of the exclusion needs to be explained (EC 2013). Characterisation factors express the relative ability of individual substances to contribute to an impact category and enable calculation of aggregated impact results for each of the multiple impact categories. It provides no support for the aggregation or comparison of scores for different impact categories. Therefore, if a trade-off between impact categories is created in a comparison of products, the final decision on which choice is preferable requires the use of some type of value judgment, i.e. a weighting process based on the perceived importance of the impact categories to the decision maker. How to weigh or balance the impact scores is left to the practitioner and the stakeholders involved in the study. It is important to recognise that the absence of assigning weights to impact scores results in equal weighting by default.

2 Historical Development of Weighting Methods

Life Cycle Impact Assessment (LCIA) dates back to the latter half of the 1980s, when the Society of Environmental Toxicology and Chemistry (SETAC) adopted LCA as a theme for studies and began to discuss LCA studies regularly at its annual meeting. The Code of Practice (Consoli et al. 1993) issued by SETAC in 1993 included valuation as a step in the framework for LCIA, and weighting was thus recognised as a step of LCIA. Research activities aimed at developing a weighting method were further intensified. Through the early 1990s, pioneering institutions such as CML (Heijungs et al. 1992) and countries in Northern Europe (Lindfors et al. 1995; Wenzel et al. 1997) issued LCIA guidelines one after another, aiming to establish a framework for LCIA with consecutive steps of **characterisation**, **normalisation** and **weighting**. This work along with others was reflected in the ISO framework which evolved in parallel during the second half of the 1990.

In almost the same period, the Swiss Ecoscarcity method (Ahbe et al. 1990; Braunschweig and Müller-Wenk 1993) and the Swedish EPS method (Steen and Ryding 1992) were proposed. These methods had no explicit characterisation step but directly correlated substance emissions of concern with valuation without the process of characterisation.

In the latter half of the 1990s, methods for integrating characterisation with valuation were proposed. Most notable was the distance-to-target method included in the Eco-Indicator 95 method (Goedkoop 1995) and EDIP97 (Wenzel et al. 1997) which attracted much attention. Development of weighting factors with this method requires two values, desired and actual, which differ among countries. Therefore, studies for developing weighting factors reflecting standards, etc. of each country (e.g., Hauschild and Wenzel 1998; Lee 1999; Itsubo 2000; Matsuno et al. 1999) were performed in various countries.

At the time it was recognised that while a single index obtained through valuation makes it easy to interpret the result, there were issues concerning reliability and representativeness of the assessment results. In ISO's tasks of establishing an international standard for LCIA, there was a great amount of discussion on whether or not to recognise weighting as a formal step in LCIA.

From the latter half of the 1990s until the 2000, development of damage assessment methods was intensified after their importance was pointed out by Müller-Wenk (1997) and Hofstetter (1998). Müller-Wenk developed a damage factor for assessing the impact of traffic noise on health. Hofstetter (1998) and Krewitt et al. (1999) developed a factor for assessing the impact of air pollutants on health. Jolliet and Crettaz (1997) developed one for assessing the damage toxic chemicals have on health. All these applied a damage index based on lost life expectancy. Lindeijer (2000) developed a damage factor for assessing the impact of land use on the growth of plants. Van de Meent (1999) developed a damage factor for assessing the impact of chemical substances on loss of species. Goedkoop et al. in the EI99 method developed a damage factor for assessing the impact of acidification and eutrophication on loss of plant species (Lindeijer 2000; van de Meent 1999). Given the progress in the research and development concerning damage assessment, development of a method of weighting by comparing end-points attracted attention. Several methods went on to support obtaining a single index by weighting and aggregating endpoint scores, including the revised version of the Eco-indicator (Goedkoop and Spriensma 1999), the Life-Cycle Impact assessment Method based on Endpoint modelling (LIME; Itsubo and Inaba 2005), the revised version of the EPS (Steen 1999), and the revised version of ExternE (EC 2005), all of which were developed in the above mentioned period. Many of these methods were also developed through studies in which economic indices were used in attempts to calculate external costs.

Historically, the development of weighting methods was undertaken mainly in industrialised countries such as Japan and European countries. Currently, weighting factors are also being developed for emerging countries such as China (Wang et al. 2011). The hope is that assessment methods covering the entire world instead of particular regions will be developed in the future.

3 Purpose of Weighting

Figure 15.1 shows characterisation results from an LCA of beverage containers (aluminium cans, plastic bottles, and standing pouches, one 350 ml unit of each) (Yoshimura et al. 2011). The figure shows the trade-off between environmental impacts for the three containers. The standing pouches allow reduction of resource consumption; especially the use of fossil resources. The global warming impact was also found to be smaller for standing pouches due to reduced energy used in production, but the difference is not as great as for resource consumption, because the material recycling rate of standing pouches is relatively lower, while it permits reduction of carbon dioxide (CO_2) emission in the production of its materials, and because the rate of thermal treatment is high, which means that the amount of CO_2 emitted in disposing of standing pouches is larger than that for the other containers. With regard to a third impact category, photochemical oxidants, the value for standing pouches is the largest among all three types of containers because volatile organic compounds (VOCs) are emitted from solvents when films consisting of multiple layers are pressure-bonded to each other.

Figure 15.2 shows the results of weighting across the same three impact categories shown in Fig. 15.1, and others. Thus, weighting allows aggregation of different environmental impacts into a single score. We can say it is a useful step because it resolves trade-offs in an explicit and transparent way in support of decision-making.

Weighting based on valuation is used to make it easier to transmit information to general consumers as well as for decision-making by product designers and for other purposes. For example, the French supermarket chain Casino calculated a single index of beverages by valuating impacts related to global warming, water contamination, and water consumption. Puma, a sportswear manufacturer, valuated five types of environmental impacts, including global warming and water use, in monetary terms and disclosed its aggregated yearly environmental impact in a report (Puma 2010; PwC World Watch Issue 2011). In recent years, we have seen cases where the results of valuation are used for calculation of an environmental index that takes into account functions and values of products, presented as an environmental efficiency. Toshiba used weighting factors of LCA for comprehensive environmental efficiency calculation called Factor T (Toshiba 2009). BASF suggests and uses a method of showing environmental efficiency by

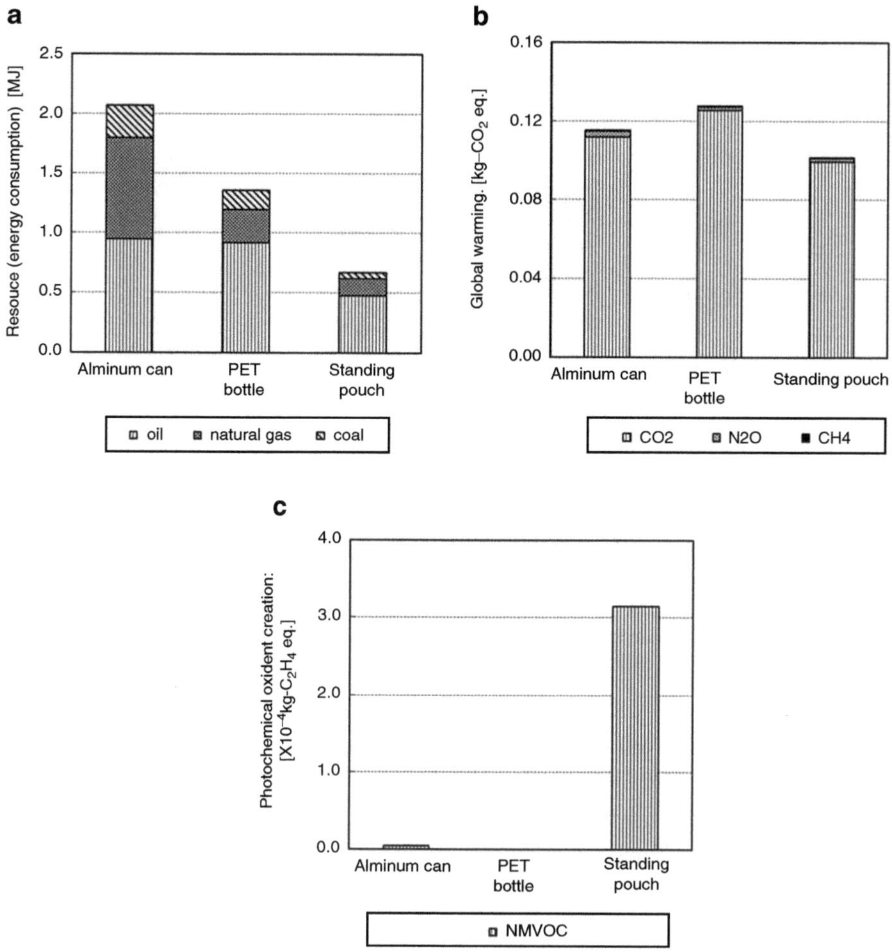

Fig. 15.1 Characterisation results of an LCA on beverage containers (aluminium cans, plastic bottles, and standing pouches, one 33 cl 350 ml unit of each) (**a**) energy consumption, (**b**) global warming, (**c**) photochemical oxidant

diagrammatically indicating relative relations between costs and environmental impacts (Saling et al. 2002). In this way, weighting is used by many companies to derive an environmental index from a comprehensive set of impact indicators.

Weighting environmental impacts offers the following benefits:

(a) Any trade-offs between the included impact category results are resolved in an explicit and transparent way and the results are prepared to be shown in a single index.
(b) It permits easy interpretation and communication of results so it is useful when transmitting information via environmental reports, etc.

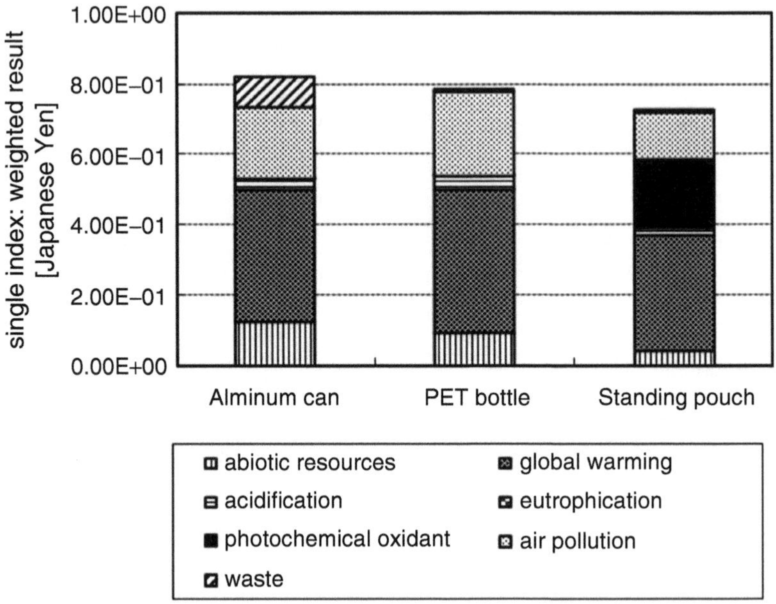

Fig. 15.2 Calculated results of a weighting across impact categories including energy consumption, global warming and photochemical oxidants

(c) It is highly applicable for tools for other environmental assessment tools such as environmental accounting (Bringezu et al. 1998) and environmental efficiency.

Common to all of these examples multiple environmental impacts are weighted and aggregated to be shown with a single index. However, the method for doing this process differs.

4 Restrictions on the Use of Weighting in International Standards

As previously described, weighting is a convenient method for resolving trade-off situations, but the ISO standard for LCA (ISO 2006) puts certain restrictions on its use. This is because value judgment is unavoidable in weighting. Environmental impacts include impacts on human health, ecosystems, and a host of other subjects. It is not possible to summarise such diverse impacts and show them with a single index only based on scientific criteria, a value judgment is inevitable in such a process.

Values differ among individuals and societies. Therefore, different stakeholders will often have different values and require different weighting factors, which could result in different conclusions. LCA results are used either to support improving products without disclosing the results (internal use) or for supporting assertions that the given products are eco-friendly by disclosing them to the public (external use). Above all, weighting could cause problems when applied in LCAs where results are intended for external use. Suppose that weighting is applied in a comparison between a company's product and a competing product from another producer, and the environmental impact of the company's product is found to be less than that of its competitor's when a particular set of weighting factors is applied. If use of another set of plausible weighting factors results in a different conclusion, and the company only discloses the result that is advantageous for itself, the competitor whose product was compared with the company's may suffer in a market that gives priority to environmental performance of the products.

To prevent such inappropriate use of weighting, and in general the influence of value-based choices in comparative assertions, ISO 14044 prescribes the following. First, the elements of LCIA are divided into mandatory elements and optional elements. Weighting is positioned as an optional element along with normalisation and grouping. While it is mandatory to conduct characterisation in LCIA, whether or not to perform weighting is determined on a case-by-case basis in accordance with the goal of the study and the target audience of the report. In other words, even when use of weighting is the main purpose of LCA, it is required to at the same time practice characterisation. Second, ISO 14044 also restricts use of weighting. Above all, it prohibits comparison of a company with the goal of the study and the target audience of the report. In other words, there are restrictions for when use of the results are for comparative assertion to be disclosed to the public.

On the other hand, ISO 14044 does not restrict the use of weighting in comparisons for other purposes, such as informing internally or disclosing to the public results of comparison between a company comparison to the As an alternative to quantitative weighting, the ISO standard also mentions the possibility of grouping or ranking the impact categories according to their importance. In this way, an alternative that scores best in all the most important impact categories may be identified as the best without resorting to a weighting, even if there are trade-offs to some of the other impact categories.

5 Different Approaches to Weighting

Generally, weighting means procedures for obtaining a single index based on subjective evaluations of different environmental impacts. The various methods of assessment with the use of weighting factors can be classified as follows.

1. Proxy method – weighting factor is directly applied to the inventory data.

$$I_1 = \sum_s (Inv(X) \times WF_1(X)) \tag{15.1}$$

I_1 indicates the result of valuation, Inv(X) is the inventory data of substance X, and $WF_1(X)$ is the weighting factor of substance X. This method enables obtaining the valuation result by directly multiplying the inventory data by the weighting factor. The weighting factor is set for each substance.

2. Midpoint method – the value obtained by multiplying the characterisation factor of the midpoint by the inventory data is converted to a non-dimensional figure or expressed in a common unit across the different impact categories and then multiplied by the weighting factor.

$$I_2 = \sum_{\text{Impact}} \sum_{X} \left(\frac{\text{Inv}(X) \times \text{CF}^{\text{Impact}}(X)}{\text{NV}^{\text{Impact}}} \times \text{WF}_2{}^{\text{Impact}} \right) \tag{15.2}$$

I_2 indicates the result of valuation, $CF^{Impact}(X)$ is the midpoint characterisation factor of substance X in the impact category (Impact), NV^{Impact} is the **normalisation reference** for the impact category (Impact), and WF_2 is the weighting factor of the impact category (Impact).

The result obtained by multiplying the characterisation factor of midpoint type by the inventory data is normalised to remove any bias caused by the different dimensions of the impact categories. Weighting is then performed by multiplying the value thus obtained by a non-dimensional weighting factor. A weighting factor is set for each impact category. The yearly amount of environmental impact in the subject country or region is used as the **normalisation value** in many cases (see this volume, Chap. 14 on normalisation by Alexis Laurent and Michael Hauschild).

3. Endpoint method (type1) – the value obtained by multiplying the characterisation factor of the endpoint by the inventory data is converted to a non-dimensional figure or expressed in a common unit across the different endpoint categories and then multiplied by weighting factor.

$$I_3 = \sum_{\text{Impact}} \sum_{\text{Endpoint}} \sum_{X} \left(\frac{\text{Inv}(X) \times \text{CF}^{\text{Impact}}(\text{Endpoint}, X)}{\text{NV}(\text{Endpoint})} \times \text{WF}_3(\text{Endpoint}) \right) \tag{15.3}$$

I_3 indicates the result of valuation, CF^{Impact} (Endpoint, X) is the endpoint characterisation factor of substance X for the endpoint (Endpoint) in the impact category (Impact), NV (Endpoint) indicates the **normalisation value** of Endpoint, and WF_3 (Endpoint) is the weighting factor of the Endpoint.

In this case, the inventory data is multiplied by the characterisation factor of the endpoint type and then divided by the **normalisation reference** that has the same dimension as the characterisation result. Weighting is then performed by multiplying this value by a non-dimensional weighting factor. A weighting factor is set for each endpoint.

4. Endpoint method (type 2) – the value obtained by multiplying the characterisation factor of the endpoint by the inventory data is multiplied by a weighting factor.

$$I_4 = \sum_{\text{Impact}} \sum_{\text{Endpoint}} \sum_{X} \left(\text{Inv}(X) \times \text{CF}^{\text{Impact}}(\text{Endpoint}, X) \times \text{WF}_4(\text{Endpoint})\right) \tag{15.4}$$

I_4 indicates the result of valuation and WF_4 is the weighting factor of the endpoint (Endpoint). In this case, the result of valuation is obtained by multiplying the inventory data by the endpoint characterisation factor and then multiplying by the weighting factor. In contrast to the previous approaches to weighting, **no normalisation is performed** and the endpoint score is expressed in amount of damage to the endpoint. WF_4 is expressed as the value per unit of the amount of damage (e.g., willingness to pay; WTP). A weighting factor is set for each endpoint.

These calculation methods make it possible to consolidate diverse environmental impacts into a single index.

Table 15.1 summarises the features of the various weighting methods introduced in this chapter.

6 Weighting Methods

This section presents details of the characteristics of the weighting methods developed thus far, by following the classification made in the previous section.

6.1 Proxy Method

With the proxy method, the actual environmental impact is not assessed through an explicit characterisation, but the inventory flow is converted into some preselected index, which is taken to express the environmental impact as a proxy parameter. Examples include MIPS (Material Input Per Service; Schmidt-Bleek 1994), CED (Cumulative Energy Demand; VDI-Richtlinie 1997), TMR (Total Material Requirement; Wuppertal Institute 1996), Ecological Footprint (Wackernagel and Rees 1996), and CExD (Cumulative Exergy Demand; Finnveden and Östlund 1997; Bösch et al. 2007). These methods are based on the premise that assessment of the actual environmental impact is difficult. They are also based on the assumption that the chosen index, e.g. the amount of energy consumption or the total amount of

Table 15.1 Summary of existing weighting methods in LCIA

	Proxy method	Midpoint method	Endpoint method (type1)	Endpoint method (type2)
Existing studies	MIPS TMR CED Ecological footprint	Eco-indicator 95 EDIP Eco scarcity etc.	Eco-indicator 99 LIME ReCiPe	EPS ExternE LIME ReCiPe etc.
Advantages	Indices are easy to understand Physical indices can be used Calculations are relatively easy	Full compliance with ISO 14044 Characterization results are used Calculations of weighting factors are relatively easy	Permits weighting among small number of items (endpoints), which reduces burden on respondents Methodologies of social sciences can be used	Permits weighting among small number of items (endpoints), which reduces burden on respondents Methodologies of social sciences and economics can be used Can be used for cost-benefit analysis
Problems	Low compliance with ISO 14044 Environmental impacts are not assessed	Weighting is difficult because of the large number of environmental issues Weighting is difficult for general consumers; highly representative weighting factors are difficult to obtain	Normalisation is required; it is necessary to tell the normalised contents to respondents Number of existing studies is small A large-scale survey is required for obtaining highly representative results	Number of existing studies is small Discussions on ethical issues are necessary A large-scale survey is required for obtaining highly representative results

substances used, will serve as an acceptable proxy for the actual environmental impact. The proxy parameter of environmental impact applied by MIPS and TMR is the total amount of substances related to the raw materials used throughout the lifecycle of the subject product. The proxy parameter applied by CED is the total amount of energy consumed either directly or indirectly through the life cycle. For the Ecological Footprint, the area of land used directly or required to assimilate the CO_2-emission is used as the alternative indicator, and this method was applied for assessing the sustainability of contemporary society. For the Ecological Footprint, there are cases where it has actually been used as a macro index for assessing environmental impact on a countrywide or global level (Kitzes et al. 2007).

Advantages of proxy methods include that the simple concept of the method makes it easy for practitioners to understand and apply, and it is easy to develop a weighting factor. On the other hand, issues of those methods include the points that the actual environmental impact is not analysed or assessed, that because of this it is impossible to verify the precision of the impact assessment results. They do not include the characterisation as a step and therefore are not compliant with ISO rules, and many findings that would be obtained through natural scientific analyses, such as the impact of global warming on temperature rise and human health, are not taken into account. Because of these issues, proxy methods are not used frequently for case studies of LCIA. They can be used for obtaining rough indicators of a country's or an organisation's environmental performance given that a chosen index is considered relevant for the purpose (e.g. ecological footprint).

6.2 Midpoint Method

In a midpoint-type impact assessment, the environmental impact is assessed for a range of different environmental problems, such as global warming, resulting in a profile of midpoint impact scores and then a single index is obtained through weighting among the problems. Representative methods include Eco-indicator 95 (Goedkoop 1995), EDIP (Hauschild and Wenzel 1998), and ones developed by Huppes et al. (1997), Walz et al. (1996), Lindeijer (1996), Nagata et al. (1995), Itsubo (2000), Matsuno et al. (1999), and Yasui (1998). The midpoint method uses the result of characterisation as the basis of integration, and therefore it is highly compliant with international standards. Its other advantages include the point that its integration concept is easy to understand and the development of the weighting factors themselves is relatively easy, depending on the choice of weighting principle.

Midpoint methods are largely divided according to the weighting principle they are based on, which is either a panel method and or a distance-to-target method. With panel methods, weighting factors for the midpoint impact categories are determined based on the level of importance assigned to each environmental problem by sampled subjects or a panel of experts. Nagata et al. (1995) directly asks the level of importance of impact categories to particular groups of respondents (including students, members of industrial associations, and people related to LCA). Yasui (1998) obtains weighting factors by asking respondents the length of the remaining period before the onset of a critical situation and the seriousness of the crisis in the impact categories. Huppes et al. (1997) calculates a factor for determining the weight of each impact category based on discussions by a panel of policymakers, while Walz (1996) and Lindeijer (1996) turn to panels of environmental specialists for the calculation. However, the panel approach is problematic: the task of comparing more than ten impact categories places an excessive burden on respondents; the statistical significance of weighting factors obtained from answers to questionnaires is not examined in many cases; and information provided to the respondents as the basis for weighting is limited, which limits the transparency of the obtained factors. Because of these shortcomings, researchers have not developed midpoint methods based on the panel method in recent years.

The distance-to-target (DtT) approach was applied in the Eco-scarcity method (Frischknecht et al. 2006), Eco-indicator 95 (Goedkoop 1995), EDIP (Hauschild and Wenzel 1998), and methods developed by Matsuno et al. (1999) and Itsubo (2000), respectively. The Eco-scarcity method was developed by Müller-Wenk for assessing the ecological balance of companies. Equation 15.5 is used for applying this method to assess the impact. With this method, the ratio between the actual level of each substance and its pre-determined desired level is calculated. The larger the difference between the two is, the greater the resulting Eco-scarcity score of the substance becomes and this is accentuated compared to other methods by the squaring of the ratio in the expression.

$$SI = \sum_{S} (Inv._S \times IF_S) = \sum_{S} \left(Inv._S \times \frac{N_S}{{T_S}^2} \right) = \sum_{S} \left(\frac{Inv._S}{N_S} \times \left(\frac{N_S}{T_S} \right)^2 \right) \tag{15.5}$$

The actual level (such as the environmental concentration of a substance) and desired level (such as the environmental standard) will vary depending on the location and local political priorities. Therefore, eco-scarcity methods compliant with the environmental standards, etc. have been developed in European countries. JEPIX (Miyazaki et al. 2003) is the Japanese equivalent of the Eco-scarcity method.

Because the Eco-scarcity method determines the weighting factor of each substance without a characterisation to calculate its environmental impacts, it is not classified as a midpoint-type assessment method in the strict sense.

Eco-indicator 95, which was developed by Goedkoop and co-workers, integrates the environmental impacts in Europe in ten impact categories. Equation 15.6 is applied for this method and other distance to target weighting methods like EDIP (Wenzel et al. 1997).

$$\mathrm{SI} = \sum_{\mathrm{Impact}} \left(\frac{\mathrm{CI}^{\mathrm{Impact}}}{\mathrm{NV}^{\mathrm{Impact}}} \times \frac{\mathrm{NV}^{\mathrm{Impact}}}{\mathrm{T}^{\mathrm{Impact}}} \right) = \sum_{\mathrm{Impact}} \left(\frac{\mathrm{CI}^{\mathrm{Impact}}}{\mathrm{NV}^{\mathrm{Impact}}} \times \mathrm{W}^{\mathrm{Impact}} \right) \tag{15.6}$$

In this formula, SI is the single index (non-dimensional) and $\mathrm{CI}^{\mathrm{Impact}}$, $\mathrm{NV}^{\mathrm{Impact}}$, $\mathrm{T}^{\mathrm{Impact}}$, and $\mathrm{W}^{\mathrm{Impact}}$ indicate the **characterisation result**, **normalisation value**, **target value**, and **weighting factor**, respectively, in the impact category (Impact). This method is the same as the Eco-scarcity method in the basic idea of obtaining the weighting factor based on the ratio between the desired flow and actual flow. However, it differs in that a characterisation is performed and the weighting factors are set for impact categories. To enable comparison among impact categories, the bias inherent in the different metrics of the impact categories is removed from the result of characterisation by normalisation, which is conducted as a preparation of weighting.

The DtT method is based on the assumption that the importance of an environmental impact is represented by the difference between the desired value and actual value. It attracted a great deal of attention especially in the latter half of the 1990s, when many DtT methods were suggested. One advantage is that the grounds for weighting is easy to understand because the national environmental standard or emission-reduction level is used as the desired value. The desired value is set based on information authorised by the national government, etc., instead of being determined subjectively by individuals, so involvement of the practitioner's subjective view can be prevented, and generic weighting factors can be developed. Another advantage is that it permits weighting factors to be determined relatively easily because only two parameters – the desired flow and actual flow – are used for weighting. On the other hand, the method has problems from the standpoint that the results may differ greatly depending on how the desired value is determined. For example, the weighting factor for global warming will differ completely between a case where the Kyoto Protocol target is used as the desired value and one where a level that allows virtually no impact of global warming to be generated is used. The weighting factor for eutrophication may be the environmental standard set for each lake or a stricter standard, where the latter will increase the weighting factor considerably and it will be difficult to ensure consistency in the ideas behind the desired values of all impact categories.

Thus, there exist many potential desired values, from among which methodology developers may choose in order to find the ones they feel are most appropriate, these chosen values then determine the weighting factors. This means that weighting factors are determined in a highly arbitrary (meaning they can vary from user to user) and potentially biased manner. To avoid such arbitrariness as much as possible, discussion is needed for setting common targets in advance; but this discussion has not occurred and agreement has not yet been reached. For these reasons, there has been little development research concerning this method in recent years. Instead, the prevailing attitude has been to leave the final valuation step up to the users.

A critique raised against DtT weighting methods is that they are really not weighting methods but rather a sort of **normalisation method** where the targeted level of impact is used as the normalisation reference instead of the current level of impact, as seen in the first part of the expression in Eq. 15.7, where NV^{Impact} cancels out, leaving the characterisation result normalised by the target value:

$$SI = \sum_{Impact} \left(\frac{CI^{Impact}}{NV^{Impact}} \times \frac{NV^{Impact}}{T^{Impact}} \right) = \sum_{Impact} \left(\frac{CI^{Impact}}{T^{Impact}} \right) \tag{15.7}$$

An inherent assumption is thus that the further away from the target, the worse, no matter which impact is studied – exceeding the target by 50 % is equally important for climate change and for acidification. With all targets inherently equally important to reach in a DtT approach, opponents criticise that an explicit weighting of the targets is missing (Finnveden 1997).

6.3 *Endpoint Method*

With an endpoint method, environmental impacts are integrated by weighting among the endpoints that are damaged by the modelled impacts. Advantages of methods classified as endpoint-type include:

- They make it possible to clearly distinguish between the specialised area based on natural scientific knowledge (until the damage assessment of the endpoint) and the specialised area based on social scientific analysis (from the endpoint until the single index),
- Transparency is improved by clarifying items to be included in the assessment (types of disease, types of species), and
- Burden on respondents is small because the number of items to compare is small.

On the other hand, the following problems are included:

- Studies for assessing the amount of damage to the endpoints, which become the premises of integration, have yet to be mature, because of which a large amount of work is required for developing assessment methods.
- The assessable range (substances or endpoints) may be limited.

Endpoint methods are divided into methods where normalised midpoint scores are weighted among endpoints (type 1) and methods where weighting is performed by multiplying the value per unit of damage to the endpoint by the result of endpoint characterisation (type 2).

6.3.1 Endpoint Method (Type 1)

With the endpoint method (type 1), specialists, general consumers, etc. determine the values of environmental impacts in questionnaires or through group discussions. With Eco-indicator 99 (Goedkoop and Spriensma 1999), weighting among three predefined endpoints (human health, ecosystem quality, and resources) was performed by LCA specialists. Table 15.2 shows the weighting factors obtained with the method.

Normalisation was performed in a way that allowed the sum of the weighting factors to be 1. Because the number of endpoints for weighting is limited to three,

Table 15.2 Rounded weighting factors per cultural perspective in Eco-indicator 99 (Goedkoop and Spriensma 1999)

	Average (%)	Individualist (n = 10) (%)	Egalitarian (n = 14) (%)	Hierarchist (n = 5) (%)
Ecosystem quality	40	25	50	40
Human health	40	55	30	30
Resources	20	20	20	30

Table 15.3 Weighting factors of LIME1 and LIME2

	LIME1 (N = 400, Kanto region)	LIME2 (N = 1,000, Japan)
Human health	0.31	0.28
Social assets	0.21	0.15
Biodiversity	0.26	0.36
Primary productivity	0.23	0.21

the burden on the respondents is relatively small. This method is further characterised by the point that weighting is undertaken for each of three perspectives or lines of environmental thought (hierarchist, egalitarian, and individualist). This permits the practitioner to make analysis based on his or her own environmental thinking by determining the group to which subjects belong. On the other hand, the number of samples from which the weighting factors are determined is small, and therefore the weighting factors lack representation, which makes this method unsuitable for general use.

For LIME (Itsubo and Inaba 2005, 2012), weighting factors were calculated for comparison among four endpoint items (human health, social assets, biodiversity, and primary productivity). The first version based the factors on interviews with 400 people from the Kanto region, but the latest version, LIME2, is aimed at obtaining factors that represent the environmental thinking of the Japanese people, and interviews of 1,000 general consumers, who were selected by applying the random sampling method, were conducted for that purpose. Dimensionless weighting factors were obtained by multiplying the result of conjoint analysis (willingness to pay per unit of damage) by the **normalisation value**. (Explanations about willingness to pay per unit of damage are given in the next section.) Table 15.3 shows weighting factors of LIME1 and LIME2.

Here again, the weighting factors were scaled to give sum up to 1. The number of samples is smaller for LIME1 than for LIME2 and the areas surveyed for LIME1 are limited to particular regions. Thus, the respondents differ between these two surveys, and direct comparison between them is impossible. As the table shows, the items weighted relatively heavily are human health in LIME1 and biodiversity in LIME2. The method of calculating these values is explained in the next section.

6.3.2 Endpoint Method (Type 2)

With this method, the result of integration is obtained by calculating the value per unit of damage to an endpoint and multiplying the value thus obtained by the result of characterisation. EPS, ExternE, and LIME are classified as belonging to this approach. In all of these, the economic value per unit of endpoint is calculated and results of assessments made by using these methods are expressed in economic metrics. Much effort in environmental economics has been put into translation of environmental impact into economic value.

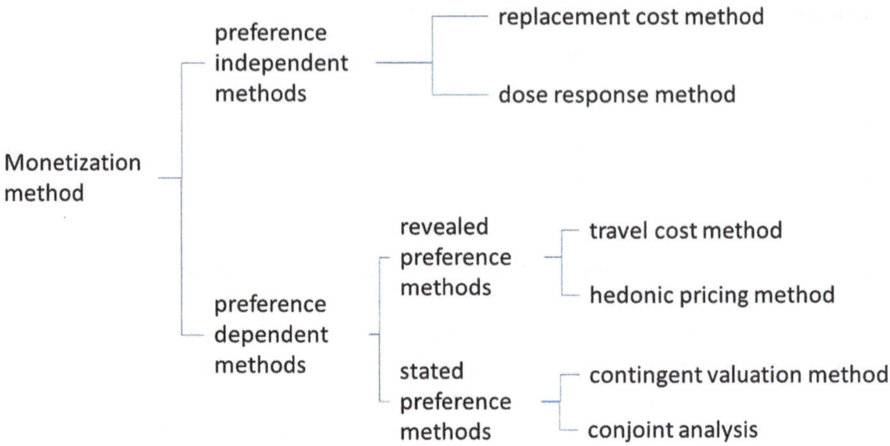

Fig. 15.3 Classification of assessment methods based on environmental economics

Using economic assessment means that the results are expressed in monetary values. This not only makes the results easy to understand and communicate but also allows them to be applied for cost-effectiveness analysis, for example. On the other hand, methods for converting environmental impacts resulting in health damage and ecosystem decline into economic values are still in the development phase, and results may be strongly biased or incomplete and their use often inappropriate depending on how the information is conveyed. It has also been pointed out that conversion of people's health, etc. into economic value is ethically problematic, which is another issue regarding this method.

The assessment methods based on environmental economics are classified mainly as shown in Fig. 15.3. First classification goes on whether environmental value is assessed independently from individual preferences (preference-independent methods) or based on individual preferences (preference-dependent methods).

Preference-independent methods include replacement cost methods and dose–response methods. A replacement cost method is an approach with which an environmental value is substituted by the cost needed for producing a substitute for the subject environment. For example, if the water-retaining function of a forest is lost due to e.g. clearing of the forest, it could be provided by the building of a dam. The cost for constructing and maintaining the dam is then regarded as the value of the forest's water-retaining function. Other examples in this vein could be the use of repair or mitigation costs associated with cleaning a pollution that has occurred to bring the environment back to the state it had before the impact was caused. With a dose–response approach, if any monetary value is created by deteriorating an environment this value is regarded as the value of the lost environment. Weidema applied annual income as the value of avoidance of the loss of one life-year (Weidema 2009). Preference-independent methods excel in that the calculations are relatively easy to perform, but the replacement cost methods are strongly dependent on the chosen technological solution. Furthermore, the dose–response methods mean that if no value is generated by damaging the environment, then the resulting damage has no value. These problems with the above approaches make their societal acceptance difficult. Currently, there are few cases in which a preference-independent method is used.

Preference-dependent methods are divided into the revealed preference method and the stated preference method. The former are applied for assessing the individual preferences on the environment indirectly based on the actual amount of money spent by individuals, while with the latter methods environmental values are assessed based on individual preferences revealed by asking the individuals directly. The revealed preference methods include the travel cost method and hedonic pricing method. With the travel cost method the value of the subject environment is assessed based on the expenses people incur in visiting the subject site. The hedonic pricing method is used to assess values of various living environments based on real estate prices, including those of land and houses. The advantage of these methods is that the results are highly reliable because values found for different individuals are estimated by the amount of money they actually spent. On the other hand, a problem with these methods is that subjects of the assessment are limited to fields where application of individuals' payment behaviour can be observed as compensatory behaviour, so these methods are restricted in scope, and accordingly, there are few cases where they have been used for weighting in LCA.

Stated preference methods excel in that they permit measuring not only of use values but also non-use values such as bequest value and existence value. Therefore, in many cases these days these methods are used for assessments based on environmental economics. Typical methods that fall into the stated preference methods include the contingent valuation method (CVM) and conjoint analysis. CVM is an approach where the value of a certain environment is measured by directly asking people related to the environment how much they would be willing to pay for the said ecosystem or environmental service (willingness to pay, WTP) or how much they would be willing to accept to give up the ecosystem or environmental service (willingness to accept, WTA), making estimates based on the distribution of the answers, and extrapolating the results to the entire relevant population. This is the most widely used method for assessing values of ecosystems including existence value because it is highly flexible, allowing assessors to tell their respondents the characteristics of the subject environment by using the questionnaire forms they prepare in advance. Diverse creative measures are applied for obtaining respondents' true willingness to pay, and there exist guidelines on how to make questionnaire forms (Arrow et al. 1993). CVM has been used for a variety of subjects. In the United States, it is used to provide bases for calculations of compensation costs in court procedures, for example. In this way, assessments made with CVM are introduced in the real world for purposes other than cost-benefit analyses. With regard to LCIA, endpoint values converted into monetary values with CVM are applied for integration with ExternE and EPS.

Conjoint analysis is similar to CVM in that individuals are directly asked about their intentions. However, it greatly differs from CVM in that it not only integrates the environment as a whole but also enables identification of differences in the strength of people's preferences among various attributes of the subject environment. If a natural environment is regarded as a single attribute, CVM is the method to be used. However, there are often cases where a natural environment should be regarded as having multiple attributes. For example, when the value of a tideland is measured, analysis made in consideration of tradeoffs among various factors (attributes) such as the natural landscape, protection of species, and cost for conserving the tideland may be advantageous for finding the direction of policy decision-making. Advantages of conjoint analysis are exerted in such a case. With regard to LCIA, conjoint analysis has been used for development of weighting factors in LIME.

7 Examples of Midpoint Methods

As part of the EDIP midpoint LCIA methodology, a weighting step was developed relying on political reduction targets and applying a distance-to-target approach. For each of the midpoint impact categories covered by the EDIP methodology weighting factors were calculated by

1. Identifying politically set reduction targets for elementary flows that contribute to the impact category. Only stated and binding targets set e.g. as part of an international treaty or national action plan are considered.
2. Harmonising the reduction targets to a common format across impact categories. Politically set targets are typically stated as a targeted reduction in the emission level in the reference year that must be attained in the target year. Target years and reference years differ between agreements and elementary flows, and a harmonisation is therefore performed by linear inter- or extrapolation to represent the corresponding reduction over a 10 year period.
3. Application of the harmonised reduction targets to the inventory of society's current emissions, which in EDIP is also used for calculation of the normalisation reference, to arrive at the inventory of targeted annual emissions.
4. Characterisation of the targeted annual emission inventory, applying the EDIP characterisation factors, to arrive at the targeted level of impact after 10 years in accordance with the political reduction targets. Both the current level of impact applied in the normalisation, and the targeted level of impact calculated for the weighting, are expressed on a per capita basis as current and targeted Person Equivalents respectively.
5. Calculation of the weighting factor as the current level of impact (the normalisation reference) divided by the targeted level of impact after 10 years

Table 15.4 Midpoint weighting factors of the EDIP methodology based on distance to politically set environmental targets

Impact category	Weighting factor
Global warming	1.3
Ozone depletion	23
Photochemical ozone formation	1.2
Acidification	1.3
Nutrient enrichment	1.2
Human toxicity	2.8
Ecotoxicity	2.3

Excerpt from Wenzel et al. 1997

$$\begin{aligned} \mathit{Weighting\ factor} &= \frac{\mathit{Current\ level\ of\ impact\ in\ reference\ year}}{\mathit{Targeted\ level\ of\ impact\ ten\ years\ after\ reference\ year}} \\ &= \frac{\mathit{Person\ Equivalent\ in\ reference\ year}}{\mathit{Targeted\ Person\ Equivalent\ ten\ years\ after\ reference\ year}} \end{aligned} \tag{15.8}$$

Table 15.4 shows an excerpt of EDIP weighting factors.

When applying these weighting factors to the normalised impact scores, the impacts of the product system are expressed in the metric of targeted person equivalents:

$$\begin{aligned} \mathit{Weighted\ impact\ score} &= \mathit{Weighting\ factor} \cdot \mathit{normalised\ impact\ score} \\ &= \frac{\mathit{Person\ Equivalent\ in\ reference\ year}}{\mathit{Targeted\ Person\ Equivalent\ ten\ years\ after\ reference\ year}} \cdot \frac{\mathit{Characterised\ impact\ score\ of\ product\ system}}{\mathit{Person\ Equivalent\ in\ reference\ year}} \\ &= \frac{\mathit{Characterised\ impact\ score\ of\ product\ system}}{\mathit{Targeted\ Person\ Equivalent\ ten\ years\ after\ reference\ year}} \end{aligned} \tag{15.9}$$

The weighting using EDIP's distance to target weighting factors may thus also be seen as a normalisation using the targeted level of impact as reference information.

As visible from Table 15.4, the weighting factors based on political reduction targets show a modest variation across the impact categories (apart from ozone depletion where a nearly complete phase out of the contributing elementary flows was the target).

As an alternative to political targets, EDIP also investigated the use of environmental carrying capacity or sustainability targets for calculation of weighting factors. With these targets the resulting person equivalents represent the environmental space that is available on average to each of us in a sustainable society or a society that respects nature's carrying capacities. They are generally far below the politically based targeted person equivalents and show a considerably larger variation in the weight put on the different midpoint categories, but they are also more uncertain due to the ambiguity of the setting of carrying capacities and sustainability targets (Hauschild and Wenzel 1998).

8 Examples of Endpoint Methods

ExternE applies a method called impact pathway analysis. With this method, as with the endpoint-type methods, monetary value of the environmental impact is obtained by estimating the amount of potential damage on endpoints and multiplying the estimated value by WTP per unit of amount of damage. Endpoints are subdivided into death, disorders, and others, and the monetary value is set for each of the subdivided cases. Different methods are applied for different WTP endpoints. For example, with regard to death, WTP for reduction of health risk was calculated with CVM for subjects from multiple countries, and the value thus obtained was converted into WTP for one life year. As a result, a value of life year (VOLY) at 50,000 euro/year was obtained. Concerning disorders, WTP per case was set for each level of severity. The value for disorders was obtained from the sum of (1) the resource cost including insurance cost and cost of treatment at hospital, etc., (2) the opportunity cost including productivity lowered by reduction of work hours, and (3) disutility including nuisance and pain. In addition to these, weighting factors related to noise, view, cultural heritage, buildings, ecosystems, and crops were also defined. Table 15.5 shows major weighting factors obtained with the method.

In addition to the values for endpoints shown in Table 15.5, the ExternE report shows integration factors for LCA obtained by multiplying these values by the amount of damages. For example, where 1 kg of PM2.5 is emitted by road traffic in a suburb, the average value is 15.2 €/kg. Here, uncertainty analysis was made and confidence intervals also shown.

Table 15.5 Different ways of determining WTP per unit of endpoint by applying ExternE

Endpoint	Approach	Value (€) per unit
Death	Contingent valuation method	50,000 per VOLY
Morbidity	Resource cost + opportunity cost + disutility	2,000 per admission (hospital admission), 670 per case (emergency room visit for respiratory illness)
Cultural and historical heritage	Expenditures for renovation of historical buildings	
Crop loss	Prices	Prices per ton of each crop
Ecosystem	Abatement costs	WTP per hectare protected ecosystem

Table 15.6 Examples of weighting factors of EPS method (Steen 1999)

Category	Unit	WF (Euro)
Life expectancy	Person-years	85,000
Severe morbidity	Person-years	100,000
Morbidity	Person-years	10,000
Severe nuisance	Person-years	10,000
Nuisance	Person-years	100

With this method, WTP per case of disorder was determined by type, such as hospitalisation and use of an emergency room. It differs from LIME and Eco-indicator 99 in that conversions to monetary values are made without consolidation to the Area of Protection (AoP). Therefore, in the strict sense, weighting among AoPs was not conducted. For setting monetary values, the method applies different approaches for different endpoints by quoting results obtained with CVM for the impact on health while applying abatement cost concerning the impact on ecosystem, for example. Discussions are needed for determining whether or not the weighted results can be summed up because conditions for integration, groups of respondents, and period when the survey was made differ among the endpoints.

With the EPS method, four endpoints – human health, production capacity of ecosystem, non-biological resources, and biodiversity – were set, and then a weighting factor was defined for each of five types such as death and disorder for human health, each of five types including crops, fish, and irrigation water for production capacity of ecosystem, each type of resource such as oil, coal, and iron for non-biological resources, and one type for biodiversity. Table 15.6 shows examples of the weighting factors thus defined. In the case of impact on health, the weighting factor per case differs greatly depending on the severity of the disorder. For example, the weighting factor concerning death was obtained by using the value for Value of Statistical Life (VSL) of ExternE.

As for ExternE, endpoints that have yet to be consolidated into AoPs, such as human health and production capacity, are converted into economic indices with EPS, and therefore weighting among AoPs in the strict sense of the term was not conducted.

For LIME, conjoint analysis was used to develop weighting factors concerning four types of AoP. Economic value for the amount of damage per unit of each AoP was calculated by statistical analysis such as logit model and random parameter logit model. The results are shown in Table 15.7 below.

This method differs from ExternE and EPS in that values are given only to four types of AoPs. Respondents were asked to choose from the hypothetical profiles including four types of environmental attributes and monetary attribute. Their choices are analysed to obtain WTP for avoidance of a unit of damage on each AoP such as biodiversity. LIME2, the latest version, is characterised by the point that highly representative weighting factors were obtained from results of a survey of 1,000 households. On the other hand, it must be noted that LIME2 does not reflect environmental thinking of other countries with different cultures and different economic situations.

Table 15.7 Weighting factors estimating external cost in LIME1 and LIME2

AoP	Unit	LIME1 (N = 400, Kanto region)	LIME2 (N = 1,000, Japan)
Human health	1 DALY	1.42E + 7	9.70E + 6
Social assets	10,000 yen	1.00E + 4	1.00E + 4
Primary productivity	1 ton	3.78E + 4	2.02E + 4
Biodiversity	1 species	1.27E + 13	4.80E + 12

9 Comparison Between Endpoint-Type Methods and Research Needs

Table 15.8 shows a comparison of endpoint-type weighting methods. All of these methods share the same assessment framework but differ on various points, such as assessment objects in terms of substances or impact categories. Important differences and future problems concerning weighting methods are described below.

9.1 *Area of Protection and Damage Indicators*

All of the above methods define human health as an AoP. However, though the ecosystem is commonly included in the objects of assessment, the methods differ in what part of the ecosystem should receive attention. The object of EPS is the degree of contribution (ratio) to extinction of species in a year, while that of Eco-indicator 99 is the ratio of vanished species (vascular plant species) and LIME uses the expected number of extinct species.

The methods also differ in how to consider the impact on human society, such as resources, materials, and agricultural products. LIME has established ‘social assets’ as a concept comprehensively covering what is treated as valuable things in human society (non-biological resources, agricultural products, marine resources, and forest resources). In addition to this, EPS includes cations, which is used as a buffer for soil acidification, and divides AoPs into resources and production capacity. Eco-indicator 99 does not include agricultural products or marine resources in the AoP, but includes resources, limited to mineral resources and fossil fuels. ExternE includes cultural properties, materials, and agricultural products for calculations, but has no clear definition of the area of protection.

LIME and EPS include the impact on primary production (plant production) in the objects of calculation, whereas Eco-indicator 99 and ExternE do not. In addition, though LIME defines ‘primary production’ as an area of protection, EPS considers it a part of the ‘production capacity of the ecosystem’. Therefore, the two methods differ in their range of AoP.

Table 15.8 Comparison among five endpoint methods of integration of environmental impacts

Method	EPS		ExternE	Eco indicator'99		ReCiPe		LIME2	
Country of development, year of publication	Sweden (Revised in 2000)		EC (Revised in 2005)	Netherlands (Revised in 2000)		Netherlands		Japan (2008)	
Assessable steps	Weighting, integration		Weighting, integration	Damage assessment, normalisation, weighting, integration		Characterisation, damage assessment, weighting, integration		Characterisation, damage assessment, weighting, integration	
Region subject to impact assessment	Sweden		Europe	Europe		Europe		Japan	
AoP and damage index	Human health		YOLL, etc.	No definition (Damage to human health, ecosystem, and materials are taken into account)		Human health		DALY	
AoP and damage index Impact categories	Biodiversity	NEX (ratio to extinc-tion of species)	No definition (Damage to human health, ecosystem, and materials are taken into account)	Quality of ecosystem	PDF (Ratio of disappeared species)	Human health	DALY	Human health	DALY
	Production capacity	kg	10 categories including resources, global warming, ozone depletion, and carcinogenic substances	Quality of ecosystem Resources	PDF (Ratio of disappeared species) Excess energy	Quality of ecosystem	PDF (Ratio of disappeared species)	Biodiversity	EINES (increase in number of extinct species)
	Resources	Amount of money						Primary production	tons
	Sensuousness	Not measured		Resources 18 impact categories	Excess energy 15 categories including global warming, ozone depletion, and urban area air pollution	Resources	Surplus cost (€)	Social assets	Damage cost (yen)
	The above five items have been defined as impact categories.	Air pollution, global warming							

(continued)

Table 15.8 (continued)

Method	EPS	ExternE	Eco indicator'99	ReCiPe	LIME2
Assessment process	Inventory → category endpoint → single index	Inventory → category endpoint → single index	Inventory → area of protection → normalisation → single index	Inventory → midpoint → area of protection → normalisation → single index	Inventory → characterisation → category endpoint → area of protection → single index
Weighting method	Market value where it exists, otherwise citation from CVM, etc.	Market value where it exists, otherwise citation from CVM, etc.	Panel method	Panel method, damage cost, prevention cost	Conjoint analysis
Unit of single index	Damage cost	Damage cost	Non-dimensional index (3 types: hierarchist, egalitarian, individualist)	Non-dimensional index, damage cost, prevention cost	Damage cost, non-dimensional index
Number of samples for weighting survey and survey method	Citation only; no field survey	Unknown	80 people (collection rate: 20 %), mail survey	Panel weighting, citation from literatures (damage cost, prevention cost)	1,000 households (collection rate: 48 %), visiting interview survey

9.2 Methods of Indicating Weighting and Integration Result

Approaches to weighting can be roughly divided into economic assessment methods (ExternE, EPS, LIME2) and panel methods (Eco-indicator 99). The integration result is expressed in amount of money (Euro or Yen) under the former methods, whereas it is expressed in a non-dimensional index under the latter. An advantage of expressing the environmental impact in an amount of money is that the results are easy to interpret, and therefore can be compared with costs or used for cost-benefit analyses, for example. However, indicating human health or the ecosystem in economic value involves an ethical problem. Indicated values for developed countries may also be higher than those of developing countries, which is another problem pertaining to the economic assessment methods. Expression in a non-dimensional index will reduce these problems but will make results difficult to understand for general consumers and policy decision-makers, which limits their external use.

9.3 Individual Differences and Variation of Weighting

Weighting differs among individuals. It would be important to express such differences as a range of weighting factors. ExternE shows the range of integration factors for each substance, but it is unknown where there is uncertainty of the amount of damage or variation in weighting. Eco-indicator 99 does not show variation in weighting factors but sets a weighting factor for each type of cultural perspective. LIME2 expresses individual differences in weighting quantitatively through an analysis that uses the random parameter logit model (Itsubo et al. 2012), showing that individuals constituting a group differ in value judgment but have a certain distribution. It is desirable for ensuring transparency of weighting factors to show the level of variation of weighting factors as an indicator of individual differences in value judgment.

The result of weighting is subject to influences of various factors such as culture, income, age, gender, religion, and educational background. Currently, there has been no study that looks at how much influence is given to weighting by differences in these factors.

9.4 Representativeness of Weighting Factors

Many of the methods shown above were developed based on the assumption they would be used for general purposes, regardless of what the products are and who the users are. Accordingly, they require confirmation that the value judgment represents the relevant population. CVM and conjoint analysis, which are normally used in environmental economics and other fields, are based on a merger between economics and inferential statistics. With inferential statistics, a statistical model is applied to a result of a survey of samples selected from a specific population by applying the random sampling method, and social preferences of the population are inferred through mathematical analysis. Results of the inference are verified for determining whether or not they are statistically significant. It is also verified whether or not the statistical model used for the regression to the population represents the social preferences of the population. Only results that pass these verifications may be used for general purposes. Eco-indicator 99 is based on weighting by environmental specialists in Europe. The number of samples is small and their representativeness is unclear. With regard to CVM applied in ExternE and EPS, it is unknown whether or not multiple results of CVM were obtained from the same group of respondents. LIME represents Japanese people's views but cannot be used for weighting in other countries.

Most of the existing methods calculate their weighting factors with the preferences of current generation, because their responses to questionnaire would be analysed. None of the weighting factor systems considers the preferences or interests of future generations, although weighting would be changed by time transition. Only ExternE method took into account discount rate in their monetisation.

As described above, existing methods of weighting differ in diverse points and leave a number of problems. Research and development for improving the problems while making use of the advantages of these methods will be required in the future.

10 Outlook

Integrating a wide range of environmental impacts means allocating weights to midpoint impact categories or to the objects that receive impacts of environmental changes, such as human health, biodiversity, agricultural products, and marine products industry, regardless of whether or not they are explicitly shown. Results of such comparisons cannot be obtained from knowledge based on natural science but are determined by subjective views of the assessors or practitioners or by how the given group views the environment. How people value the environment differs according to their social background, such as cultural background, educational background, and economic conditions. Methods of integration suggested thus far differ in the population subject to weighting (e.g., population in Europe or Japan), and therefore results of LCA are often inconsistent with each other due to the integration methods applied. In addition, even when the population subject to weighting is identical, use of different integration methods may lead to different conclusions (Itsubo 2000).

This leads to a concern over abuses of weighting, such as: manipulating weighting factors in a way that the assessment results of a company's products will be better than those of its competitors', and only disclosing results obtained by using assessment methods that are advantageous for a company's products. To restrict such abuses, ISO 14044 positions weighting as an optional element and prohibits its use in comparative assertions disclosed to the public.

Among practitioners some have negative attitudes towards the practice of weighting and integration because of the above reasons. However, a substantial number of companies make use of integration by placing greater emphasis on the positive features of integration, that is, easy-to-understand assessment results and the wide range of application, based on the recognition that integration comes into effect on the premise of ethical, social, and economical elements. Many pioneering companies make assessments in support of corporate evaluation, environmental accounting, and environmental efficiency assessments by using their own weighting factors or existing ones, and disclose the results in their environmental reports or on their websites. These are examples of using integration indices as tools for communication.

More companies may in the future use integration by making use of its advantages, that is, ease of interpretation and high applicability of assessment results. For this to happen, there is an urgent need to develop an integration method that can be used for such general purposes.

11 Conclusions

This chapter presents the different features of weighting and integration approaches applied in LCIA. Methods of weighting are classified into proxy methods, midpoint methods, and endpoint methods. **With an endpoint method, weighting is conducted after normalisation, or values for results of characterisation are multiplied without normalisation.**

An advantage of a **proxy method** is that assessment results are easy to understand because physical quantities such as weight and energy are used. On the other hand, this method has problems such as incompliance with the ISO standard's requirement that a characterisation be performed as part of the impact assessment, and the point that the environmental impacts are not assessed directly.

Advantages of a **midpoint method** include compliance with the ISO framework and the point that it permits weighting that uses characterisation results. On the other hand, it has problems such as that weighting is difficult because of the large number of impact categories and that it is difficult with the DtT method to set desired values that are truly equivalent among all impact categories.

Advantages of an **endpoint method** include the points that allocating weights to areas of protection rather than midpoints reduces the number of subject items and therefore reduces burden on respondents, and that furthermore it permits use of an assessment method of environmental economics. On the other hand, problems of this method include the limited number of studies conducted thus far and the need for a large-scale interview survey, which is costly.

In this way, advantages and problems differ among approaches. In recent years, however, weighting with the endpoint-type methods has been attracting attention due to the advancement of characterisation methodologies of this type.

References

Ahbe S, Braunschweig A, Müller-Wenk R (1990) Method for ecobalancing based on ecological optimization. Bundesamtfür Umwelt, Wald und Landschaft, Bern, p 39

Arrow K, Solow R, Portney PR, Learner EE, Radner R, Schuman H (1993) Report of the NOAA panel on contingent valuation. Fed Regist 58(10):4602–4614

Bösch M, Hellweg S, Huijbregts M, Frischknecht R (2007) Applying cumulative exergy demand (CExD) indicators to the ecoinvent database. Int J Life Cycle Assess 12(3):181–190

Braunschweig A, Müller-Wenk R (1993) Ökobilanzen für Unternehmungen. Eine Wegleitung für die Praxis. Bern/Stuttgart

Bringezu S, Behrensmeier R, Schutz H (1998) Material flow accounts indicating environmental pressure from economic sectors. In: Uno K, Bartelmus P (eds) Environmental accounting in theory and practice. Kluwer Academic Publishers, Dordrecht/Boston/London, pp 213–227

Consoli F, Allen D, Boustead I, de Oude N, Fava J, Franklin W, Quay B, Parrish R, Perriman R, Postlethwaite D, Seguin J, Vigon B (eds) (1993) Guidelines for life-cycle assessment: a 'Code of Practice', 1st edn. SETAC-Europe, Brussels

European Commission (2005) In: Bickel P, Friedrich R (ed) ExternE, externalities of energy, methodology 2005 update. ISBN 92-79-00423-9

European Commission (2013) 2013/179/EU: Commission recommendation of 9 April 2013 on the use of common methods to measure and communicate the life cycle environmental performance of products and organisations. Off J Eur Union 56

Finnveden G (1997) Valuation methods within LCA – where are the values? Int J Life Cycle Assess 2(3):163–169

Finnveden G, Östlund P (1997) Exergies of natural resources in life-cycle assessment and other applications. Energy 22:923–931

Frischknecht R, Steiner R, Jungbluth N (2006) The ecological scarcity method – Eco-factors: a method for impact assessment in LCA. 2009, Federal Office for the Environment FOEN: Zürich und Bern. Retrieved from www.bafu.admin.ch/publikationen/publikation/01031/index.html?lang=en. Accessed 29 Mar 2014

Goedkoop M (1995) The Eco-indicator 95. Final report and manual for designers. PRé Consultants, Amersfoort

Goedkoop M, Spriensma R (1999) The Eco-indicator 99. A damage oriented method for life cycle impact assessment. PRé Consultants, Amersfoort

Hauschild MZ, Wenzel H (1998) Environmental assessment of products, vol 2, Scientific background. Chapman & Hall/Kluwer Academic Publishers, London/Hingham, p 565

Heijungs R, Guinée J, Huppes G, Lankreijer RM, Udo de Haes HA, Sleeswijk AW, Ansems AMM, Eggels PG, Van Duin R, De Goede HP (1992) Environmental life-cycle assessment of products-guide. Center of Environmental Science (CML), Leiden Univ, Leiden

Hofstetter P (1998) Perspectives in life cycle impact assessment. A structured approach to combine models of the technosphere, ecosphere and valuesphere. Ph.D. Thesis. Kluwer Academic Publishers, Dordrecht

Huppes G, Sas H, Haan E, Kuyper J (1997) Efficiënte milieu investeringen. Milieu 3:126–133

ISO (2006) ISO 14044:2006 – environmental management – life cycle assessment – requirements and guidelines. International Standards Organization, Geneva

Itsubo N (2000) Screening life cycle impact assessment with weighting methodology based on simplified damage functions. Int J Life Cycle Assess 5(5):273–280

Itsubo N, Inaba A (2005) Life cycle impact assessment method, LIME—a methodology and database for LCA, environmental accounting and ecoefficiency. Japan Environmental Management Association for Industry (in Japanese)

Itsubo N, Inaba A (eds) (2012) LIME2 life-cycle impact assessment method based on endpoint modeling. Japan Environmental Management Association for Industry. http://lca-forum.org/english/. Accessed 17 Dec 2012

Itsubo N, Sakagami M, Kuriyama K, Inaba A (2012) Statistical analysis for the development of national average weighting factors—visualization of the variability between each individual's environmental thoughts. Int J Life Cycle Assess 17:488–498

Jolliet O, Crettaz P (1997) Critical surface-time 95. A life cycle impact assessment methodology including fate and exposure. Swiss Federal Institute of Technology, Institute of Soil and Water Management, Lausanne

Kitzes J, Peller A, Goldfinger S, Wackernagel M (2007) Current methods for calculating national ecological footprint accounts. Sci Environ Sustain Soc 4(1):1–9

Krewitt W, Trukenmüller A, Friedrich R (1999) Site dependent LCIA impact indicators for human health from integrated air quality and exposure modeling. Poster presented at the 9th annual meeting of SETAC-Europe, Leipzig

Lee KM (1999) A weighting method for the Korean Eco-Indicator. Int J Life Cycle Assess 4 (3):161–164

Lindeijer EW (1996) Normalisation and valuation, Part VI. In: Guinée JB (ed) Handbook on life cycle assessment. Operational guide to the ISO standards. Kluwer

Lindeijer EW (2000) Impact assessment of resources and land use. Report of the SETAC WIA-2 taskforce on resources and land. 6th draft

Lindfors L-G, Christiansen K, Hoffman L, Virtanen Y, Juntilla V, Hanssen OJ, Rönning A, Ekvall T, Finnveden G (1995) Nordic guidelines on life-cycle assessment, vol 1995:20, Nord. Nordic Council of Ministers, Copenhagen

Matsuno Y, Inaba T, Mizuno T (1999) Development of site-and source-specific life cycle impact assessment methodology for local impact categories. 9th annual meeting of SETAC-Europe, Leipzig

Miyazaki N, Siegenthaler C, Kumagai S, Shinozuka E, Nagayama A (2003) JEPIX – Japan Environmental Policy Priorities Index, Japan Science and Technology Inc/Sustainable Management Forum Japan, Tokyo, in Japanese

Müller-Wenk R (1997) Safeguard subjects and damage functions as core elements of life-cycle impact assessment, vol 42, IWÖ-Diskussionsbeitrag, Institut für Wirtschaft und Ökologie, Universität St. Gallen (IWÖ-HSG)

Nagata K, Fujii Y, Ishikawa M (1995) Proposing a valuation method based on panel data, preliminary report, Tokyo

PUMA (2010) PUMA's environmental profit and loss account for the year ended 31 December 2010. http://about.puma.com/wp-content/themes/aboutPUMA_theme/financial-report/pdf/EPL080212final.pdf. Accessed 29 Mar 2014

PwC World Watch Issue (2011) Puma's reporting highlights global business challenges. http://www.pwc.com/gx/en/corporate-reporting/sustainability-reporting/pumas-reporting-highlights-global-business-challenges.jhtml. Accessed 10 Feb 2010

Saling P, Kicherer A, Dittrich-Krämer B, Wittlinger R, Zombik W, Schmidt I, Schrott W, Schmidt S (2002) Eco-efficiency analysis by BASF: the method. Int J Life Cycle Assess 7(4):203–218

Schmidt-Bleek F (1994) Wieviel Umwelt braucht der Mensch – MIPS, das Maß für ökologisches Wirtschaften. Birkhäuser, Berlin

Steen B (1999) A systematic approach to environmental priority strategies in product development (EPS) version 2000—models and data of the default method. Chalmers University of Technology, Technical Environmental Planning, Gothenburg

Steen B, Ryding SO (1992) The EPS enviro-accounting method. An application of environmental accounting principles for evaluation and valuation of environmental impact in product design. IVL Report B 1080. IVL, Gothenburg

Toshiba (2009) Advancing together with factor T 2009. http://www.toshiba.eu/eu/Environmental-Management/Factor-T/. Accessed 14 Dec 2011

van de Meent D (1999) Potentieel aangetaste tractie als maatlat voor toxische druk op ecosystemen. RIVM rapport nr. 60750400, RIVM, Bilthoven

VDI-Richtlinie (1997) Cumulative energy demand, terms, definitions, methods of calculation. Band 4600 VDI-Richtlinien, Beuth

Wackernagel M, Rees WE (1996) Our ecological footprint: reducing human impact on the earth. Press New Society Publishing, Gabriola

Walz R, Herrchen M, Keller D, Stahl B (1996) Impact category ecotoxicity and the valuation procedure. Ecotoxicological impact assessment and the valuation step within LCA Pragmatic approaches. Int J Life Cycle Assess 4:193–198

Wang H, Hou P, Zhang H, Weng D (2011) A novel weighting method in LCIA and its application in Chinese policy context. In: Finkbeiner M (ed) Towards life cycle sustainability management. Springer, Dordrecht, pp 65–72

Weidema BP (2009) Using the budget constraint to monetarise impact assessment results. Ecol Econ 68:1591–1598. doi:10.1016/j.ecolecon.2008.01.019

Wenzel H, Hauschild MZ, Alting L (1997) Environmental assessment of products, vol 1, Methodology, tools and case studies in product development. Chapman & Hall/Kluwer Academic Publishers, London/Hingham, p 544

Yasui I (1998) A new scheme of life cycle impact assessment method based on the consumption of time. 3rd Int. Conf. on Ecobalance, Tsukuba

Yoshimura Y, Komatsu I, Itsubo N (2011) Life cycle impact assessment and approaches for reducing the environmental impact of each type of container, Technical Report. J Life Cycle Assess, Japan 7(3):264–273

Index

M.Z. Hauschild, M.A.J. Huijbregts (eds.), *Life Cycle Impact Assessment*, LCA Compendium – The Complete World of Life Cycle Assessment, DOI 10.1007/978-94-017-9744-3

MIX
Papier aus verantwortungsvollen Quellen
Paper from responsible sources
FSC® C105338

If you have any concerns about our products,
you can contact us on
ProductSafety@springernature.com

In case Publisher is established outside the EU,
the EU authorized representative is:
Springer Nature Customer Service Center GmbH
Europaplatz 3, 69115 Heidelberg, Germany

Printed by Libri Plureos GmbH
in Hamburg, Germany